Microarray Biochip Technology

Other BioTechniques® Books Titles

Gene Cloning and Analysis by RT-PCR
P.D. Siebert and J.W. Larrick (Eds.)

Apoptosis Detection and Assay Methods
L. Zhu and J. Chun (Eds.)

Protein Staining and Identification Techniques
R.C. Allen and B. Budowle

Immunological Reagents and Solutions: A Laboratory Handbook
B.B. Damaj

Affinity and Immunoaffinity Purification Techniques
T.M. Phillips and B.F. Dickens

Yeast Hybrid Methods
L. Zhu and G. Hannon (Eds.)

Viral Vectors: Gene Therapy and Basic Science
A. Cid-Arregui and A. García Carrancá (Eds.)

Ribozyme Biochemistry and Biotechnology
G. Krupp and R. Gaur (Eds.)

Antigen Retrieval Techniques
S.-R. Shi, J. Gu, and C.R. Taylor (Eds.)

Bioinformatics: A Biologist's Guide to Biocomputing and the Internet
S. Brown

Microarray Biochip Technology

Mark Schena, PhD
TeleChem International, Inc.
Sunnyvale, CA, USA

A BioTechniques Books Publication
Eaton Publishing

Mark Schena
TeleChem/arrayit.com
524 E. Weddell Drive
Suite 3
Sunnyvale, CA 94089-2115, USA

Library of Congress Cataloging-in-Publication Data

Schena, Mark.
Microarray biochip technology / Mark Schena.
p. cm.
Includes bibliographical references and index.
ISBN 1-881299-37-6
1. Biochips. 2. DNA microarrays. I. Title.

R857.B5 S35 2000
660'.65--dc21

99-088465

ISBN 1-881299-37-6

Printed in the United States of America

9 8 7 6 5 4 3

Eaton Publishing
BioTechniques Books Division
154 E. Central Street
Natick, MA 01760
www.BioTechniques.com

Francis W. Eaton: *Publisher and President*
Stephen Weaver: *Director and Editor-in-Chief*
Christine McAndrews: *Managing Editor*
Sandy Lamont: *Production Manager*
Paul Haje: *Cover Designer*

Contributors

Karl Adler
NEN Life Science Products, Inc.
Boston, MA, USA

David Barker
Molecular Dynamics, Inc.
Sunnyvale, CA, USA

Trent Basarsky
Axon Instruments, Inc.
Foster City, CA, USA

Kate Bechtol
Molecular Dynamics, Inc.
Sunnyvale, CA, USA

Jeffrey Broadbent
NEN Life Science Products, Inc.
Boston, MA, USA

Jing-Ying Chen
BioDiscovery, Inc.
Los Angeles, CA, USA

Ronald W. Davis
Department of Biochemistry
Beckman Center
Stanford University
Medical Center
Stanford, CA, USA

Suzanne Dee
Affymetrix, Inc.
Santa Clara, CA, USA

David Englert
Packard Instrument Company
Meriden, CT, USA

Elisabeth Evertsz
Incyte Pharmaceuticals, Inc.
Palo Alto, CA, USA

Russell Garlick
NEN Life Science Products, Inc.
Boston, MA, USA

Robert Gupta
Incyte Pharmaceuticals, Inc.
Palo Alto, CA, USA

Paul Haje
arrayit.com
Sunnyvale, CA, USA

David Hanzel
Molecular Dynamics, Inc.
Sunnyvale, CA, USA

John Hartwell
Gene Logic, Inc.
Gaithersburg, MD, USA

Glenn Hoke
Gene Logic, Inc.
Gaithersburg, MD, USA

Richard Joseph
NEN Life Science Products, Inc.
Boston, MA, USA

Peter Kalocsai
BioDiscovery, Inc.
Los Angeles, CA, USA

Anis Khimani
NEN Life Science Products, Inc.
Boston, MA, USA

Jeffrey Killian
NEN Life Science Products, Inc.
Boston, MA, USA

Ants Kurg
Institute of Molecular and
Cell Biology
University of Tartu
Estonian Biocentre, Tartu, Estonia

Elin Lõhmussaar
Institute of Molecular and
Cell Biology
University of Tartu
Estonian Biocentre, Tartu, Estonia

Todd Martinsky
TeleChem International, Inc.
Sunnyvale, CA, USA

Myles L. Mace, Jr.
Genetic MicroSystems, Inc.
Woburn, MA, USA

Greg McGuinness
Genetic MicroSystems, Inc.
Woburn, MA, USA

Andres Metspalu
Institute of Molecular and
Cell Biology
University of Tartu
Estonian Biocentre
and Asper Ltd., Tartu, Estonia

Alvydas Mikulskis
NEN Life Science Products, Inc.
Boston, MA, USA

Jean Montagu
Genetic MicroSystems, Inc.
Woburn, MA, USA

Sharron Penn
Molecular Dynamics, Inc.
Sunnyvale, CA, USA

Peter Rapiejko
AlphaGene, Inc.
Woburn, MA, USA

David Roach
Molecular Dynamics, Inc.
Sunnyvale, CA, USA

Don Rose
Cartesian Technologies, Inc.
Durham, NC, USA

Stanley D. Rose
Genetic MicroSystems, Inc.
Woburn, MA, USA

Soheil Shams
BioDiscovery, Inc.
Los Angeles, CA, USA

Mark Schena
TeleChem International, Inc.
and arrayit.com
Sunnyvale, CA, USA

Pascual Starink
Incyte Pharmaceuticals, Inc.
Palo Alto, CA, USA

Adam Steel
Gene Logic, Inc.
Gaithersburg, MD, USA

Nan Ting
Gene Logic, Inc.
Gaithersburg, MD, USA

Neeme Tõnisson
Institute of Molecular and
Cell Biology
University of Tartu
Estonian Biocentre
and Asper Ltd., Tartu, Estonia

Matt Torres
Gene Logic, Inc.
Gaithersburg, MD, USA

Mary Trounstine
Molecular Dynamics, Inc.
Sunnyvale, CA, USA

Mark Trulson
Affymetrix, Inc.
Santa Clara, CA, USA

Damian Verdnik
Axon Instruments, Inc.
Foster City, CA, USA

Janet A. Warrington
Affymetrix, Inc.
Santa Clara, CA, USA

Drew Watson
Incyte Pharmaceuticals, Inc.
Palo Alto, CA, USA

David Wellis
Axon Instruments, Inc.
Foster City, CA, USA

Jennifer Worley
Molecular Dynamics, Inc.
Sunnyvale, CA, USA

Hongjun Yang
Gene Logic, Inc.
Gaithersburg, MD, USA

Yong-Yi Yu
Gene Logic, Inc.
Gaithersburg, MD, USA

Jack Ye Zhai
Axon Instruments, Inc.
Foster City, CA, USA

Yi-Xiong Zhou
BioDiscovery, Inc.
Los Angeles, CA, USA

PREFACE

"Science knows only one commandment—contribute to science."

- Bertolt Brecht, *The Life of Galileo*

Microarray technology has catapulted into the limelight, promising to accelerate genetic analysis in much the same way that microprocessors have sped up computation. Microarrays are miniature arrays of gene fragments attached to glass chips. These biochips are used to examine gene activity and identify gene mutations, using a hybridization reaction between the sequences on the microarray and a fluorescent sample. After hybridization, the chips are read with high-speed fluorescent detectors and the intensity of each spot is quantified. The location and intensity of each spot reveals the identity and amount of each sequence present in the sample. The data are then mined and modeled using the tools of computational biology. Because thousands or tens of thousands of gene fragments can be present on a single microarray, data for entire genomes can be acquired in a single experiment.

This book covers all of the main areas of microarray technology, including theory, sample preparation and labeling, manufacturing methods, fluorescent imaging, and data analysis and mining. A wide spectrum of applications and approaches are presented, and readers may read the book from beginning to end or focus on specific chapters for their content. *Microarray Biochip Technology* contains material that is not available from any other source and therefore is a "must read" for anyone interested in biochips. Many of the tools and technologies described in this book will allow researchers to examine biological questions that were entirely refractory a few years ago; thus, this compendium is very forward-looking in its scope. At the same time, to add some balance and perspective to this explosive new technology, I think it is valuable to place DNA microarrays in a broader historical context. As we stand on the brink of a new millennium and look to the future with much enthusiasm and optimism, a quick digression into the history of science is as prudent as it is entertaining.

Had Lippershey not invented the telescope in Holland in 1608, Galileo Galilei would have probably remained a respected but relatively obscure academician. But after improving on Lipperhseys device in 1609 and using it to examine the nighttime sky, Galileo embarked on a set of experiments that were as monumental as any in the history of science; indeed, by some accounts, Galileo actually *invented* science. By examining nature in a pragmatic way, rather than according to the edicts of the Church, Galileo forged a way of thinking that provides the foundation for modern science. His accounts of the moon and Jupiter launched him into science "stardom," and place him among the greatest of all early empirical investigators.

It is important to remember the rather sobering state-of-the-art in Renaissance science. The predominant view was that earth represented the human world and that all other bodies including the moon, planets, and stars were heavenly bodies; by definition, the moon was thought to be perfectly round because of its "heavenly origin." But owing in part to advances in Italian glass and lens manufacturing, Galileo had the resources to build a telescope that allowed approximately 30× magnification (instead of the 3× power of the Lippershey instrument) and with such an instrument observed irregularities on the surface of the moon that he later described as mountains and valleys. The descriptions of the moon as "earthlike" were so shocking that they landed Galileo under house arrest until his death in 1642.

A second startling account involved the planet Jupiter. Owing again to the power of his telescope, Galileo was able to observe three novel bodies arranged in an elliptical configuration around Jupiter. He described them as "stars" in his notes, but their variable location with respect to Jupiter gave reason to pause and became the subject of intense interest and curiosity. The key observation, made in a series of successive nightly observations, was that the three small bodies (now known of course to be three of Jupiter's moons) dramatically changed position with respect to Jupiter on successive evenings. Because Jupiter's orbital path was firmly established, Galileo correctly concluded that the only reasonable explanation for the behavior of these bodies was that they were revolving around Jupiter and not earth, calling into serious question the predominant notion of earth as the center of the universe. Galileo's observations concerning the moon and Jupiter touched off heated debates between practitioners of religion and science, a conflict that has persisted in general form to the present.

Although the Galilean approach to astronomy was decidedly different from hypothesis-driven research, the most common style of research as it is conducted today, his exploratory approach was noticeably similar to much present day work in genomics. Though Galileo was aware of the commonly held views of the cosmos in the 17th century—the earth as the center of the universe, the moon as perfectly round and so forth—his research approach to astronomy was more query-based than hypothesis-driven. Before pointing his telescope at the stars, Galileo was probably more likely to ask, "What does the cosmos look like?" than to hypothesize, "I think the moon is a perfect sphere, and do the accumulated data support or refute this?" Because so little astronomical data were available at the time, it was probably more useful for Galileo to observe and record than to hypothesize. The "Columbus of the Stars" explored the cosmos in much the same way that contemporary microarray researchers are exploring the human genome.

A common style of research is one similarity between early astronomy and early genomics, but there are others. Galileo's work also underscores how a singular technology advance can enable remarkable discoveries. A more powerful telescope provided an entirely unprecedented view of the moon, planets, stars, and galaxies. Other factors being equal, better astronomical tools provide better data. The same is true in biology. DNA microarrays provide a quantum technology advance for functional genomics, and with this advance, the mysteries of biology are being unraveled.

It is also intriguing that many of the same theoretical issues that challenged Galileo present themselves to microarray biochip technologists. The central data-gathering tools in early astronomy and microarray technology, the telescope and the fluorescent scanner, respectively, utilize lenses to magnify and focus incoming light. Signal-to-noise ratios, resolution, depth of focus, and field size are common theoretical challenges in both endeavors.

With much the same excitement that captivates astronomers exploring deep space, microarray technologists are using the tools of functional genomics to probe the mysteries of genes and genomes. As we stand poised to undercover the intricacies of genomic function with our colleagues worldwide, it is difficult to imagine a more timely book than *Microarray Biochip Technology*. As the editor of this wonderful compilation, it is also imponderable to have been selected for more rewarding opportunity. In this regard, so many people along the way have contributed to making my involvement in this project possible. Thanks go out to my parents for their early nurturing (and genes); my sisters Kim and Amy for their laughter and kind support; my scientific mentors Dan Koshland, Keith Yamamoto, and Ron Davis for a steady flow of careful guidance and brilliant ideas, all of which I prudently absorbed; my business manager Todd Martinsky and publicist Paul Haje for showing me the ropes in the real world; Steve Weaver for offering me this editorship and Christine McAndrews for her diligent copyediting; my family and friends for their constant support; and my wife (and boss) René Schena for all her love and affection.

Microarray Biochip Technology is of course entirely predicated on the spectacular contents provided by all the contributors. The book contains chapters from many of the top scientists, academicians, programmers, engineers, and businesspeople in the world and provides a timeless collection of theory and practice, much of which is not available in the primary literature. A veritable gold mine of ideas and insights awaits the reader.

Microarray biochips are moving us forward quickly in terms of understanding disease, developing new diagnostics and drugs, helping us design better crops, and expanding the horizons of basic research. Microarray technology is also serving to catalyze much activity in the private sector, stimulating investment and providing employment opportunities in the form of an explosive new industry for the new millennium. Because of the unprecedented power of the technology, microarrays may ultimately tell us more about ourselves and each other than we ever wanted to know. And while the latter point is unlikely to trigger any investigations by the Inquisition, chances are that some of the data afforded by the technologies described in this book will be as controversial as they are exciting.

Mark Schena
January 2000

Further Reading:

1. **Bolles, E.B.** 1999. Galileo's Commandment. W.H. Freeman, New York, NY.
2. **Bragg, M.** 1998. On Giants' Shoulders. John Wiley & Sons, Inc., New York, NY.
3. http://es.rice.edu:80/ES/humsoc/Galileo/

Contents

1 Technology Standards for Microarray Research

Mark Schena[1] and Ronald W. Davis[2]
[1] TeleChem International, Inc. and arrayit.com, Sunnyvale, CA; [2] Department of Biochemistry, Beckman Center, Stanford University Medical Center, Stanford, CA, USA

INTRODUCTION

Microarray technology is emerging as a powerful platform for biological exploration. The rapid development of assays, tools, methods, and technologies in laboratories worldwide suggests that guidelines and standards should be established for this exciting new field. Provided herein are some ideas for the standardization of experimental design, microarray formats, detection systems, data outputs, and software tools. The proposal is not intended to be comprehensive or binding, rather the hope is that it will serve as a catalyst for discussions focused on the need to unify microarray experimentation. Adopting thoughtful standards and procedures and open platform architectures will expedite the proliferation of microarray technology by providing an increasingly powerful international "tool kit" that will better enable researchers to gather information and share discoveries with their colleagues in the scientific community.

VALUE OF STANDARDS—PERSONAL COMPUTERS AS A CASE STUDY

The enormous success of the personal computer in recent years provides a vivid example of why open platforms with common and interchangeable components are so powerful and valuable. Personal computers occupy nearly every desk in every research laboratory in the world. Personal computing has been successful because it was made affordable and available to the individual in a set of

Microarray Biochip Technology
Edited by Mark Schena

formats that were powerful as well as fun and easy to use. The microarray industry should share the similar goals of bringing genomic analysis to the benchtop of every biochemist, developmental biologist, immunologist, cell biologist, structural biologist, geneticist, population biologist, and molecular biologist in the world. But how will this be accomplished?

Using desktop computers as a general model for the microarray industry makes sense. The issue of instrument size is one important consideration. The microarray industry should strive to create arraying devices, sample analyzers, hybridization stations, detection systems, and so forth that fit on common laboratory benches. Benchtop microarray instruments, like desktop computers, will stimulate interest in microarray experimentation by intermingling these tools with the scientists that perform the experiments. Except in production or industrial situations, "mainframe"-sized microarray instruments should be avoided if possible; indeed, many advanced computational workstations including the Sun and Silicon Graphics products, fit nicely on the desktop. The efficiency and fun of microarray experimentation will be compromised if the machines that are used to make and read the microarrays are unwieldy.

Another important issue for the microarray industry is the physical format or configuration of the microarrays themselves. Data storage technology in the computer industry provides some instructive guidance. Portable data storage devices such as the 3.5-in (floppy) disc and compact disc (CD) are made by many vendors, and yet these standard data storage devices can be used in virtually every personal computer (PC) and Macintosh (Mac) computer in the world. The size and shape of floppies and CDs is arbitrary, and yet many vendors tailor their products to these arbitrary but important standards. The fact that floppies and CDs are made by many vendors for use in many different brands of computer does not reduce the value of these products, but rather increases it. The microarray industry should follow this example and choose one or a small number of microarray formats that can be used on as many supporting pieces of hardware as possible (see below).

Microarray experiments generate a huge amount of biological data. The field of computation biology (bioinformatics) arose to manage and model biological data generated by microarray and other high-throughput methods. In making software tools for microarrays, bioinformaticists should borrow lessons from software developers in the computer industry. The openness and flexibility of personal computing software such as the Microsoft® and Adobe® products as well as the search engines for the internet provide wonderful examples of the value of unified file formats (e.g., JPEG, GIF, TIFF) and fun and easy-to-use software tools. Microsoft Word software, for example, allows the user to import multiple file formats into any body of text, and Adobe Photoshop® allows interconversion between all the main file types. Many desktop computing programs also allow direct conversion of files of interest into the internet Hypertext Markup Language (HTML). Bioinformaticists, in developing motion control software and data mining tools for the microarray industry, should strive to make software packages that are open, flexible, internet-ready, and compatible with PCs, Macs, and other popular computing platforms.

The fact that the internet was developed in the 1970s but lay dormant for 20 years until Netscape® and Microsoft developed reasonable user interfaces makes a compelling case for the value of reducing barriers to enter a given field. Now that browsing is easy, fun, and affordable, the internet has rapidly become the most powerful communication tool in human history. The power of microarray technology is indisputable, yet the goal of putting chips in every biological research laboratory in the world has yet to be achieved, and the speed at which the transition from filters to chips takes place will depend in large part on the quality of the commercial tools. Every microarray tool must be based on superior science, although tool makers might also include a racing stripe or two, especially considering the success of the not overly powerful but enormously successful iMac.

BIOLOGICAL QUESTIONS AND THE MICROARRAY LIFE CYCLE

As a graduate student at UCSF, one of the authors (M.S.) was provided a lasting memory by Ira Herskowitz during his genetics lectures, in which he would encourage his students to formulate a biological question prior to initiating an experiment in the research laboratory. Posing a question, it was reasoned, helps to focus the experimentalist and assists in interpreting the results. An important question in yeast genetics at that time was: "Which cellular genes are involved in mating type switching in *Saccharomyces cerevisiae*?" With question in hand, Herskowitz reasoned, the experimentalist could then go off and initiate a mutant hunt to identify yeast genes involved in mating for example.

In retrospect, fewer pieces of methodological advice have proved more useful. Formulating a biological question encourages the researcher to construct models, examine experimental approaches, consider pitfalls, and identify "show stoppers" *prior to* embarking on the arduous, time-consuming, and expensive process of laboratory research. Question-oriented research seems subtly different from approaches that do not involve a question, but the distinction is actually quite large. Subjecting yeast cells to DNA-damaging agents for the sake of simply making yeast mutants has little value. A thoughtful, meditative period on the upstream side of bench work generally improves the chances of experimental success and is a practice well worth the time. Are DNA microarrays undermining query-based biology?

To the contrary, microarray analysis not only preserves the practice of query-based biology but has actually made biological questioning more important than ever. For this reason, a recent proposal places the biological question as the first step in the microarray life cycle (66). Because of the enormous power of microarray assays as data-generating devices, a single set of microarray experiments often generates an amount of data that requires several months to analyze (1,2). This places a premium on the soundness of the experimental design. It is important to emphasize that microarray analysis is fundamentally different from traditional assays that employ gels and nylon membranes. In traditional assays, the ratio of analysis time to experimental time is small, and in microarray assays this ratio is enormous. It is therefore suggested that a biological (or technical) question precede the other four basic steps of the life cycle of every microarray experiment (Figure 1). If the experimentalist adheres to this basic methodological principle, the microarray data glut is more easily managed (1,2).

Microarray experimentation should begin with a question, but need not begin with a hypothesis. The distinction seems subtle, but again the difference is actually substantial. Starting with a question means that the focus, scope, and intent of a study has been formulated prior to undertaking an experiment. Starting with a hypothesis means that the study has been undertaken with an a priori understanding of what the outcome is likely to be. Because microarray analysis allows a totally unprecedented look at biological systems (3,4,7,9,10–20,22,25, 28–30,32,35,38–50,52,55,56,58,61,64,65,67–70,73,78,80–82,86,87,89–91), it is often impossible to formulate a hypothesis a priori because so little quantitative information is available about biological systems at the genomic level. Adhering to a query-based paradigm, but not necessarily requiring hypothesis-driven research, provides a careful guide for microarray experimentation without quashing the discovery-oriented aspects of working with chips (6).

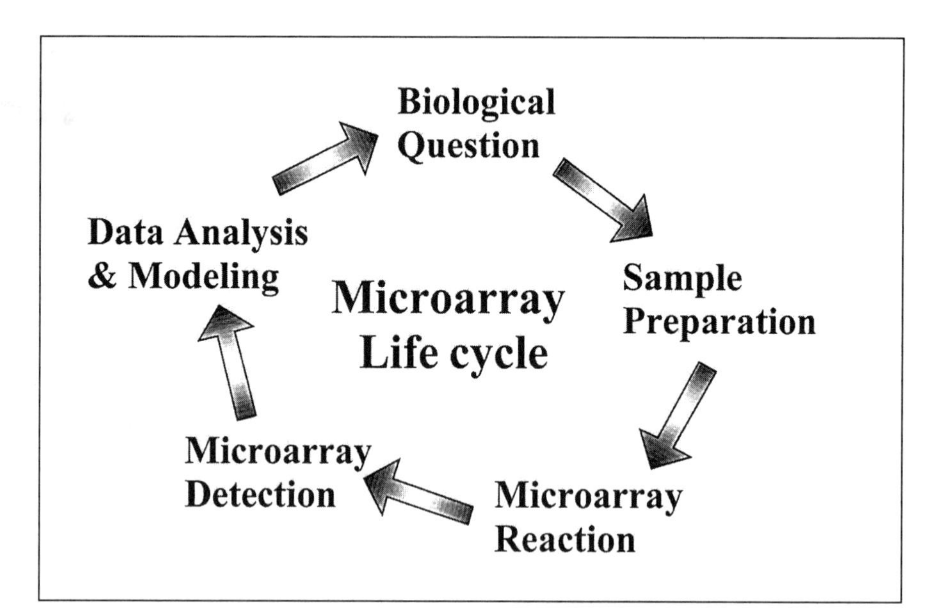

Figure 1. Schematic of the microarray assay life cycle with the five main steps. Every microarray experiment should begin with a biological question or query and proceed stepwise. Successive rounds of experimentation provide an increasingly detailed view of biological function.

SAMPLE PREPARATION, mRNA AMPLIFICATION, AND LABELING

Microarray assays generate a large data flow (1,2,24,88), placing a premium on high-quality nucleic acid samples. Particularly in gene expression experiments, differential degradation of RNA can lead to erroneous conclusions about both relative and absolute mRNA levels in specimens of interest. Simply grinding up a tissue and hoping for the best is not good enough. Poor sample preparation will lead to inferior labeled samples and poor data output. Assembling databases of erroneous results can rapidly contaminate a microarray expression or genotyping database. The microarray field should therefore endeavor to develop methods for sample preparation that provide quantitative results of the highest quality.

A variety of methods that have been used for RNA and DNA isolation from plants (62), yeast (17,29,48,69,73,89), bacteria (19), mouse (50), and human cells (17,42,64) and human tissues (41,81) give satisfactory results with microarray formats. The main consideration with RNA is the rapid and complete inactivation of ribonucleases. With DNA, the most important consideration is probably the efficiency of isolating viscous, high-molecular-weight molecules. Protocols for RNA isolation should take into account a large spectrum of seemingly mundane considerations including incubator temperature, CO_2 concentration, growth media, and buffering components, hormones and other additives, reagent lot numbers, rotor temperatures, and the like. For comparative analyses, it is essential that the biological specimens be processed in an identical fashion. Sample processing artifacts that are invisible at the level of single gene analyses can become glaring when analyzing thousands of genes in a single experiment. Different commercial RNA isolation kits, for example, have been shown to yield twofold or greater differences between kits for as many as 1% of the human transcripts analyzed by microarray (M. Schena and R.W. Davis, unpublished data).

Although microarray experiments involving cultured cells have been extremely informative, there is an increasing interest in analyzing small amounts of nucleic acid from limited quantities of tissue. Some of the applications include diagnostics with blood samples or needle biopsies, analysis of transgenic plants for which the DNA and RNA content is extremely low, examination of gene expression patterns in specific cell types, and detection of bacterial and viral infections in humans and other organisms. The development of laser capture microdissection (LCM) technology (23), coupled with a T7-based RNA amplification scheme (79), allows functional analysis of single human cells (21). Combining LCM, RNA amplification, and microarray analysis (Figure 2) has permitted analysis of as few as 500 rat neural cells (51,59). Further refinements in the technique will probably allow microarray analysis of single cells, opening the door to the assembly of anatomical three-dimensional databases of gene expression profiles from whole animals. Adding a time component would provide four-dimensional data, thereby allowing the action of drugs, hormones, and other stimuli to be monitored quantitatively on a cell-by-cell basis in intact

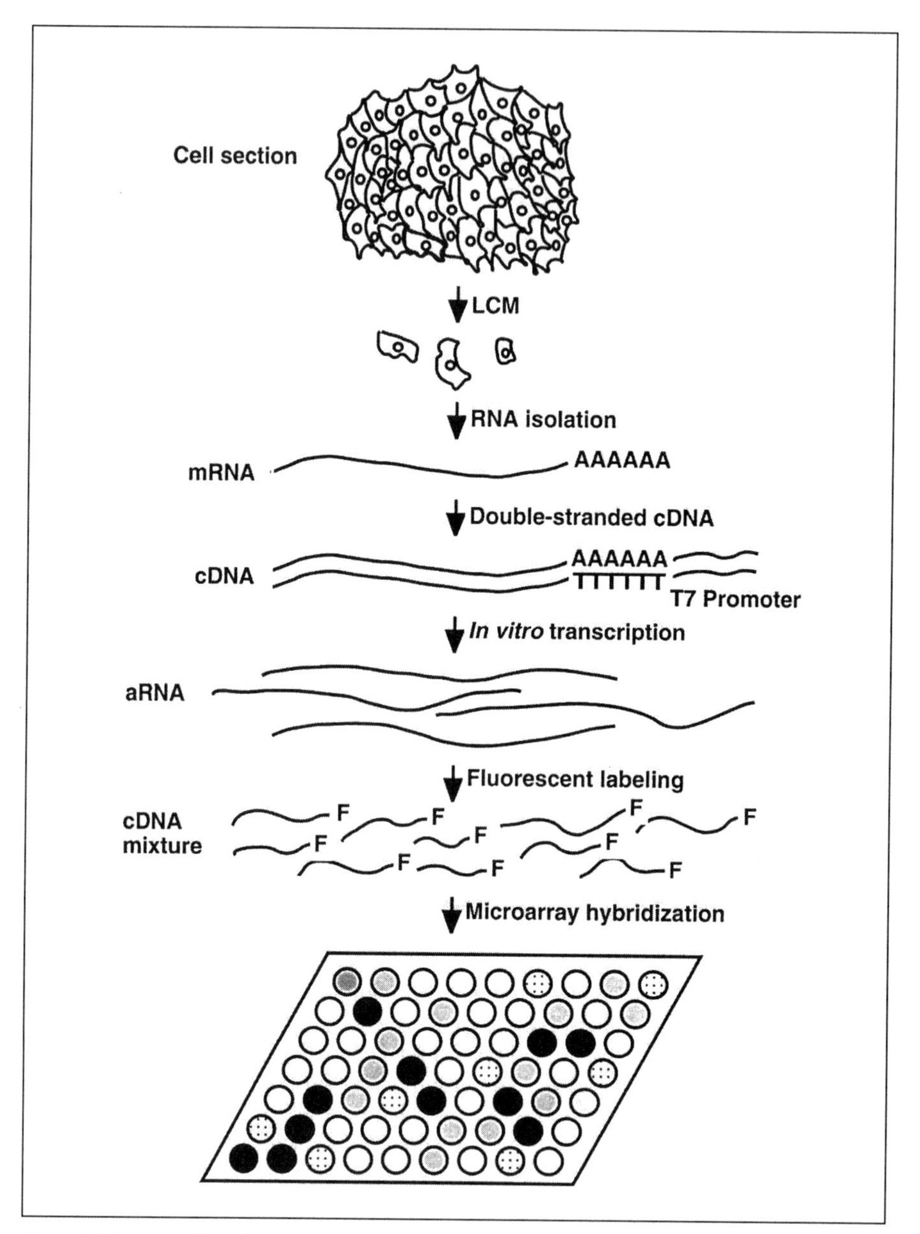

Figure 2. Schematic of how laser capture microdissection (LCM) is used in concert with microarray analysis to analyze gene expression in a small number of cells. LCM is used to capture cells from a cell section, and the isolated RNA is converted to double-stranded complementary DNA (cDNA) using primers that contain T7 promoter sequences. The double-stranded cDNA is transcribed in vitro with T7 polymerase to yield a large amount (approximately 100-fold amplification per round) of amplified RNA (aRNA). The aRNA is then labeled with reverse transcriptase in the presence of fluorescent nucleotides to yield a fluorescent cDNA mixture that is hybridized to a DNA microarray. The location and extent of fluorescence provides a quantitative measure of gene expression in the sample. The use of multicolor schemes allows sensitive differential expression analysis of multiple biological samples.

mammals. Such analyses would provide unprecedented insights into pharmacodynamics, disease progression, and aging.

There are two important advantages of T7 amplification over other methods such as the polymerase chain reaction (PCR). First, because amplification with T7 polymerase is a linear process, skewing of the resultant amplified RNA (aRNA) population relative to the input mRNA is minimized. The T7-based scheme therefore maintains quantitation in the assay, whereas the PCR with two primers biases heavily for small inserts. The second advantage of the T7 scheme is that aRNA is readily labeled with reverse transcriptase, which incorporates fluorescent nucleotide analogs much more readily than the DNA polymerases. These and other advantages of the T7 scheme suggest that the microarray field should probably adopt this strategy as the method of choice when mRNA amplification is required (Table 1).

Methods for labeling nucleic acids for microarray analysis are numerous but fall into two broad categories known as direct and indirect labeling (Table 1). Direct labeling schemes, which were used in many of the first microarray experiments (17,50,60,62,64,69,73), incorporate fluorescent tags directly into the nucleic acid probe mixtures that are hybridized to the chip. Fluorescent tags are incorporated into probes by enzymatic synthesis in the presence of either labeled nucleotides or PCR primers. Direct labeling schemes are simple to implement, provide strong hybridization signals, and make use of fluorescent dye families that share similar chemical structures but absorb and emit light at different wavelengths (Table 2). The latter point is particularly important for comparative analyses because fluorescent moieties that differ widely in structure produce hybridization and detection artifacts that are sequence dependent, thereby producing false positives in differential expression experiments. The basis of this effect is poorly understood but can be obviated by a reciprocal labeling approach in which dyes are "swapped" with respect to the mRNA sources in two separate experiments (64). To maximize the precision of direct labeling schemes involving multiple fluors for comparative analysis, the microarray field should try to use groups of chemically related dyes such as the cyanine and alexa analogs (Table 2).

The indirect labeling schemes, developed after the direct labeling approaches, incorporate epitopes into the nucleic acid probe mixture that is hybridized to the microarray. Following hybridization of the epitope-tagged nucleic acid, the microarray is stained with a protein that binds to the epitopes, thereby providing a fluorescent signal by virtue of the conjugate that is tethered to the staining protein. One common method of indirect labeling involves the use of a biotin epitope and a fluorescent streptavidin-phycoerythrin conjugate (see Chapter 6). A second, more recent, approach (see Chapter 10), known as Tyramide Signal Amplification (TSA™), utilizes biotin and dinitrophenol (DNP) epitopes as well as streptavidin and antibody conjugates linked to horseradish peroxidase (HRP). In the presence of hydrogen peroxide, the HRP enzyme catalyzes the deposition of cyanine-tyramide compounds onto the microarray surface. Because tyramide is available as

Table 1. Microarray Tools and Technologies

Tool/Technology	Platform of Choice	Comments
Assay substrate (two-dimensional)	Glass	Low intrinsic fluorescence, affordable
Research platform	Microscope slides (1 × 25 × 76 mm)	Optically flat substrates give superior results
Production platform	Affymetrix cassettes (12.7 × 12.7 mm)	Enclosed chip enables high-quality control
Direct nucleic acid labeling technique	Enzymatic incorporation of fluorescent nucleotides	Easy and rapid protocols, Cy and Alexa dyes
Indirect nucleic acid labeling technique	Biotin and dinitrophenol (DNP)	Allows amplification
RNA amplification	Eberwine T7 polymerase method	Linear amplification minimizes skewing of mRNA population
Fluorescent signal amplification	Tyramide signal amplification (TSA)	Provides 10–100-fold amplification
Fluorescent detection	Scanning or charge-coupled device imaging	10 pixels/feature diameter preferred
Personal computing operating system	Windows NT	Fast, reliable and will run on any PC

both Cy3 and Cy5 derivatives, the TSA approach can be used for comparative analysis of two samples on a single microarray using Cy3/Cy5 fluorescence detection. The main advantage of the indirect labeling methods is that they provide 10- to 100-fold signal amplification relative to the direct labeling approaches (Table 1). The main disadvantages are that they are somewhat more difficult to implement and generally somewhat less precise for comparative analysis because the epitopes have different labeling efficiencies and protein binding affinities.

A simple uniformity test for both the direct and indirect multicolor labeling schemes is a concordance comparison in which a single source of nucleic acid is labeled separately with two fluors or epitopes and then hybridized to a single microarray. Fluorescent signals for all the elements on the microarray are then determined at both emission wavelengths and plotted as ratios as a function of signal intensity. Ideally, a ratio of 1.0 should be obtained for each microarray element such that the data cluster tightly along the "sameness" line. Deviations from 1.0 in either the upward or downward direction suggest discordance or imbalance between the two channels that may be due to differences in incorporation or staining of the fluors or epitopes, or to inaccuracies in detection and quantitation. For Cy3 and Cy5 epitopes, for example, the concordance of a

Table 2. Fluorescent Dyes for Microarray Assays*

Fluorophore	Absorbance (nm)	Emission (nm)	Structural Partner	Comments
FITC	494	518		5-FAM derivative use for DNA sequencing
Fluor X	494	520		Less bright than FITC
Alexa 488	495	519	Alexa 532, 546, 568, and 594	
Oregon Green 488	496	524		
JOE	522	550		6-JOE used for DNA sequencing
Alexa 532	531	554	Alexa 488, 546, 568, and 594	
Cy3	550	570	Cy2, -3.5, -5, and -5.5	
Alexa 546	556	573	Alexa 488, 532, 568, and 594	
TMR	555	580		6-TAMRA used for DNA sequencing
Alexa 568	578	603	Alexa 488, 532, 546, and 594	
ROX	580	605		6-ROX used for DNA sequencing
Alexa 594	590	617	Alexa 488, 532, 546, and 568	
Texas Red	595	615		
Bodipy 630/650	625	640	Bodipy series	
Cy5	649	670	Cy2, -3, -3.5, and -5.5	Less soluble in aqueous than Cy3

*Absorbance and emission values are approximate and are given in nanometers (nm).

FITC, fluorescein isothiocyanate; TMR, tetramethylrhodamine; ROX, X-rhodamine. Additional electronic information available at http://www.probes.com.

single source of mRNA is generally <1.8 for >99% of the gene sequences present on the microarray (see Chapter 7). A concordance test provides a simple means of evaluating existing and new labeling methods, and the field should probably employ this test whenever possible.

MICROARRAYS AND MICROARRAY SURFACES

Derivatized 1 × 3-in microscope slides (1 × 25 × 76 mm) were used for the first microarray experiments (62); this is the microarray format most commonly used in laboratories at present. Advantages of this convenient platform include the following: compatibility with many different fluorescent detection and robotic fabrication systems; large surface area (approximately 19 cm^2), which allows arrays of >100 000 features to be made using microspotting and ink-jetting technologies (Figure 3); compatibility with many existing microscopes and microscopy tools; ease of surface modification, hybridization, and washing; coverslips allow thin film hybridization, providing low hybridization volumes (5–20 µL), high probe concentration, and rapid kinetics; low intrinsic fluorescence; and low cost. The main disadvantage of the microscope slide format is that the open microarray surface is susceptible to contamination with dust and other airborne particles, and mishandling can lead to scratches and other forms of surface damage, although contamination at the manufacturing stage can be entirely eliminated by use of a cleanroom (see Chapter 9). Another disadvantage is that the quality of the glass used for microscope slides is rather low, although recent developments with ultraflat solid substrates provide a high-quality "microscope slide" at an affordable price (see Chapter 9).

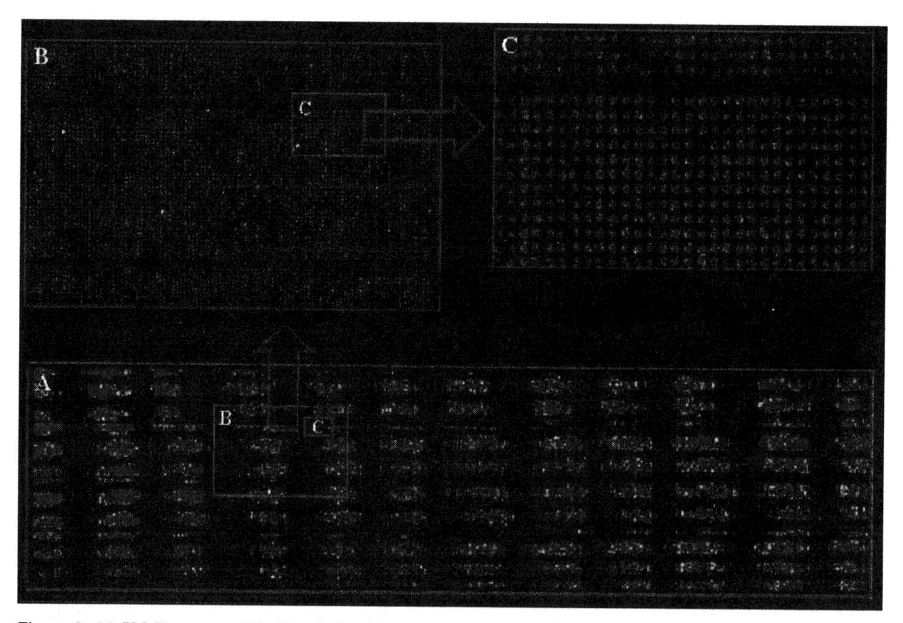

Figure 3. 82-K Microarray. TeleChem's Stealth printing technology was used to deposit 0.4-nL volumes (approximately 100-µm diameter spots) of Cy3-labeled oligonucleotide at 120 µm spacing on an aldehyde-derivatized glass substrate (CEL). The subarray configuration of 36 × 36 × 72 gives a total of 82 944 features. The areas represented by regions A, B, and C correspond to 12.96, 0.25, and 0.07 cm^2, respectively. **(See color plate A1.)**

The proprietary Affymetrix cassette is another widely used microarray format (9,11,19,36–38,50,80,89). The Affymetrix cassette is a plastic enclosure that seals the glass chip from the environment, protects the chip from user damage, provides a fixed reagent volume to ensure consistency in reactions, has greater ease of handling, and allows automated loading. Some disadvantages of this platform are increased cost and lack of compatibility with all but the Hewlett Packard scanning device. At the moment, the Affymetrix cassette is the microarray of choice for high-density production data mining with oligonucleotide arrays.

For technical reasons, Protogene has developed a 4 × 4-in microarray format that is currently under testing. This square format is symmetrical and allows the chips to be spun during manufacturing, thereby allowing reagents to be washed off the surface by centrifugal force during each cycle of phosphoramidite coupling. This platform greatly expedites manufacturing and has a large surface area, although like the Affymetrix cassette, requires a custom detection platform. Product literature suggests that the Virtek ChipReader™ scanning device will accommodate the Protogene substrate.

There need not be one size and shape of microarray, but it is prudent to limit the formats to a small number, especially when organizations are pondering the development of new products. The "microscope slide," the Affymetrix cassette, and the square Protogene substrates are the standards at the moment, and the field should probably embrace these formats to bolster unity in the field. Substrate dimensions and configurations are inherently arbitrary, but similar to computer data storage devices (e.g., floppy, CD, and Zip), the microarray field should try to focus on as small a number of different formats as possible and design hardware and software around these devices.

FLUORESCENCE DETECTION DEVICES

Detection is the fourth step in the microarray life cycle (Figure 1), and this critical step is being addressed by an explosion in fluorescence detection technology. High-quality instruments from GSI Lumonics (Watertown, MA, USA), Amersham Pharmacia Biotech/Molecular Dynamics (Sunnyvale, CA, USA), Axon (Foster City, CA, USA), HP (Palo Alto, CA, USA), Virtek (Woburn, MA, USA), and others are making fluorescence detection powerful, affordable, and user friendly. Most of the commercial systems use some form of scanning technology with photomultiplier tube (PMT) detection (see Chapters 3 and 4). There is a trend toward detection systems with three-, four-, and five-color capabilities utilizing either gas or solid state lasers and a multitude of different filter sets to provide spectral separation into each of the different channels. Multiple detection channels allow analysis of multiple samples in a single experiment, as well as additional channels for hybridization and surface DNA controls. Some general guidelines for making detection devices are given below.

Pixel Size

The spatial resolution of detection instruments is typically given as the pixel size, the physical "bin" in which a single datum is acquired. A reasonable guideline is that the pixel size should be approximately 1/10 the diameter of each spot or feature on the chip. Microarrays containing 100-μm features require fluorescent detectors with approximately 10-μm pixel size, 50-μm features require approximately 5-μm pixels, and so forth. Most charge-coupled device (CCD)-based detection instruments are equipped with 1-megapixel cameras that contain 1000 × 1000 chips for a pixel size of approximately 20 μm when imaging a 20 × 20-mm area. At the moment, fluorescent scanners are probably best suited for imaging fairly large arrays, and CCD-based systems are most compatible with arrays $\leq$1.0 cm^2.

Photobleaching

Microarray dyes emit fluorescent photons in the presence of excitation light. If the light intensity is excessive, however, the incoming photons can damage the dyes and reduce the fluorescent readout of the microarray during successive scans. Photobleaching is particularly problematic if users excite a portion of the array and then interrupt the imaging to adjust laser or PMT settings before scanning the remaining array area. Photobleaching can greatly degrade array data when portions of arrays are selectively exposed to excitation light. Most commercial scanners use 5- to 35-mW light sources with scanning speeds of 5 to 25 lines per second. More powerful light sources or longer pixel dwell times can lead to considerable photobleaching. A guideline is that photobleaching should be <1% per imaging cycle.

It should be noted that dyes differ considerably in their photostabilities, with fluorescein isothiocyanate (FITC) being more susceptible to photobleaching than the Cy dyes. Differential photostability can be particularly problematic in experiments involving multiple fluors and ratio measurements in that photobleaching in one channel (but not the other) can lead to errors in quantitation. Photobleaching should be minimized as much as possible.

Cross-Talk

Cross-talk refers to emitted light from one channel being detected in another channel. This optical "contamination" primarily occurs because emission spectra are typically rather broad, with as much as 10% of the emitted signal observable at wavelengths 100 nm longer than the peak emission wavelength. In dual-labeling and detection experiments, fluorescence intensity from the Cy3 channel, for example, can contaminate the Cy5 channel if care is not taken to minimize cross-talk. As the Stokes shift dictates, the emission wavelength is always longer than the absorbance wavelength, and therefore Cy3 cross-talk into the Cy5 channel is a practical problem but not vice versa. There are many ways to minimize cross-talk, although the most common and least expensive involves the use of emission filters that reject light outside the desired wavelengths. Optical cross-talk should be kept to <0.1%/channel, with lesser amounts desired for high-sensitivity applications such as gene expression analysis.

DATA FILES AND MINING

Microarray experiments generate an enormous data stream, placing bioinformatics at the core of the microarray industry. Attention must be paid to data output and modeling. The most common graphical output format from fluorescence imaging instruments is the 16-bit tiff (.tif) file. The 16-bit format allows intensity readings from 0 to 65 536 (2^{16}) to be recorded, thereby permitting a theoretical dynamic range in excess of four orders of magnitude. Instrument makers should probably embrace the 16-bit tiff file format as the format of choice for the present time, with a 16-bit bitmap (.bmp) format as a common alternative.

Output graphical files in 16-bit format are, in essence, two-dimensional number sets with values ranging from 0 to 65 536. The intensity values corresponding to the regular pattern of each microarray element are generally higher than the background fluorescence values corresponding to the area between the spots. Extracting the data is basically a task of summing the pixel values of each spot and dividing by the total quantified area. Tiff files allow a single microarray image to be mined with multiple quantitation packages. A 16-bit output also allows data sets from different instruments or laboratories to be compared with reasonable precision. Output files of 12 and 8 bits reduce the storage space required for the images, but these formats only allow 4096- and 256-fold theoretical dynamic ranges, which is insufficient for gene expression applications.

Data quantitation packages such as QuantArray™ from GSI Lumonics, ImaGene™ from BioDiscovery (Los Angeles, CA, USA), the software on the Axon GenePix™ system, and other commercial tools perform this function well. Typically, a user-defined gridding pattern is overlaid on the image, and the areas defined by the regular pattern of circles or squares are subjected to data extraction. The computational theory underlying data quantitation is presented in detail in Chapter 8.

Once values are quantified from the microarray, the data are commonly displayed in a variety of ways. One common depiction is known as a scatter plot, in which all the data points from a two-color experiment are plotted as a function of ratio and signal intensity, with those ratios greater than 1.0 plotted above the diagonal and those with ratios less than 1.0 plotted below the diagonal. The scatter plot provides a means of viewing very large datasets such as those involving expression profiles of thousands of genes. A mouse click on a particular data point usually provides information about the gene sequence present at that position on the microarray. Scatter plots are typical capabilities of most commercial quantitation packages and should be included as a feature for products that are in the planning stages.

Once the data are quantified and genes of interest are identified, a common form of data mining allows genes to be grouped or "clustered" according to expression profile. The biological basis for clustering derives from the fact that gene expression correlates well with biological function (11,12,17,18,22,41, 42,48, 59,62–64,89). Clustering tools help identify functionally related genes such as those in common signaling pathways and those that share common biochemical activity. Stingray™ from the Molecular Applications Group (Palo Alto, CA, USA) and GeneSpring™ 2.2 from Silicon Genetics (San Carlos, CA, USA) are among the commercial clustering tools that work well for this purpose.

CONCLUSIONS

Microarray technology allows an unprecedented view of biological systems by allowing genomes to be examined on glass chips. What used to require several weeks per gene with traditional assays, using nylon membranes and radioactivity, can now be performed in less than a day for 10 000 genes. The pillars of the microarray industry are similar to those that have served as the foundation for the computer chip industry: parallelism, miniaturization, and automation. The Human Genome Project and other large sequencing programs are assembling massive sequence databases that provide a wealth of sequence information for microarray manufacture and analysis. It is noteworthy that PE Celera (Rockville, MD, USA) has decided to release primary genomic sequence data into the public domain, with the idea that the greatest value of the fly, human, and other genomes will be realized by functional genomics.

The rapid and continued success of microarray technology development (5, 8,26,27,31,33,34,53,54,57,71,72,74–77,83–85) rests in part on the extent to which microarray tools are affordable, fun, and easy to use. The microarray technology standards put forth in this chapter will hopefully serve as a source of ideas and help to catalyze discussion about the importance of having guidelines and standards for this explosive new field. Open platforms based on solid microarray fundamentals will allow users to navigate seamlessly between commercial tools and share their data efficiently with their colleagues. The commercial tools will also provide a way for basic research laboratories to explore their favorite biological question without the burden of expensive and time-consuming technology development. The power of microarray technology renders organisms such as bacteria, fungi, marine species, flowering plants, and others not currently on the list of "model organisms" immediately tractable. Microarray technology provides a wonderful meeting place for the practitioners of biology, chemistry, physics, mathematics, and computer science. Strong alliances with social scientists and clergy will ensure that the genomic data are utilized according to the highest ethical standards.

REFERENCES

1. **Bassett, D.E. Jr., M.B. Eisen, and M.S. Boguski.** 1999. Gene expression informatics—it's all in your mine. Nat. Genet. *21(1 Suppl)*:51-55.
2. **Bellenson, J.L.** 1999. Expression data and the bioinformatics challenges, p. 139-164. *In* M. Schena (Ed.), DNA Microarrays: A Practical Approach. Oxford University Press, Oxford.
3. **Bowtell, D.D.** 1999. Options available—from start to finish—for obtaining expression data by microarray. Nat. Genet. *21(1 Suppl)*:25-32.
4. **Braxton, S. and T. Bedilion.** 1998. The integration of microarray information in the drug development process. Curr. Opin. Biotechnol. *9*:643-649.
5. **Brignac, S.J., Jr., R. Gangadharan, M. McMahon, J. Denman, R. Gonzales, L.G. Mendoza, and M. Eggers.** 1999. A proximal CCD imaging system for high-throughput detection of microarray-based assays. IEEE Eng. Med. Biol. *18*:120-122.
6. **Brown, P.O. and D. Botstein.** 1999. Exploring the new world of the genome with DNA microarrays. Nat. Genet. *21(1 Suppl)*:33-37.
7. **Bubendorf, L., J. Kononen, P. Koivisto, P. Schraml, H. Moch, T.C. Gasser, N. Willi, M.J. Mihatsch, G. Sauter, and O.P. Kallioniemi.** 1999. Survey of gene amplifications during prostate cancer progression by high-throughout fluorescence in situ hybridization on tissue microarrays. Cancer Res. *59*:803-806.
8. **Case-Green, S.C., K.U. Mir, C.E. Pritchard, and E.M. Southern.** 1998. Analysing genetic information with DNA arrays. Curr. Opin. Chem. Biol. *2*:404-410.
9. **Chee, M., R. Yang, E. Hubbell, A. Berno, X.C. Huang, D. Stern, J. Winkler, D.J. Lockhart, M.S. Morris, and S.P. Fodor.** 1996. Accessing genetic information with high-density DNA arrays. Science *274*:610-614.
10. **Cheung, V.G., M. Morley, F. Aguilar, A. Massimi, R. Kucherlapati, and G. Childs.** 1999. Making and reading microarrays. Nat. Genet. *21(1 Suppl)*:15-19.
11. **Cho, R.J., M..J. Campbell, E.A. Winzeler, L. Steinmetz, A. Conway, L. Wodicka, T.G. Wolfsberg, A.E. Gabrielian, D. Landsman, D.J. Lockhart, and R.W. Davis.** 1998. A genome-wide transcriptional analysis of the mitotic cell cycle. Mol. Cell *2*:65-73.
12. **Cho, R.J., M. Fromont-Racine, L. Wodicka, B. Feierbach, T. Stearns, P. Legrain, D.J. Lockhart, and R.W. Davis.** 1998. Parallel analysis of genetic selections using whole genome oligonucleotide arrays. Proc. Natl. Acad. Sci. USA *95*:3752-3757.
13. **Cole, K.A., D.B. Krizman, and M.R. Emmert-Buck.** 1999. The genetics of cancer—a 3D model. Nat. Genet. *21(1 Suppl)*:38-41.
14. **Cronin, M.T., R.V. Fucini, S.M. Kim, R.S. Masino, R.M. Wespi, and C.G. Miyada.** 1996. Cystic fibrosis mutation detection by hybridization to light-generated DNA probe arrays. Hum. Mutat. *7*:244-255.
15. **Debouck, C. and P.N. Goodfellow.** 1999. DNA microarrays in drug discovery and development. Nat. Genet. *21(1 Suppl)*:48-50.
16. **Der, S.D., A. Zhou, B.R. Williams, and R.H. Silverman.** 1998. Identification of genes differentially regulated by interferon alpha, beta, or gamma using oligonucleotide arrays. Proc. Natl. Acad. Sci. USA *95*:15623-15628.
17. **DeRisi, J, L. Penland, P.O. Brown, M.L. Bittner, P.S. Meltzer, M. Ray, Y. Chen, Y.A. Su, and J.M. Trent.** 1996. Use of a cDNA microarray to analyze gene expression patterns in human cancer. Nat. Genet. *14*:457-460.
18. **DeRisi, J.L., V.R. Iyer, and P.O. Brown.** 1997. Exploring the metabolic and genetic control of gene expression on a genomic scale. Science *278*:680-686.
19. **de Saizieu, A., U. Certa, J. Warrington, C. Gray, W. Keck, and J. Mous.** 1998. Bacterial transcript imaging by hybridization of total RNA to oligonucleotide arrays. Nat. Biotechnol. *16*:45-48.
20. **Duggan, D.J., M. Bittner, Y. Chen, P. Meltzer, and J.M. Trent.** 1999. Expression profiling using cDNA microarrays. Nat. Genet. *21(1 Suppl)*:10-14.
21. **Eberwine, J., H. Yeh, K. Miyashiro, Y. Cao, S. Nair, R. Finnell, M. Zettel, and P. Coleman.** 1992. Analysis of gene expression in single live neurons. Proc. Natl. Acad. Sci. USA *89*:3010-3014.
22. **Eisen, M.B., P.T. Spellman, P.O. Brown, and D. Botstein.** 1998. Cluster analysis and display of genome-wide expression patterns. Proc. Natl. Acad. Sci. USA *95*:14863-14868.
23. **Emmert-Buck, M.R., R.F. Bonner, P.D. Smith, R.F. Chuaqui, Z. Zhuang, S.R. Goldstein, R.A.**

Weiss, and L.A. Liotta. 1996. Laser capture microdissection. Science *274*:998-1001.
24.**Ermolaeva, O., M. Rastogi, K.D. Pruitt, G.D. Schuler, M.L. Bittner, Y. Chen, R. Simon, P. Meltzer, J.M. Trent, and M.S. Boguski.** 1998. Data management and analysis for gene expression arrays. Nat. Genet. *20*:19-23.
25.**Falus, A., A. Váradi and I. Raskó.** 1998. The DNA-chip, a new tool for medical genetics. Orvosi Hetilap *139*:957-960.
26.**Ferguson, J.A., T.C. Boles, C.P. Adams, and D.R. Walt.** 1996. A fiber-optic DNA biosensor microarray for the analysis of gene expression. Nat. Biotechnol. *14*:1681-1684.
27.**Gentalen, E. and M. Chee.** 1999. A novel method for determining linkage between DNA sequences: hybridization to paired probe arrays. Nucleic Acids Res. *27*:1485-1491.
28.**Geschwind, D.H., J. Gregg, K. Boone, J. Karrim, A. Pawlikowska-Haddal, E. Rao, J. Ellison, A. Ciccodicola, M. D'Urso, R. Woods, G.A. Rappold, R. Swerdloff, and S.F. Nelson.** 1998. Klinefelter's syndrome as a model of anomalous cerebral laterality: testing gene dosage in the X chromosome pseudoautosomal region using a DNA microarray. Dev. Genet. *23*:215-229.
29.**Giaever, G., D.D. Shoemaker, T.W. Jones, H. Liang, E.A. Winzeler, A. Astromoff, and R.W. Davis.** 1999. Genomic profiling of drug sensitivities via induced haploinsufficiency. Nat. Genet. *21*:278-283.
30.**Gingeras, T.R., G. Ghandour, E. Wang, A. Berno, P.M. Small, F. Drobniewski, D. Alland, E. Desmond, M. Holodniy, and J. Drenkow.** 1998. Simultaneous genotyping and species identification using hybridization pattern recognition analysis of generic Mycobacterium DNA arrays. Genome Res. *8*:435-448.
31.**Ginot, F.** 1997. Oligonucleotide micro-arrays for identification of unknown mutations: how far from reality? Hum. Mutat. *10*:1-10.
32.**Graves, D.J.** 1999. Powerful tools for genetic analysis come of age. Trends Biotechnol. *17*:127-134.
33.**Graves, D.J., H.J. Su, S.E. McKenzie, S. Surrey, and P. Fortina.** 1998. System for preparing microhybridization arrays on glass slides. Anal. Chem. *70*:5085-5092.
34.**Gunderson, K.L., X.C. Huang, M.S. Morris, R.J. Lipshutz, D.J. Lockhart, and M.S. Chee.** 1998. Mutation detection by ligation to complete n-mer DNA arrays. Genome Res. *8*:1142-1153.
35.**Hacia, J.G., L.C. Brody, M.S. Chee, S.P. Fodor, and F.S. Collins.** 1996. Detection of heterozygous mutations in BRCA1 using high density oligonucleotide arrays and two-colour fluorescence analysis. Nat. Genet. *14*:441-447.
36.**Hacia, J.G., S.A. Woski, J. Fidanza, K. Edgemon, N. Hunt, G. McGall, S.P. Fodor, and F.S. Collins.** 1998. Enhanced high density oligonucleotide array-based sequence analysis using modified nucleoside triphosphates. Nucleic Acids Res. *26*:4975-82.
37.**Hacia, J.G., K. Edgemon, B. Sun, D. Stern, S.P. Fodor, and F.S. Collins.** 1998. Two color hybridization analysis using high density oligonucleotide arrays and energy transfer dyes. Nucleic Acids Res. *26*:3865-3866.
38.**Hacia, J.G., B. Sun, N. Hunt, K. Edgemon, D. Mosbrook, C. Robbins, S.P.A. Fodor, D. Tagle, and F.S. Collins.** 1998. Strategies for mutational analysis of the large multiexon ATM gene using high density oligonucleotide arrays. Genome Res. *8*:1245-1258.
39.**Hacia, J.G., W. Makalowski, K. Edgemon, M.R. Erdos, C.M. Robbins, S.P. Fodor, L.C. Brody, and F.S. Collins.** 1998. Evolutionary sequence comparisons using high-density oligonucleotide arrays. Nat. Genet. *18*:155-158.
40.**Hacia, J.G.** 1999. Resequencing and mutational analysis using oligonucleotide microarrays. Nat. Genet. *21(1 Suppl)*:42-47.
41.**Heller, R.A., M. Schena, A. Chai, D. Shalon, T. Bedilion, J. Gilmore, D.E. Woolley, and R.W. Davis.** 1997. Discovery and analysis of inflammatory disease-related genes using cDNA microarrays. Proc. Natl. Acad. Sci. USA *94*:2150-2155.
42.**Iyer, V.R., M.B. Eisen, D.T. Ross, G. Schuler, T. Moore, J.C.F. Lee, J.M. Trent, L.M. Staudt, J. Hudson, Jr., M.S. Boguski, D. Lashkari, D. Shalon, D. Botstein, and P.O. Brown.** 1999. The transcriptional program in the response of human fibroblasts to serum. Science *283*:83-87.
43.**Jelinsky, S.A. and L.D. Samson.** 1999. Global response of Saccharomyces cerevisiae to an alkylating agent. Proc. Natl. Acad. Sci. USA *96*:1486-1491.
44.**Khan, J., R. Simon, M. Bittner, Y. Chen, S.B. Leighton, T. Pohida, P.D. Smith, Y. Jiang, G.C. Gooden, J.M. Trent, and P.S. Meltzer.** 1998. Gene expression profiling of alveolar rhabdomyosarcoma with cDNA microarrays. Cancer Res. *58*:5009-5013.
45.**Khan, J., M.L. Bittner, Y. Chen, P.S. Meltzer, and J.M. Trent.** 1999. DNA microarray technology:

the anticipated impact on the study of human disease. Biochim. Biophys. Acta *1423*:M17-28.
46.**Kononen, J., L. Bubendorf, A. Kallioniemi, M. Bärlund, P. Schraml, S. Leighton, J. Torhorst, M.J. Mihatsch, G. Sauter, and O.P. Kallioniemi.** 1998. Tissue microarrays for high-throughput molecular profiling of tumor specimens. Nat. Med. *4*:844-847.
47.**Kozal, M.J., N. Shah, N. Shen, R. Yang, R. Fucini, T.C. Merigan, D.D. Richman, D. Morris, E. Hubbell, M. Chee, and T.R. Gingeras.** 1996. Extensive polymorphisms observed in HIV-1 clade B protease gene using high-density oligonucleotide arrays. Nat. Med. *2*:753-759.
48.**Lashkari, D.A., J.L. DeRisi, J.H. McCusker, A.F. Namath, C. Gentile, S.Y. Hwang, P.O. Brown, and R.W. Davis.** 1997. Yeast microarrays for genome wide parallel genetic and gene expression analysis. Proc. Natl. Acad. Sci. USA *94*:13057-13062.
49.**Lipshutz, R.J., S.P. Fodor, T.R. Gingeras, and D.J. Lockhart.** 1999. High density synthetic oligonucleotide arrays. Nat. Genet. *21(1 Suppl)*:20-24.
50.**Lockhart, D.J., H. Dong, M.C. Byrne, M.T. Follettie, M.V. Gallo, M. Chee, M. Mittmann, C. Wang, M. Kobayashi, H. Horton, and E.L. Brown.** 1996. Expression monitoring by hybridization to high-density oligonucleotide arrays. Nat. Biotechnol. *14*:1675-1680.
51.**Luo, L., R.C. Salunga, H. Guo, A. Bittner, K.C. Joy, J.E. Galindo, H. Xiao, K.E. Rogers, J.S. Wan, M.R. Jackson, and M.G. Erlander.** 1999. Gene expression profiles of laser-captured adjacent neuronal subtypes. Nat. Med. *5*:117-122.
52.**Marra, M., L. Hillier, T. Kucaba, M. Allen, R. Barstead, C. Beck, A. Blistain, M. Bonaldo, Y. Bowers, L. Bowles, et al.** 1999. An encyclopedia of mouse genes. Nat. Genet. *21*:191-194.
53.**McGall, G., J. Labadie, P. Brock, G. Wallraff, T. Nguyen, and W. Hinsberg.** 1996. Light-directed synthesis of high-density oligonucleotide arrays using semiconductor photoresists. Proc. Natl. Acad. Sci. USA *2693*:13555-13560.
54.**Milner, N., K.U. Mir, and E.M. Southern.** 1997. Selecting effective antisense reagents on combinatorial oligonucleotide arrays. Nat. Biotechnol. *15*:537-541.
55.**Moch, H., P. Schraml, L. Bubendorf, M. Mirlacher, J. Kononen, T. Gasser, M.J. Mihatsch, O.P. Kallioniemi, and G. Sauter.** 1999. High-throughput tissue microarray analysis to evaluate genes uncovered by cDNA microarray screening in renal cell carcinoma. Am. J. Pathol. *154*:981-986.
56.**Nuwaysir, E.F., M. Bittner, J. Trent, J.C. Barrett, and C.A. Afshari.** 1999. Microarrays and toxicology: the advent of toxicogenomics. Mol. Carcinog. *24*:153-159.
57.**Pastinen, T., A. Kurg, A. Metspalu, L. Peltonen, and A.C. Syvänen.** 1997. Minisequencing: a specific tool for DNA analysis and diagnostics on oligonucleotide arrays. Genome Res. *7*:606-614.
58.**Ramsay, G.** 1998. DNA chips: state-of-the art. Nat. Biotechnol. *16*:40-44.
59.**Salunga, R.C., H. Guo, L. Luo, A. Bittner, R.C. Joy, J.R. Chambers, J.S. Wan, M.R. Jackson, and M.G. Erlander.** 1999. Gene expression analysis via cDNA microarrays of laser capture microdissected cells from fixed tissue, p. 121-136. *In* M. Schena (Ed.), DNA Microarrays: A Practical Approach. Oxford University Press, Oxford.
60.**Sapolsky, R.J. and R.J. Lipshutz.** 1996. Mapping genomic library clones using oligonucleotide arrays. Genomics *33*:445-456.
61.**Sapolsky, R.J., L. Hsie, A. Berno, G. Ghandour, M. Mittmann, and J.B. Fan.** 1999. High-throughput polymorphism screening and genotyping with high-density oligonucleotide arrays. Genet. Anal. *14*:187-192.
62.**Schena, M., D. Shalon, R.W. Davis, and P.O. Brown.** 1995. Quantitative monitoring of gene expression patterns with a complementary DNA microarray. Science *270*:467-470.
63.**Schena, M.** 1996. Genome analysis with gene expression microarrays. Bioessays *18*:427-431.
64.**Schena, M., D. Shalon, R. Heller, A. Chai, P.O. Brown, and R.W. Davis.** 1996. Parallel human genome analysis: microarray-based expression monitoring of 1000 genes. Proc. Natl. Acad. Sci. USA *93*:10614-10619.
65.**Schena, M., R.A. Heller, T.P. Theriault, K. Konrad, E. Lachenmeier, and R.W. Davis.** 1998. Microarrays: biotechnology's discovery platform for functional genomics. Trends Biotechnol. *16*:301-306.
66.**Schena, M. and R.W. Davis.** 1999. Parallel Analysis with Biological Chips: PCR Applications. Academic Press, San Diego.
67.**Schena, M. and R.W. Davis.** 1999. Genes, genomes and chips, p. 1-16. *In* M. Schena (Ed.), DNA Microarrays: A Practical Approach. Oxford University Press, Oxford.
68.**Service, R.F.** 1998. Microchip arrays put DNA on the spot. Science *282*:396-399.

69.**Shalon, D., S.J. Smith, and P.O. Brown.** 1996. A DNA micro-array system for analyzing complex DNA samples using two-color fluorescent probe hybridization. Genome Res. *6*:639-645.
70.**Shalon, D.** 1998. Gene expression micro-arrays: a new tool for genomic research. Pathol. Biol. *46*:107-119.
71.**Shchepinov, M.S., I.A. Udalova, A.J. Bridgman, and E.M. Southern.** 1997. Oligonucleotide dendrimers: synthesis and use as polylabelled DNA probes. Nucleic Acids Res. *25*:4447-4454.
72.**Shchepinov, M.S., S.C. Case-Green, and E.M. Southern.** 1997. Steric factors influencing hybridisation of nucleic acids to oligonucleotide arrays. Nucleic Acids Res. *25*:1155-1161.
73.**Shoemaker, D.D., D.A. Lashkari, D. Morris, M. Mittmann, and R.W. Davis.** 1996. Quantitative phenotypic analysis of yeast deletion mutants using a highly parallel molecular bar-coding strategy. Nat. Genet. *14*:450-456.
74.**Silzel, J.W., B. Cercek, C. Dodson, T. Tsay, and R.J. Obremski.** 1998. Mass-sensing, multianalyte microarray immunoassay with imaging detection. Clin. Chem. *44*:2036-2043.
75.**Southern, E.M.** 1996. DNA chips: analysing sequence by hybridization to oligonucleotides on a large scale. Trends Genet. *12*:110-115.
76.**Southern, E.M., N. Milner, and K.U. Mir.** 1997. Discovering antisense reagents by hybridization of RNA to oligonucleotide arrays. Ciba Found. Symp. *209*:38-44.
77.**Southern, E., K. Mir, and M. Shchepinov.** 1999. Molecular interactions on microarrays. Nat. Genet. *21(1 Suppl)*:5-9.
78.**Spellman, P.T., G. Sherlock, M.Q. Zhang, V.R. Iyer, K. Anders, M.B. Eisen, P.O. Brown, D. Botstein, and B. Futcher.** 1998. Comprehensive identification of cell cycle-regulated genes of the yeast Saccharomyces cerevisiae by microarray hybridization. Mol. Biol. Cell *9*:3273-3297.
79.**Van Gelder, R.N., M.E. von Zastrow, A. Yool, W.C. Dement, J.D. Barchas, and J.H. Eberwine.** 1990. Amplified RNA synthesized from limited quantities of heterogeneous cDNA. Proc. Natl. Acad. Sci. USA *87*:1663-1667.
80.**Wang, D.G., J.B. Fan, C. Siao, A. Berno, P. Young, R. Sapolsky, G. Ghanour, N. Perkins, E. Winchester, J. Spencer, L. Kruglyak, et al.** 1998. Large-scale identification, mapping, and genotyping of single-nucleotide polymorphisms in the human genome. Science *280*:1077-1082.
81.**Wang, K., L. Gan, E. Jeffery, M. Gayle, A.M. Gown, M. Skelly, P.S. Nelson, W.V. Ng, M. Schummer, L. Hood, and J. Mulligan.** 1999. Monitoring gene expression profile changes in ovarian carcinomas using cDNA microarray. Gene *229*:101-108.
82.**Wang, Y., T. Rea, J. Bian, S. Gray, and Y. Sun.** 1999. Identification of the genes responsive to etoposide-induced apoptosis: application of DNA chip technology. FEBS Lett. *445*:269-273.
83.**Watson, A., A. Mazumder, M. Stewart, and S. Balasubramanian.** 1998. Technology for microarray analysis of gene expression. Curr. Opin. Biotechnol. *9*:609-614.
84.**Weiler, J. and J.D. Hoheisel.** 1996. Combining the preparation of oligonucleotide arrays and synthesis of high-quality primers. Anal. Biochem. *243*:218-227.
85.**Weiler, J., H. Gausepohl, N. Hauser, O.N. Jensen, and J.D. Hoheisel.** 1997. Hybridisation based DNA screening on peptide nucleic acid (PNA) oligomer arrays. Nucleic Acids Res. *25*:2792-2799.
86.**Welford, S.M., J. Gregg, E. Chen, D. Garrison, P.H. Sorensen, C.T. Denny, and S.F. Nelson.** 1998. Detection of differentially expressed genes in primary tumor tissues using representational differences 87.analysis coupled to microarray hybridization. Nucleic Acids Res. *26*:3059-3065.
87.**Winzeler, E.A., D.R. Richards, A.R. Conway, A.L. Goldstein, S. Kalman, M.J. McCullough, J.H. McCusker, D.A. Stevens, L. Wodicka, D.J. Lockhart, and R.W. Davis.** 1998. Direct allelic variation scanning of the yeast genome. Science *281*:1194-1197.
88.**Wittes, J. and H.P. Friedman.** 1999. Searching for evidence of altered gene expression: a comment on statistical analysis of microarray data. J. Natl. Cancer Inst. *91*:400-401.
89.**Wodicka, L., H. Dong, M. Mittmann, M.H. Ho, and D.J. Lockhart.** 1997. Genome-wide expression monitoring in Saccharomyces cerevisiae. Nat. Biotechnol. *15*:1359-1367.
90.**Yang, G.P., D.T. Ross, W.W. Kuang, P.O Brown, and R.J. Weigel.** 1999. Combining SSH and cDNA microarrays for rapid identification of differentially expressed genes. Nucleic Acids Res. *27*:1517-1523.
91.**Zhu, H., Cong, J.P., G. Mamtora, T. Gingeras, and T. Shenk.** 1998. Cellular gene expression altered by human cytomegalovirus: global monitoring with oligonucleotide arrays. Proc. Natl. Acad. Sci. USA *95*:14470-14475.

2 Microfluidic Technologies and Instrumentation for Printing DNA Microarrays

Don Rose
Cartesian Technologies, Inc., Durham, NC, USA

INTRODUCTION

DNA microarray technology permits the simultaneous analysis of thousands of sequences of DNA for genomic research and diagnostics applications. The power of this technology was demonstrated primarily by the work of Affymetrix (4,11,12,14,17,21,31) and of the Stanford University groups of Ronald Davis and Patrick Brown (8,9,15,18,19,25–29).

The Affymetrix approach involves the in situ synthesis of oligonucleotides on a solid substrate using photolithographic techniques. The result is a microarray or "DNA chip," an array of tens of thousands of unique oligonucleotides in an area of several square centimeters. Biological samples are tagged with a fluorescent dye and incubated with the array. Fluorescent sequences in the probe mixture that are complementary to array sequences will bind or hybridize. Interrogation of the array with a fluorescent scanning device reveals the sequences that have a complementary match.

The Stanford groups have also used a variety of printing approaches, whereby tweezers, split pins, microspotting pins, or ink jets are used to deposit presynthesized oligonucleotides or polymerase chain reaction (PCR) products onto solid substrates such as chemically treated microscope slides. After printing, the microarrays are processed to remove unbound DNA and probed with a fluorescent sample. Similar to the Affymetrix approach, the location and intensity of the fluorescent signal provide quantitative information about the sequences present in a biological sample. Several recent reviews provide a good overview of the field (1–3,5–7,10,13,20,24,28,30).

This chapter reviews the microfluidic technologies and instrumentation for

Microarray Biochip Technology
Edited by Mark Schena

printing or "spotting" microarrays and also examines some of the practical considerations in using a pin printing system. It should be noted that there has been some confusion about nomenclature in the microarray field; this review uses the terminology established in the 1970s for filter-based methods out of deference to the pioneers in that field. The *target* is the DNA spotted on the surface to form the microarray, and the *probe* is the labeled DNA that is hybridized to the surface-bound DNA of the microarray.

MICROARRAY PRINTING TECHNOLOGIES

Making a microarray by printing involves delivering a small volume of DNA sample, the target DNA, onto the solid surface. The volume delivered is typically in the nanoliter (10^{-9} L) or picoliter (10^{-12} L) range. These volumes are below the range capable of being dispensed by typical liquid handling systems. Thus, new technologies have emerged to allow the production of microarrays.

The technologies for spotting DNA material onto a substrate fall into two distinct categories: noncontact and contact dispensing. Noncontact dispensing involves the ejection of drops from a dispenser onto the surface. The most common type of noncontact dispensing uses ink-jet printing technologies, which are essentially modifications of devices used for printing ink onto paper. Contact printing involves direct contact between the printing mechanism and the solid support. Contact printing devices include solid pins, capillary tubes, tweezers, split pins, and microspotting pins or "ink stamps," all of which deliver sample spots onto the solid surface. These two types of technology are described in the following sections.

Noncontact Ink-Jet Printing

Ink-jet printing involves the dispensing of the target DNA using a dispenser derived from the ink-jet printing industry. The DNA sample is withdrawn from the source plate up into the print head and then moved to a location above the slides. The sample is then forced through a small orifice, causing the ejection of a droplet from the print head onto the surface. Two types of noncontact ink-jet technology, piezoelectric and syringe-solenoid, are currently being used to print DNA microarrays (Figure 1).

Piezoelectric printing technology. Piezoelectric printing technology uses a piezoelectric crystal (e.g., ceramic material) closely apposed to the fluid reservoir. One configuration places the piezoelectric crystal in contact with a glass capillary, which holds the sample fluid (Figure 1A). The sample is drawn up into the reservoir and the crystal is biased with a voltage, which causes the crystal to deform, squeeze the capillary, and eject a small amount of fluid from the tip. An example of an array made with a piezoelectric dispensing system is shown in Figure 2.

The fast response time of the crystal permits fast dispensing rates, on the order of several thousand drops per second. Furthermore, the small deflection of the crystal results in drop volumes on the order of hundreds of picoliters. To date, piezoelectric dispensing technology has been shown to work for making small numbers of gene expression microarrays (28), but development into a commercially viable dispensing system has been slow. The main difficulties in implementing piezoelectric dispensing include air bubbles, which reduce the reliability of the system, relatively large sample volumes, and problems with sample changing. Most of the commercial activity has been carried out by Packard Instrument Company (Meriden, CT, USA) and Incyte Pharmaceuticals (Palo Alto, CA, USA).

Syringe-solenoid printing technology. Syringe-solenoid technology combines a syringe pump with a microsolenoid valve to provide quantitative dispensing in the low nanoliter range. Shown schematically in Figure 1B, a high-resolution syringe pump is connected to both a high-speed microsolenoid valve and a reservoir through a switching valve. For printing microarrays, the system is filled with a system fluid, usually water, and the syringe is connected to the microsolenoid valve. Withdrawing the syringe causes the sample to move upward into the tip. The syringe then pressurizes the system such that opening the microsolenoid valve causes droplets to be ejected onto the surface. With this configuration, the minimum dispense volume is on the order of 4 to 8 nL. However, given the positive displacement nature of the dispensing mechanism, the reliability of the system is very high. An example of an array made with the syringe-solenoid tech-

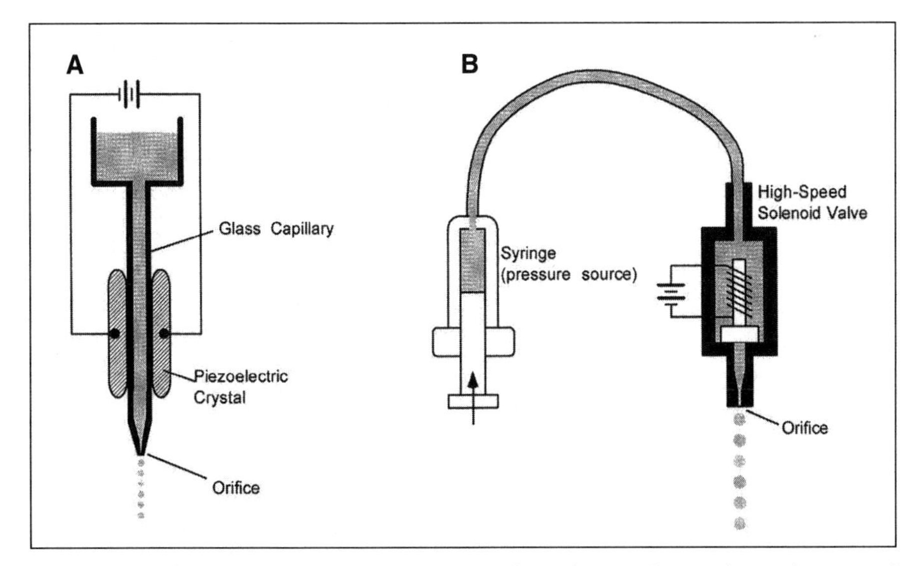

Figure 1. Types of noncontact ink-jet dispensers. (A) Piezoelectric dispenser shown with piezoelectric crystal surrounding a glass capillary. (B) Syringe-solenoid ink-jet dispenser shown with a high-resolution syringe pump coupled to a high-speed solenoid valve.

nology is shown in Figure 3. Most of the commercial activity in the syringe-solenoid area has been carried out by Cartesian Technologies (Irvine, CA, USA).

Contact Printing

Pin printing technology. The second and more common means for dispense-based printing of microarrays involves the use of rigid pin tools and related technologies that were originally developed for use in making filter arrays. In surface contact printing, the pin tools are dipped into the sample solution, resulting in the transfer of a small volume of fluid onto the tip of the pins. Touching the pins or pin samples onto the slide surface leaves a spot, the diameter of which is determined by the surface energies of the pin, fluid, and slide. The typical spot volume is in the high picoliter to low nanoliter range.

Pin printing can be done using solid pins for transferring samples from microwell plates onto microscope slides (23). The tips of the solid pins are generally flat, and the diameter of the pins determines the volume of fluid that is transferred to the substrate. A recent modification of solid pins involves the use of solid pin tips with concave bottoms, which have been shown to print more efficiently than flat pins in certain cases. Because the loading volume of both types of solid pins is relatively small, only one or a few microarrays can be printed with a single sample loading, making the overall printing process rather slow.

To permit the printing of multiple arrays with a single sample loading, several groups have developed printing capillaries, tweezers, and split pins that hold larger sample volumes than solid pins and therefore allow more than one array to be printed from a single sample loading. Capillaries and tweezers have found use in research applications (16,25,29), although the open configuration of the sample channel leads to irregular loading volumes; moreover, the need to break the meniscus of the loaded sample by tapping the tweezers on the surface reduces printing durability in a production setting (Figure 4). Split pins, developed by

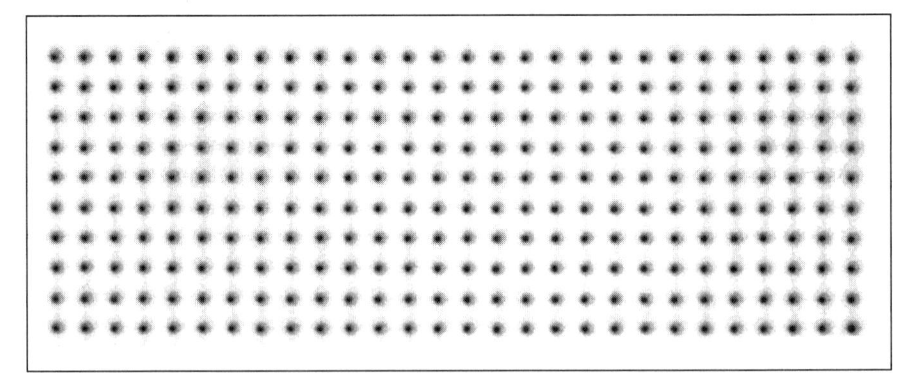

Figure 2. Array of ^{33}P-labeled material made with a single-channel piezoelectric dispenser. Spot spacing is 750 µm and spot size approximately 145 µm. (Courtesy of Packard Instrument Company.)

several groups, have a fine slot machined into the end of the pin to hold the sample. When the split pin is dipped into the sample solution, sample is loaded into the slot. Tapping the pin onto a solid surface with sufficient force deposits a small volume of sample (Figure 4).

The most widely used printing technology for microarray manufacture was developed by TeleChem International. TeleChem's microspotting pins work like ink stamps, whereby sample solution is loaded into each pin, which can be customized to hold from 0.2 to 1.0 μL of sample solution (Figure 4). The sample solution on the end of the pins is then brought in contact with the substrate. When the pins are moved away from the surface, the attractive force of the substrate on the liquid withdraws a small amount of sample from the sample channel of each pin. TeleChem pins are available in a wide range of tip dimensions (http://arrayit.com), allowing the user to obtain spot sizes from 75 to 360 μm in diameter. An example array made with the TeleChem ChipMaker™ 3 pin is shown in Figure 5.

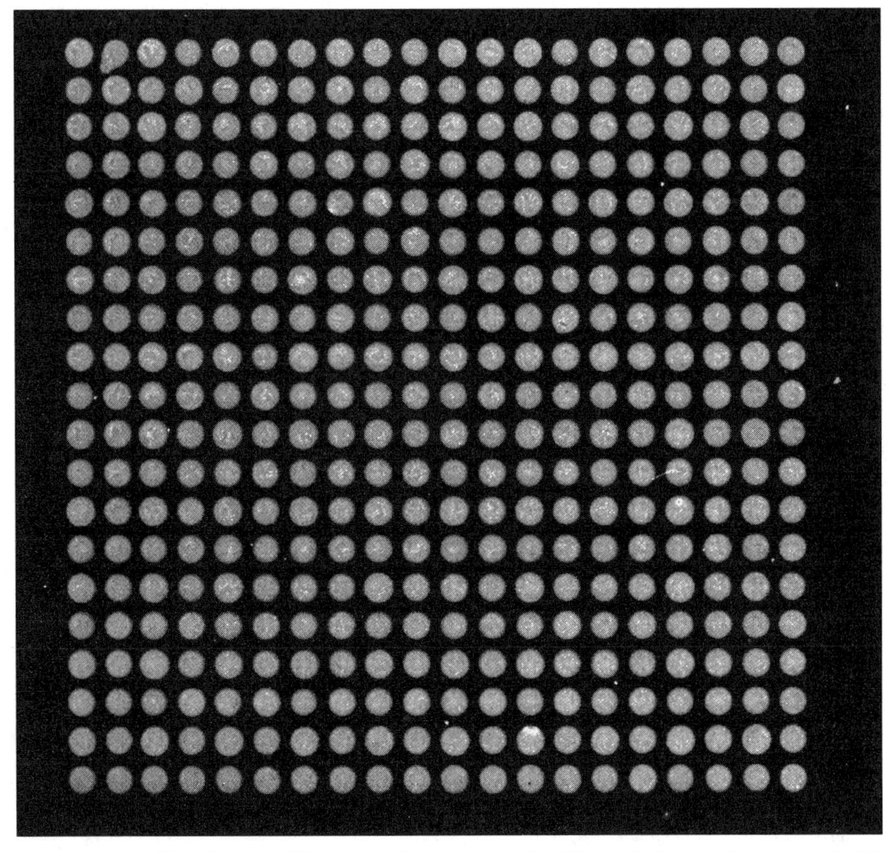

Figure 3. Array of 4.2-nL spots of Cy3 made using a syringe-solenoid type of dispenser. Spot spacing is 500 μm with spot size approximately 325 μm. **(See color plate A2.)**

A variation of the pin printing process is the Pin-and-Ring™ technique developed by Genetic MicroSystems. This technique involves dipping a small ring into the sample well and removing it to capture liquid in the ring. A solid pin is then pushed through the sample in the ring, and sample trapped on the flat end of the pin is deposited onto the surface. Complete details of this technique can be found in Chapter 3.

Summary

A comparison of the different printing technologies is shown in Table 1. Various parameters are shown relative to the different technologies. Many of the numbers are estimates for comparison purposes only and will depend on the specific application.

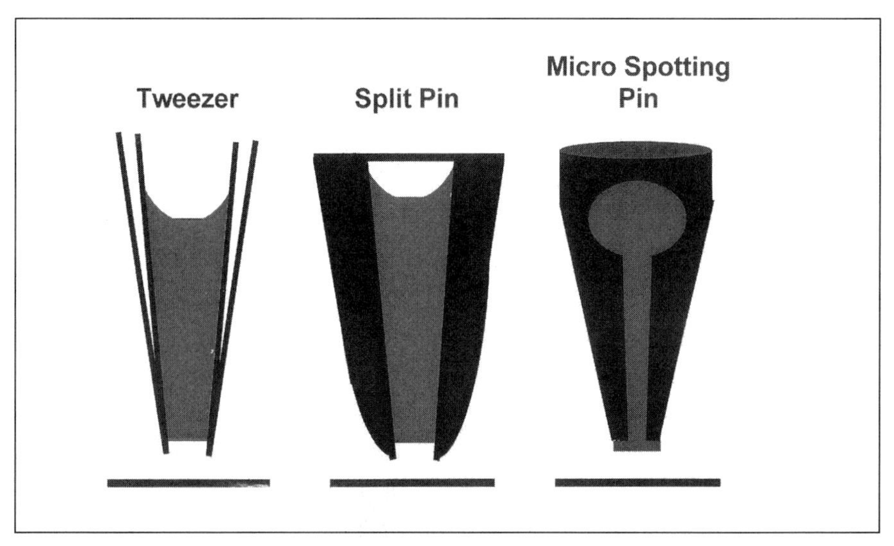

Figure 4. Schematic drawings of contact printing dispensers. The tweezer, split pin, and TeleChem's ArrayIt brand microspotting pin are shown loaded with sample solution (blue). Tweezers and split pins require a tapping force to move sample down the channel onto the surface. TeleChem pins work more like ink stamping devices, whereby sample on the end of the pin is brought in contact with the surface, and the substrate pulls a small amount of sample out of the fluid reservoir when the pin is moved away from the surface.

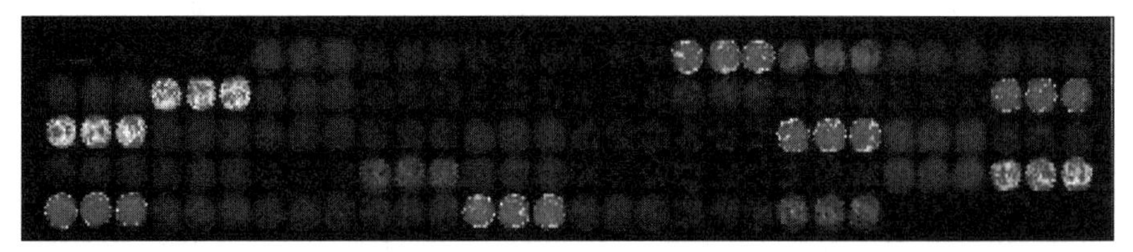

Figure 5. Example of a printed and hybridized microarray made with TeleChem's microspotting pin technology (ChipMaker 3 pins). Spot spacing, 140 µm; spot size, approximately 125 µm, with targets spotted in triplicate. Fluorescent image generated with a ScanArray 3000 (GSI Lumonics). **(See color plate A2.)**

Table 1. Comparison of Printing Technologies

Parameter	Printing Technology: Piezoelectric	Syringe-Solenoid	Microspotting Pin
Minimum sample volume (μL)[a]	20–50	20–50	5
Loading volume (μL)[b]	5–10	5–10	0.2–1.0
Print volume (nL)	0.05–10	4–100	0.5–2.5
Spot size (μm)	125–175	250–500	75–360
Spot density (spots/cm^2)	500–2500	200–400	400–10 000
Programmable volume	Yes	Yes	No
Number of nozzles or pins	4–8	8–16	1–64
Delivery speed (spots/s)	100–500	10–50	64
Simplicity	✔	✔	✔✔✔
Robustness	✔	✔✔	✔✔✔
Cost per spot	$$$	$$	$

[a]Volume of sample in the 384-well source microplate
[b]Sample volume of the dispensing device

THE MICROARRAY PRINTING SYSTEM

General Requirements

One or more of the above printing technologies can be used in an instrument for making microarrays. Putting a printing technology into a system requires both hardware and software infrastructures. There are several general requirements:

- **Robust:** Must function for an extended period (>24 hours) without user intervention due to hardware or software errors.
- **Automated:** Must process a large number of samples (3000–10 000) with little or no user intervention.
- **High precision:** The microarray spots need to be highly regular, with tight tolerances, and duplicate slides need to have high reproducibility.

Hardware

Print head. The print head is the core of the system, being the device that transfers sample from the microplate to the slides. For noncontact, ink-jet type printing, the dispense head consists of 4 to 16 channels or lines for dispensing. Each line is connected to a syringe pump for aspiration of the sample, prior to dispensing to the slide. For pin-based dispensing, the pins are held such that they can move freely up and down in the print head during contact with the glass surface. Some configurations have the pin spring-loaded, although the preferable configuration allows the pins to move along the Z-axis using gravity to return the pins to the resting position in the print head.

The spacing of pins or nozzles is determined by the source microplate configuration. Since most microplates contain either 96 or 384 wells, the center-to-center spacing is 9.0 and 4.5 mm, respectively. As will be discussed later, since the spacing of the nozzles or pins in the print head is much larger than the spacing of spots on the microarray, the spots typically do not directly map from the source plate to the array.

Plate and substrate handling. The microarraying system should have accommodations for both the source plates, 96- or 384-well plates, and the microarray substrate, typically 1 × 3 × 0.039-in (25 × 76 × 1 mm) glass microscope slides. The source plate and slides are usually placed on a platform that can be addressed by the print head. Both the source plate and each slide must be held securely to permit accurate loading and dispensing of the sample.

The number of source plates will depend on the number of spots per slide. For example, to place 10 000 spots on a slide, the system will need to accommodate 104 × 96-well plates or 26 × 384-well plates. To accommodate this number of plates, three options exist: *(i)* manually place each plate on the printing platform, *(ii)* load each plate onto the platform using an automated plate handling system, or *(iii)* arrange each plate on the platform that can be addressed by the print head. The first approach is simple but labor-intensive, especially if the printing process takes more than 8 hours, the second approach is more automated but more expensive, and the last approach results in a very large system (13.5 ft^2 for 100 microplates).

XYZ positioning stage. The XYZ positioning stage moves the print head relative to the source plate and the slides. All positioning stages have a certain level of positioning error. Generally, the smaller the error, the more expensive the positioning system. Several parameters are important when specifying the quality of a stage:

- **Repeatability:** The error in moving from position A to a series of positions and returning to position A.
- **Accuracy:** The error in moving from position A to position B, usually expressed in terms of ± microns.
- **Resolution:** The smallest step the stage is capable of moving.
- **Linear velocity:** The maximum speed (mm/s) a stage can move in one dimension and maintain its position.
- **Positioning feedback:** The means for determining the current position of the stage.

In general, there is a tradeoff between many of these parameters. For example, the maximum velocity is limited by the stage resolution; the greater the resolution, the slower the stage. The cost of the stage is related to many of these parameters. High resolution and high repeatability increase the price. Likewise, encoded stepper motors offer very high accuracy and positioning feedback but are more expensive than nonencoded motors.

Environmental Control

A variety of environmental conditions can affect microarray manufacturing. The two most important factors, humidity and dust, must be controlled to obtain the highest quality microarrays. Humidity must be controlled to prevent sample evaporation from the source plates and from the sample channels of the pins during the arraying process. For plates configured in a stack, a plate acts as a lid for the plate below to minimize sample evaporation. For the plate in use, if the printing time is long and the evaporation rate is high, then some sample evaporation will occur during the printing process. For example, if the cycle time for printing a set of spots (load sample, print on each slide, wash and dry pins) is 1 minute using 4 pins, then each 384-well microplate will require 1.6 hours to process. If evaporation is excessive during this time, the concentration of the DNA in the sample wells will increase, causing a gradual increase in the concentration of the arrayed DNA. Proper humidity control virtually eliminates sample evaporation.

Humidity control is usually achieved by use of a humidity chamber that encloses the arraying device. Humidity should be controllable over a wide range and have a feedback mechanism for maintaining the humidity at a predetermined level. Relative humidity of 65% to 75% is usually sufficient for most applications. Humidity >75% can be problematic because condensation can occur, leading to the wetting of metal parts of the positioning stage.

Dust contamination must be minimized or eliminated to create high-quality microarrays. Dust from ceiling tiles, ventilation systems, and the user can settle on the slide during the arraying process. This can lead to printing inaccuracies as well as false readings during slide scanning (dust being highly fluorescent). A humidity chamber is usually sufficient to minimize contamination due to dust and particulate matter.

Instrument Control Software

The microarray printing process involves moving the print head to the sample plate, loading sample, making spots on the substrate, and washing the dispense head prior to the next sample. The control of this operation is accomplished by software that communicates with the arraying instrument. Software of this type can span the spectrum of easy-to-use (but inflexible) to difficult-to-use but extremely flexible. For most applications and end-users, a certain amount of software flexibility is required until microarray experimentation becomes routine and stable.

Sample Tracking Software

The microarray process involves moving a sample from a source plate to the microarray, hybridizing the microarray with probes, scanning the slide, and evaluating the spots. Sample tracking software is required to track the sample through this process such that spots on the array can be readily identified.

PRACTICAL CONSIDERATIONS

This section outlines practical considerations when using microspotting pins for printing microarrays onto glass slides. Although broadly applicable, these considerations apply primarily to TeleChem ChipMaker pins used on a Cartesian PixSys PA Series Workstation.

Array Substrate

The substrate for printing the array must be rigid and amenable to surface chemistry modifications and have low background fluorescence in the region of fluorescent dye excitation wavelengths. The most commonly used substrate is the 25 × 76-mm glass microscope slide, although a number of groups have begun exploring porous polymeric surfaces (polymers coated onto glass substrates) or plastic substrates for producing microarrays.

Surface Chemistry

The surface of the microscope slide must be treated prior to use for several reasons. First, a suitable functional group must exist on the surface for attaching the target DNA to the glass. Target DNA will not attach to naked glass. Attachment prevents the target DNA from being washed away during slide processing and hybridization. Second, a hydrophobic surface will allow a higher printing density since the spotted sample (hydrophilic) will spread less on a hydrophobic surface than an untreated (hydrophilic) surface. Currently, two chemical functionalities are commonly available on glass slides. An amine- or lysine-coated slide is used for adsorption of DNA onto the glass slide (ionic interaction between the negatively charge phosphodiester backbone of DNA and the positively charged slide surface). An aldehyde-functionalized surface is used to attach amino-modified DNA onto the surface covalently via reaction with free aldehyde groups using Schiff's base chemistry. In addition to chemistries that provide a two-dimensional attachment surface, DNA attachment can also be achieved by coating the slide with polymeric reagents such as thin layers of acrylamide.

Substrate Effects

The substrate can have large effects on the overall microarray experiment.

Substrate materials with elevated intrinsic fluorescence will decrease the sensitivity of the assay. Poor surface treatment can lead to poor attachment of DNA to the slide. Nonhomogeneous surface treatment will result in variations in the amount of DNA attached. Finally, residual material from the slide treatment step can lead to background fluorescence.

Pin Characteristics

The ChipMaker and Stealth™ pins from TeleChem contain a stainless steel shaft with a fine point. Machined into the point is a narrow gap that acts as the reservoir for sample loading and spotting. The pins are mounted in a print head such that the pins float under their own weight when touched onto the substrate. Up to 64 pins, on 4.5-mm centers, can be mounted in the print head.

The pins have a loading volume of 0.2 to 0.6 μL and can produce spots ranging from 100 to 360 μm, depending on sample and surface properties. Given the fine structure of the pin, care must be used in handling the pins. Although they are robust when touched onto the surface in the Z-direction, movement of the pin across the surface in the X- or Y-directions may cause the tip to bend. Also, dragging the tip of the pin across a surface can result in clogging of the pin sample channel. In the event that a pin becomes clogged, it can be cleaned with an ultrasonic bath. It should be noted that extensive exposure to ultrasonic waves is not recommended since this may weaken the pin tips and compromise durability in a production setting.

Array Layout Options

Print area. DNA samples must be spotted or printed in an area defined by the area detectable by a fluorescent detection device. For example, the ScanArray 4000 and 5000 confocal scanners (GSI Lumonics, Watertown, MA, USA) can detect a 22 × 72-mm area, whereas the ScanArray 3000 has a 22 × 60-mm scan area.

Print head. The print head contains an array of holes, each holding a single ChipMaker or Stealth pin. The standard ChipMaker 2 print head holds up to 32 pins in a 4 × 8 array, whereas the ChipMaker 3 print head holds up to 48 pins in a 4 × 12 array. The Stealth print heads hold either 32, 48, or 64 pins. Both the ChipMaker and Stealth print heads hold pin on 4.5-mm centers.

The print head is oriented relative to the plate such that pin A1 corresponds to well A1 of a microwell plate. For a 384-well plate, pins may be placed in any of the positions for printing since the 4.5-mm pin spacing matches the 4.5-mm well spacing. For a 96-well plate (9-mm well spacing), every other position can be used (A1, A3, C1, C3, ...).

Single pin printing. Printing arrays with single pins is the most straightforward type of printing, although it is the most time-consuming. With single pin printing, a source plate can be directly mapped into an array. In other words, 384 samples from a source plate can be spotted as a 16 × 24 array such that the spot A1, B1, C1, etc. in the array corresponds to well A1, B1, C1, etc. of the source plate. This makes posthybridization analysis trivial. Using a single pin and 0.25-mm spot-to-spot spacing, 21120 spots (55 × 384-well plates) can be placed on a slide.

Multiple pin printing. Printing with more than one pin is faster than a single pin but requires more planning for array layout more sophisticated than sample tracking. To spot with multiple pins, the pins are dipped into the sample wells to load the pins and then touched in unison onto the surface to create separate spots. If the pins are on 4.5-mm spacing, the first round of spotting produces spots on 4.5-mm spacing. The next rounds of printing are done by spotting with a small offset (100–400 μm) from the previous location. This permits maximum spot density but requires deconvolution at the analysis stage to identify the sample spot. The following are possible configurations:

1. 1 × 8 pin configuration (ChipMaker 2). If 8 pins are placed in row one of the ChipMaker 2 print head (4.5-mm centers) and used to sample column-wise from a 384-well plate, then an array containing 8 subgrids will be generated. As the arraying process proceeds, spots from pin A1 will approach spots printed by pin A2 (sample B1). With the appropriate choice of center-to-center spacing, the desired number of spots can be printed between pins. For example, if spots are placed on 0.28-mm centers, then 16 spots can be placed between adjacent pins (16 × 0.28 ≈ 4.5). Sixteen is desirable since it is an even multiple of a 96- or 384-well plate.
2. 1 × 12 configuration (ChipMaker 3). If 12 pins are placed in row 1 of a ChipMaker 3 print head and used to sample row-wise from a 384-well plate, then more of the slide can be printed with spots. Using the same spot spacing as the 1 × 8 configuration, the number of spots increases to 14 976 (78 × 192 array or 39 × 384-well plates).
3. 4 × 1 pin configuration for 384-well plates. Another alternative is to place 4 pins in column A of a ChipMaker 2 or 3 print head. With this configuration, 64 spots can be placed in a column along the short edge of the slide. If one continues to spot in a left to right fashion with 0.28-mm offsets, a total of 240 columns can be printed to fill the entire print area. The resulting 64 × 240 array contains 15 360 spots (40 × 384-well plates).
4. 4 × 8 or 4 × 12 configuration. For maximum throughput, the ChipMaker 2 or 3 print head can be used with the maximum number of pins, 32 and 48, respectively. In doing so, the spots from one pin approach the other pin's two dimensions. For 250-μm spacing, each pin can print a 16 × 16 array for 8064 spots for the 4 × 8 configuration and 12 288 spots for the 4 × 12 configuration. If the spacing is decreased to 150-μm spacing, then the total features increase to 28 800 for the 4 × 8 configuration and 43 200 for the 4 × 12 configuration.
5. Pin configurations for 96-well plates. For 96-well plates, a similar strategy is

used, but the pin density in the print head is reduced by two since these plates have 9-mm well spacing.

PRINTING THE MICROARRAY

Sample Loading

The DNA sample to be printed on the slides is usually placed in a 96- or 384-well plate. For best results, a flat-bottomed plate is used with enough sample to provide a 1-mm layer (4–6 μL) on the plate bottom of the microplate. This technique uses a minimum of sample volume in each well. Dipping the pin into a larger volume of sample results in the absorption of sample onto the outside of the pin, causing printing irregularities early in the printing process and requiring preprinting for consistent printing.

Preprinting

If the sample volume exceeds >6 μL/well for 384-well plates, the pins must be spotted a number of times on a slide to create consistent spots. This preprinting is necessary to drain excess sample solution from the exterior of the pin. In the example shown in Figure 6A, three pins were loaded from wells containing 10 μL of sample per well (384-well plate) and spotted a number of times to show the change in spot size as a function of spot number. As the data show, after 10 to 20 preprint spots, the spotting becomes more consistent (Figure 6A).

Loading the pin with a larger volume of sample (i.e., dipping the pin deeply into the sample solution) results in a larger number of preprints required to achieve consistent spotting. As shown in Figure 6B, pins loaded from wells containing 35 μL of sample require a larger number of preprints to achieve uniform printing.

Reproducibility

Spot reproducibility is a measure of spot variation during the printing process. Spot variations can be due to mechanical differences between pins, slight variations in slide surface properties, and changes in the pin during the printing process (e.g., a pin becomes clogged with particulate matter). This variation was measured using four pins and printing a Cy3-labeled 31-mer printing onto five slides. The results shown in Figure 7 reveal slight variations from pin to pin and from slide to slide. As can be seen, each pin has a consistent volume, but there is a slight difference between pins, presumably due to slight mechanical differences in the pins or different surface properties of the pin. For example, the coefficient of variation of spots among each pin across the five slides ranges from 9.3% (pin 2) to 12.1% (pin 4). The variation of spots across each slide ranges from 7.1% (slide 2) to 14.4% (slide 5). The overall variation for the entire data set is 12.6% (Figure 7).

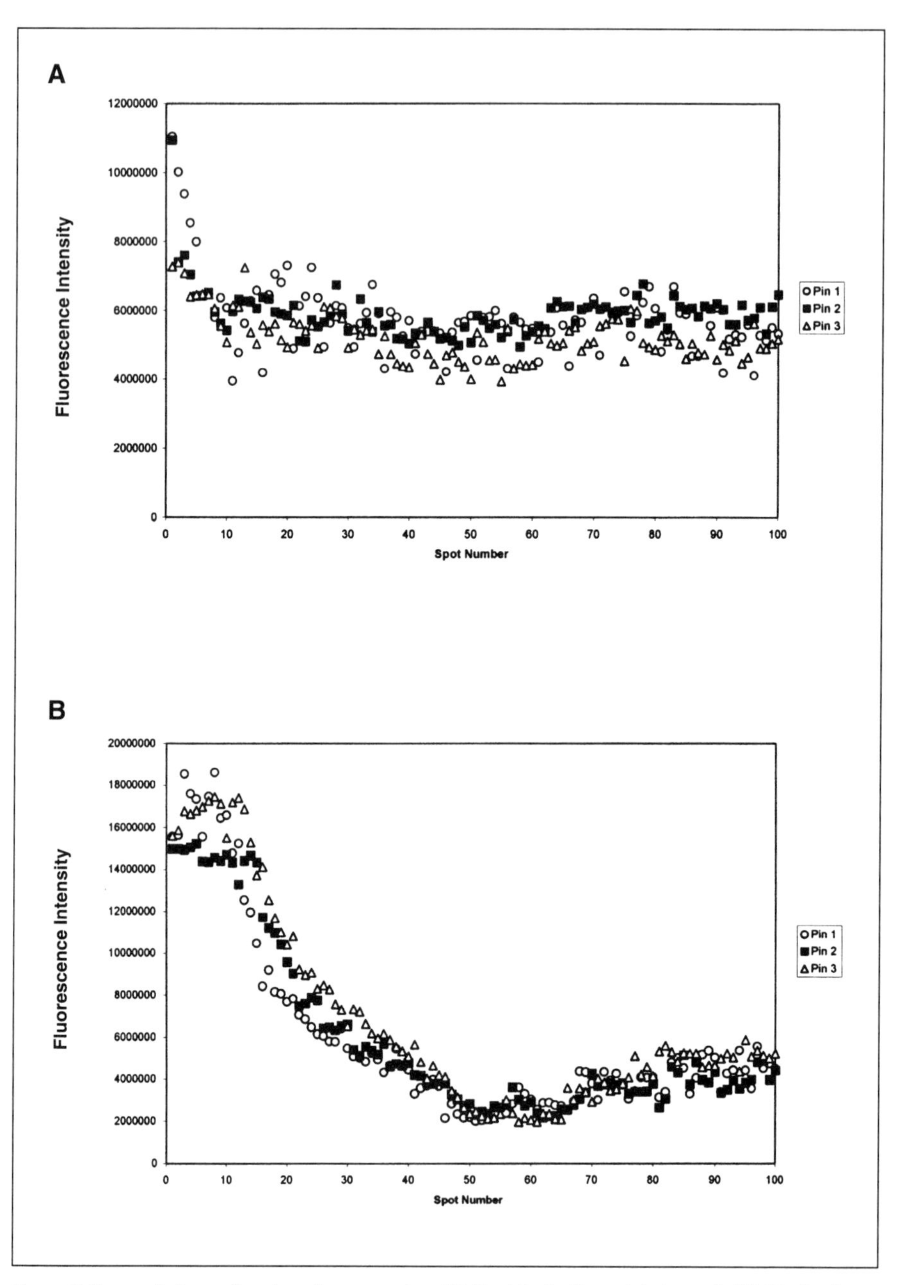

Figure 6. Spot variation as function of spot number. (A) For 10 μL of sample/microwell. (B) 35 μL of sample/microwell. The sample was a Cy3-labeled 31-mer in 1× microspotting solution (TeleChem) in square-well, flat-bottomed 384-well plate. The procedure used a PixSys 5500 gridding robot (Cartesian) equipped with four ChipMaker 2 pins (TeleChem) printing on CSS aldehyde-modified glass slides (CEL Associates) with the following steps: *(i)* dip the pin into the sample such that the pin contacts the bottom of microplate; *(ii)* print 100 spots on one slide (with no preprints); *(iii)* wash the pin with a three-water dip and vacuum wash cycles; *(iv)* scan the slide with a ScanArray 3000 (GSI Lumonics); and *(v)* calculate fluorescence intensities by integrating each spot.

Spot Size and Density

The size of the spot deposited on the glass slide determines the number of spots that can be printed on a slide. Spot size is directly related to the volume of sample deposited on the surface. The volume is determined by a number of factors:

Surface and solution properties. The properties (surface energy, viscosity) of the slide surface, pin surface, and sample determine how much sample will be deposited and how much the sample will spread once deposited. For example, an aqueous sample deposited onto a hydrophobic surface (e.g., aldehyde slide) will result in a much smaller volume and spot size than the same sample printed on a more hydrophilic surface (e.g., untreated slide).

Pin contact surface area. The surface area of the pin determines the initial contact between sample or pin and slide. The larger the area, the larger the spot size. For example, the two different ChipMaker pins (CM2 and CM3), which contain approximately 100- and 75-μm tips, respectively, were tested and shown to produce spot sizes proportional to the sample/pin tip surface area (Figure 8).

Pin velocity. Although direct contact between the pin and the substrate is not necessary for printing with microspotting pins, most users choose to touch the slide surface lightly to correct for the unevenness of the printing surface. If pin surface contact is chosen as means of calibration, the speed at which the pin strikes the surface can have an effect on spot size if excessive speeds are used. If the pins tap the surface at high velocity (>20 mm/s), fluidic inertia may force a large volume of sample out of the pin, resulting in a large spot. Tapping the pins on the surface may also lead to mechanical damage of the pin tips. Unlike tweezers and split pins, TeleChem microspotting pins *do not* require a tapping force for printing. Lightly touching the pin to the surface like an ink stamp produces small spots and extends the durability of the pins. An estimate of spot density, given a spot size and center-to-center spacing, is shown in Table 2.

Pin Washing and Sample Carryover

Efficient cleaning of the pins during the printing process is necessary to prevent sample carryover, which would complicate the hybridization results. ChipMaker and Stealth pins are cleaned by dipping the pins into distilled water and then removing the wash water from the pins with a vacuum. Repeating this procedure three times reduces sample carryover to <1 part/10 000. To measure the effect of pin washing on sample carryover, two 31-mer oligonucleotides were spotted onto aldehyde-containing microscope slides. One of the oligonucleotides (target-positive) was perfectly complementary to the fluorescent probe, whereas the second oligonucleotide (target-negative) was noncomplementary (Figure 9A). After spotting and drying, the oligos were attached to the slide using Schiff-based chemistry, hybridized overnight with a Cy3-labeled positive probe, and scanned. The resulting array is shown in Figure 9B. As can be seen, no carryover of the

Table 2. Microarray Density

Spot Size (µm)	CTC Spacing (µm)	Spots/cm^2	Total Spots (18 × 72 mm)
75	113	7901	104 296
100	150	4444	58 667
150	225	1975	26 074
200	300	1111	14 667
250	375	711	9387
300	450	494	6519
400	600	278	3667
500	750	178	2347

Shown are the densities (spots/cm^2) and total number of spots possible on a single microscope slide with an 18 × 72-mm print area, given a spot size and center-to-center (CTC) spacing given in microns (µm).

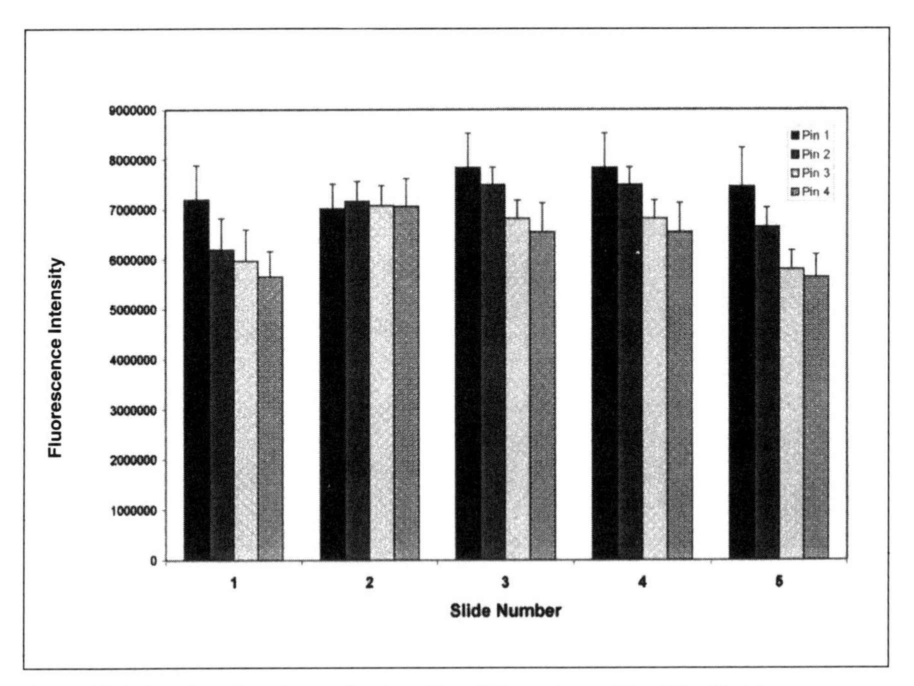

Figure 7. Variation of spot intensity as a function of four different pins and five slides. (Each bar represents an average of 20 spots; error bars = SD.) Overall CV (standard deviation divided by the mean × 100) for the data set is 12.6%. Experimental conditions: sample, Cy3-labeled 31-mer in 1× microspotting solution (TeleChem); procedure: *(i)* dip pin; *(ii)* preprint 20 spots; *(iii)* print one spot/pin/slide for five slides; *(iv)* dip and vacuum wash; *(v)* repeat with same sample to produce 20-spot array; *(vi)* scan with ScanArray 3000; *(vii)* integrate spots; and *(viii)* compute CV for 20 spots.

Table 3. Printing Time Estimates*

Step	Time		
	1 pin	8 pins	32 pins
Load pins with sample	2 s	2 s	2 s
Preprint 10 spots	4 s	4 s	4 s
Print 48 slides	48 s	48 s	48 s
Wash pins	6 s	6 s	6 s
Total time for one cycle	1 min	1 min	1 min
Total time for 1 × 384-well microplate	6.4 h	0.8 h	0.2 h
Total time for 40 × 384-well microplates (15360 spots)	256 h	32 h	8 h

*Shown are microarray manufacturing specifications for the PixSys 5500 (Cartesian) using ArrayIt ChipMaker 2 microspotting technology (TeleChem).

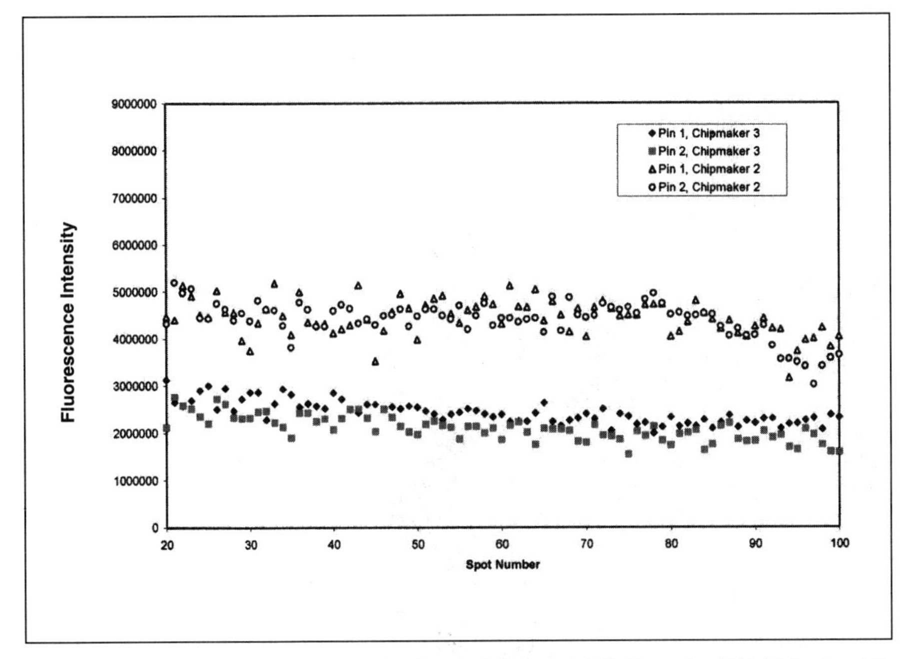

Figure 8. Spot size as a function of spot number for two different size TeleChem pins (ChipMaker 2 and 3). Experimental conditions were the same as in Figure 7, except that 30 preprints were done prior to printing. (See color plate A3.)

target-negative oligonucleotide can be detected.

Carryover can also be measured by direct spotting of Cy3- or Cy3-labeled oligonucleotide and a blank solution onto slides and comparing the results. Although a small but measurable amount of carryover is seen at the highest sensitivity settings of the scanner, this level is not directly applicable to DNA because the oily cyanine dyes are more difficult to clean from the pins than unlabeled oligonucleotides or cDNAs.

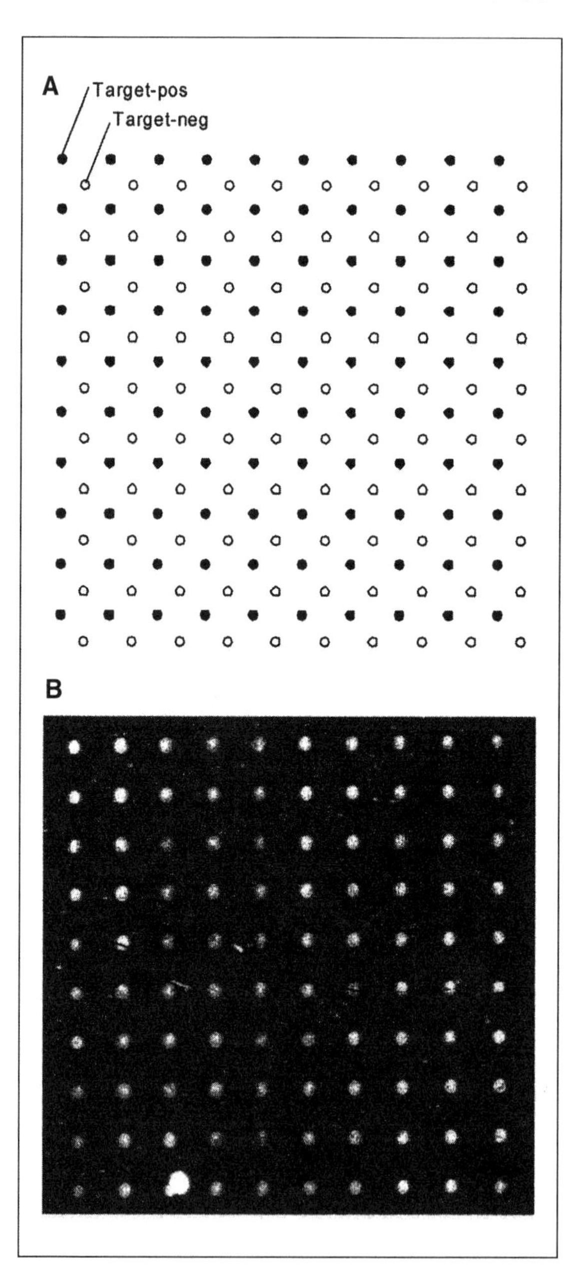

Figure 9. Effect of pin washing on sample carryover. (A) Schematic layout of two test 31-mers printed on aldehyde (silylated) slides. One oligonucleotide (target-pos) was 100% complementary to the fluorescent probe, whereas the second oligonucleotide (target-neg) was noncomplementary. The microarray was printed as follows: print 10 spots of the target-pos oligonucleotide, wash pin twice (dip into water, vacuum dry, dip, dry, dry), print 10 spots of the target-neg oligonucleotide, wash pin twice (dip into water, vacuum dry, dip, dry, dry), and repeat until a total of 400 spots are printed. (B) After spotting and drying, the oligos were attached to the slide using Schiff-based chemistry, hybridized overnight with Cy3-labeled positive probe (complimentary to the target-pos oligo), washed, and scanned three times with a ScanArray 3000 (GSI Lumonics) set at 100% laser power and 80% PMT.

Throughput

Printing throughput should be measured as total cycle time (a cycle being loading the pins, preprinting, printing arrays, and washing). Table 3 lists some printing time estimates.

REFERENCES

1. **Bassett, D.E., M.B. Eisen, and M.S. Boguski.** 1999. Gene expression informatics—it's all in your mine. Nat. Genet. Suppl. *21*:51-55.
2. **Bowtell, D.D.L.** 1999. Options available—from start to finish—for obtaining expression data by microarray. Nat. Genet. Suppl. *21*:25-32.
3. **Brown, P.O. and D. Botstein.** 1999. Exploring the new world of the genome with DNA microarrays. Nat. Genet. Suppl. *21*:33-37.
4. **Chee, M., R. Yang, E. Hubbell, A. Berno, X.C. Huang, D. Stern, J. Winkler, D.J. Lockhart, M.S. Morris, and S.P.A Fodor.** 1996. Accessing genetic information with high-density DNA arrays. Science *274*:610-614.
5. **Cheung, V.G., M. Morley, F. Aguilar, A. Massimi, R. Kucherlapati, and G. Childs.** 1999. Making and reading microarrays. Nat. Genet. Suppl. *21*:15-19.
6. **Cole, K.A., D.B. Krizman, and M.R. Emmert-Buck.** 1999. The genetics of cancer—a 3D model. Nat. Genet. Suppl. *21*:38-41.
7. **Debouck, C. and P.N. Goodfellow.** 1999. DNA microarrays in drug discovery and development. Nat. Genet. Suppl. *21*:48-50.
8. **DeRisi, J.L., L. Penland, P.O. Brown, M.L. Bittner, P.S. Meltzer, M. Ray, Y. Chen, Y.A. Su, and J.M. Trent.** 1996. Use of a cDNA microarray to analyze gene expression patterns in human cancer. Nat. Genet. *14*:457-460.
9. **DeRisi, J.L, V.R. Iyer, and P.O.Brown.** 1997. Exploring the metabolic and genetic control of gene expression on a genomic scale. Science *278*:680-686.
10. **Duggan, D.J., M. Bittner, Y. Chen, P. Meltzer, and J.M. Trent.** 1999. Expression profiling using cDNA microarrays. Nat. Genet. Suppl. *21*:10-14.
11. **Fodor, S.P.A., M.C. Pirrung, J.L. Read, and L. Stryer.** 1995. Array of oligonucleotides on a solid substrate. U.S. Patent 5,445,934.
12. **Fodor, S.P.A., L. Stryer, J.L. Read, and M.C. Pirrung.** 1998. Array of materials attached to a substrate. U.S. Patent 5,744,305.
13. **Hacia, J.G.** 1999. Resequencing and mutational analysis using oligonucleotide microarrays. Nat. Genet. Suppl. *21*:42-50.
14. **Hacia, J.G., L.C. Brody, M.S. Chee, S.P.A. Fodor, and F.S. Collins.** 1996. Detection of heterozygous mutations in *BRCA1* using high density oligonucleotide arrays and two-colour fluorescence analysis. Nat. Genet. *14*:441-447.
15. **Heller, R.A., M. Schena, A. Chai, D. Shalon, T. Bedilion, J. Gilmore, D.E. Woolley, and R.W. Davis.** 1997. Discovery and analysis of inflammatory disease-related genes using cDNA microarrays. Proc. Natl. Acad. Sci. USA *94*:2150-2155.
16. **Khrapko, K.R., A.A. Khorlin, I.B. Ivanov, B.K. Chernov, Iu.P. Yu., Lysov, S.K. Vasilenko, V.L. Florent'ev, and A.D. Mirzabekov.** 1991. Hybridization of DNA with oligonucleotides immobilized in gel: a convenient method for detecting single base replacements [in Russian]. Mol. Biol. (Mosk.) *25*:581-591.
17. **Kozal, M.J., N. Shah, N. Shen, R. Yang, R. Fucini, T.C. Merigan, D.D. Richman, D. Morris, E. Hubbell, M. Chee, and T.R. Gingeras.** 1996. Extensive polymorphisms observed in HIV-1 clade B protease gene using high-density oligonucleotide arrays. Nat. Med. *2*:793-799.
18. **Lashkari, D.A., J.L. DeRisi, J.H. McCusker, A.F. Namath, C. Gentile, S.Y. Hwang, P.O. Brown, and R.W. Davis.** 1997. Yeast microarrays for genome wide parallel genetic and gene expression analysis. Proc. Natl. Acad. Sci. USA *94*:13057-13062.
19. **Lemieux, B., A. Aharoni, and M. Schena.** 1998. Overview of DNA chip techology. Mol. Breeding *4*:277-289.
20. **Lipshutz, R.J., S.P.A. Fodor, T.R. Gingeras, and D.J. Lockhart.** 1999. High density synthetic oligonucleotide arrays. Nat. Genet. Suppl. *21*:20-24.
21. **Lockhart, D.J., H. Dong, M.C. Byrne, M.T. Follettie, M.V. Gallo, M.S. Chee, M. Mittman, C. Wang, M. Kobayashi, H. Horton, and E.L. Brown.** 1996. Expression monitoring by hybridization to high-density oligonucleotide arrays. Nat. Biotechnol. *14*:1675-1680.
22. **Macas, J., M. Nouzova, and D. Galbraith.** 1998. Adapting the Biomek 2000 laboratory automation workstation for printing DNA microarrays. BioTechniques *25*:106-110.
23. **Maier, E., S. Meier-Ewert, A.R. Ahmadi, J. Curtis, and H. Lehrach.** 1994. Application of robotic technology to automated sequence fingerprint analysis by oligonucleotide hybridisation. J. Biotech-

nol. *35*:191-203.
24.**Marshall, A. and J. Hodgson.** 1998. DNA chips: an array of possibilities. Nat. Biotechnol. *16*:27-31.
25.**Schena, M., D. Shalon, R.W. Davis, and P.O. Brown.** 1995. Quantitative monitoring of gene expression patterns with a complementary DNA microarray. Science *270*:467-470.
26.**Schena, M.** 1996. Genome analysis with gene expression microarrays. Bioessays *18*:427-431.
27.**Schena, M., D. Shalon, R. Heller, A. Chai, P.O. Brown, and R.W. Davis.** 1996. Parallel human genome analysis: microarray-based expression monitoring of 1000 genes. Proc. Natl. Acad. Sci. USA *93*:10614-10619.
28.**Schena, M., R.A. Heller, T.P Theriault, K. Konrad, E. Lachenmeier, and R.W. Davis.** 1998. Microarrays: biotechnology's discovery platform for functional genomics. Trends Biotechnol. *16*:301-306.
29.**Shalon, D., S.J. Smith, and P.O. Brown.** 1996. A DNA microarray system for analyzing complex DNA samples using two-color fluorescent probe hybridization. Genome Res. *6*:639-645.
30.**Southern, E., K. Mir, and M. Shchepinov.** 1999. Molecular interactions on microarrays. Nat. Genet. Suppl. *21*:5-9.
31.**Wodicka, L., H. Dong, M. Mittman, M-H. Ho, and D.J. Lockhart.** 1997. Genome-wide expression monitoring in *Saccharomyces cerevisiae.* Nat. Biotechnol. *15*:1359-1367.

3 Novel Microarray Printing and Detection Technologies

Myles L. Mace, Jr., Jean Montagu, Stanley D. Rose, and Greg McGuinness
Genetic MicroSystems, Inc., Woburn, MA, USA

INTRODUCTION

The Human Genome Project is generating an enormous and increasing amount of information concerning potential genetic targets for further study. At the same time, new tools of combinatorial chemistry are being used to synthesize large libraries of different molecules. Instead of thinking about individual biochemical reactions, scientists are rapidly adopting a new paradigm in which many reactions are monitored at once, enabling the study of complex pathways and systems. Investigators involved in basic genomic research and drug discovery, however, face major bottlenecks in their ability to gather and analyze all of the data effectively from all the experimentation that is now possible.

Microarray technology represents the most recent, and exciting, advance in the application of hybridization-based approaches to analysis in the biological sciences. It is now becoming generally recognized that microarray technology will be a fundamental tool used in future genomic research. As the technology becomes more widely accessible, a larger number of investigators will be able to shift their focus from a linear study of individual events, to the matrix analysis of complex systems and pathways. The technology also offers the possibility of providing similar benefits to routine analysis for clinical diagnostic or industrial analytical purposes. The obvious advantages offered by this massively parallel form of analysis include increased data per unit time, as well as significant reductions in the time of analysis, sample volumes required, and reagent and disposable consumption. Taken together, this represents considerable savings in the time and cost of associated labor. The major shortcomings of this technology, which is still in its relative infancy, include the large initial capital expense required for instrumentation, the

Microarray Biochip Technology
Edited by Mark Schena

lack of well-tested materials and protocols for analysis, and the inability to obtain truly quantitative information about the many data points acquired simultaneously. These problems are currently the intense focus of many researchers and, as the chapters in this book will attest, are becoming less problematic.

This chapter describes the approaches that Genetic MicroSystems (GMS) is using to develop and apply innovative, proprietary technologies to develop novel, enabling instrumentation for microarray-based analysis. The Pin-and-Ring™ (PAR™) spotting technology allows the creation of highly consistent array elements and offers a broad range of flexibility in terms of both the type of fluid spotted and the nature of the surface onto which these fluid spots are deposited. This technology has been incorporated into an integrated benchtop instrument for making spotted DNA microarrays that is called the GMS 417 Arrayer™. The company has also developed The Flying Objective™ scanning technology, which allows high-speed, sensitive collection of high-resolution microscopic images over large fields. This technology has been incorporated into an epifluorescent laser microscope called the GMS 418 Array Scanner™, a benchtop instrument used to capture images of fluorescent DNA microarrays.

BRIEF REVIEW OF MICROARRAY TECHNOLOGIES

Currently two approaches are used to create microarrays:

Light-Directed Synthesis

This technique, pioneered by Fodor et al. (5,6) and commercialized by Affymetrix, uses precision photolithographic masks to define the positions at which single, specific nucleotides are added to growing single-stranded nucleic acid chains. Through a stepwise series of defined nucleotide additions and light-directed chemical linking steps, high-density arrays of defined oligonucleotides are synthesized on a solid substrate. This approach is generally limited to polymers of ≤25 nucleotides in length. The fully detailed descriptions of this process are described in Chapter 6.

Microfluidic Delivery of Nucleic Acids

This section focuses on available instrumentation that is used to create microarrays via delivery techniques. Delivery technologies offer the flexibility of generating microarrays that contain a broad range of materials including plant, animal, human, fungal, and bacteria cells; viruses, peptides, antibodies, receptors, and other proteins; cDNA clones, DNA probes, peptide nucleic acids, genomic DNA, RNA probes, oligonucleotides, polymerase chain reaction (PCR) products, chemicals, and any type of small particulate in suspension. Suitable microarray substrates compatible with delivery approaches include chemically treated glass microscope slides, coverslips, plastics, membranes, gels, and other materials. The different delivery technologies offer, to a greater or lesser degree, the flexibility to deposit such a wide variety of samples, although some are more limited than others in terms of compatible sample material.

Regardless of the mechanism of microarray manufacture with a given delivery technology, certain parameters must be considered to obtain optimal results, and certain underlying factors are necessary for successful arraying and resultant data output. The key, and critical, factor is the creation of consistent and reproducible features or spots, particularly the uniformity of individual spots and of the entire array. Without spot uniformity, the data are not repeatable or reproducible, and consideration must be given to inconsistencies from a single instrument and from one instrument to another. In addition, several other experimental factors must be considered that affect successful microarray experimentation including array density, versatility, speed of operation, and environmental requirements such as humidity and temperature control.

Some of the typical parameters for current biochip arrayers are as follows:

- Number of spots for a typical array is 100 to 10 000, although 50 000 spots per microarray is possible.
- Number of simultaneous deposits is 1 to 32, depending on the configuration of pins, capillaries, or quills.
- Volume deposited per spot is 50 to 500 pL, although 10 pL to 10 nL is possible.
- Diameter of a spot is 100 to 250 μm, although 50 to 1000 μm is possible.

Certain applications can use simple yes or no answers (binary measurements) and therefore do not necessarily require a wide dynamic range of measurement or particularly consistent spots. Examples include certain genotyping, genetic mapping, and identity by descent (IBD) analyses. In general, however, high-quality microarrays provide data superior to those that have inconsistent amounts of sample at each position on the array. Gene expression and single-nucleotide polymorphism (SNP) applications, for example, benefit greatly from microarrays that have a precise and consistent amount of material at each address on the array. In addition, surface chemistry is pivotal with respect to accurate measurement of inter- and intraexperimental variation.

Quantitation of Volumes Deposited

The measurement of small liquid volumes, fractions of nanoliters, is extremely difficult. Most manufacturers of arrayers offer values for the volume deposited per spot. Piezoelectric-based spotters can deliver a continuous stream of droplets and can accurately measure the amount delivered with strobe photography. It has been reported that piezo-jetting can dispense volumes as low as 50 pL/dot with a great consistency of volume. Of course, larger volumes may be delivered with ease. The syringe technologies tend to dispense larger volumes and also suffer from material elasticity that affects repeatability at low volume settings. Contact-based dispensers (3) deliver picoliters to nanoliters of volume per spot. The solid or "replicator" pin dispensing method we used to quantify the volume deposited per spot did not correlate with the expected values. This was traced to an unexpectedly high rate of evaporative loss of water during the time between sample uptake and delivery to a nitrocellulose membrane using a procedure reported by Beckman (2).

To determine the evaporative loss using solid pins, a digital video camera was adapted to a microscope to record the volume of fluid that adheres to the pin after the pin punctures the sample in the ring, as well as the loss of sample volume on the end of the pin due to evaporation over time. Standard laboratory conditions of 35% to 40% relative humidity and 22°C were used for the experiment. Figure 1 shows the starting volume and fluid evaporation over time using a 125-µm pin that was dipped in and withdrawn from a microtiter sample plate. The results are presented as the composite of five separate experiments, from which we derived an equation relating evaporative loss to time under the experimental conditions.

The resulting equation was then used to predict the evaporative loss under conditions in which the pin and ring operate. Measurements were made with a video camera viewing the region of the meniscus in the ring under standard laboratory conditions. Figure 2 shows the extrapolated data. The large pin surface to sample volume ratio of very small pins causes extremely rapid sample evaporation, such that about 50% of a 250-pL volume is lost under standard laboratory conditions in 1 second. Independent measurements show that approximately 30 to 40 pL of liquid are displaced when the pin punctures the sample in the ring

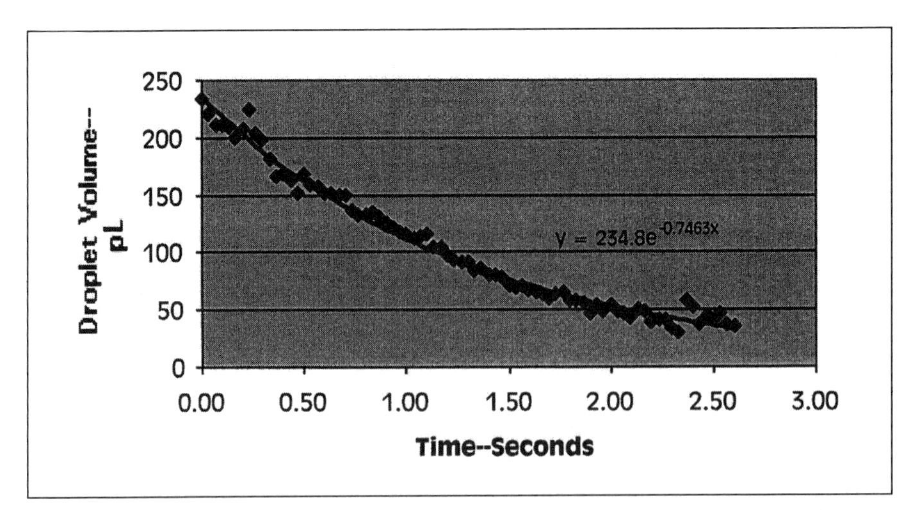

Figure 1. Droplet evaporation from a 125-µm solid pin (composite of five runs). The 240-pL volume was obtained by lifting the pin from a microplate sample using the "replicator" mode commonly used for solid pin printing tools. Volumes over time were calculated with a video camera.

assembly. Because the transfer time (duration between puncturing of the meniscus in the ring and transfer of material on the end of the pin to the surface) is only approximately 5 milliseconds, however, sample loss due to evaporation is negligible. Also, little evaporation of the sample held in the ring occurs during printing because the ring effectively encloses the aliquot and prevents evaporation. The PAR method was demonstrated to transfer a very constant volume per pin actuation.

Southern et al. (16) and Lemieux et al. (9), in recent reviews on molecular interactions on microarrays, suggest that a density of oligonucleotides of approximately 400 molecules/Å^2 on a treated glass surface is desirable and that a higher density, particularly of longer polymers, is likely to be detrimental to hybridization due to steric hindrance. This suggests that most reliable measurements are possible when the microarray is manufactured such that a tightly packed monolayer of duplex molecules would be seen if the hybridization process were taken to completion at a given site on the chip. Higher concentrations of target molecules can lead to steric hindrance, which inhibits hybridization, as well as reduced florescence resulting from shielding from the excitation laser light or emission quenching due to overlying molecules.

The first data point (Figure 1) shows the initial volume collected by the transfer pin when lifted from a microplate sample well, whereas the initial data point in Figure 2 represents the volume on the end of the pin after it has penetrated and exited the sample meniscus within the ring. Loading sample on a solid pin in the replicator mode results in approximately sevenfold more material on the end of the pin than when operating in the PAR mode; consequently, with the PAR system, a small volume of relatively concentrated material allows a nearly monolayer of molecules to be deposited, but without the drying artifacts caused by

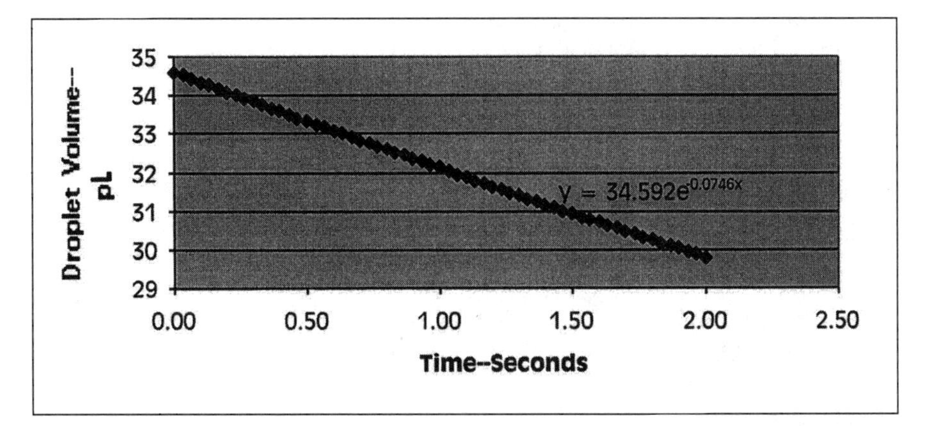

Figure 2. Predicted evaporative loss from 125-μm solid pins. Shown is the theoretical volume loss with a 125-μm solid pin used in the Pin and Ring Technology. The initial volume was obtained by puncturing the meniscus in the sample ring with the solid pin (see Figure 4A), and then examining the end of the pin with a video camera.

large volumes of solution, which frequently contain high concentrations of salt. It should be noted that the PAR system can be used in the replicator mode and that having this flexibility has some advantages for users who wish to develop or standardize protocols.

There are many important considerations for successful expression monitoring, but two critical factors are as follows:

- Consistent and reproducible spot formation and consistent performance from instrument to instrument and from array to array.
- Control of the volume and area deposited per spot.

In addition, several other experimental parameters must be met for robust expression analysis:

- Sufficient array density
- Versatility
- Speed of operation
- Environmental requirements
- Humidity and temperature control
- Cleanroom setting

Completely quantitative measurements are currently difficult to obtain using microarrays, due in part to the large number of variables and potential errors involved at each step of the experimental protocol. For example, a typical procedure might involve the following steps:

- Collection of mRNA sample
- Isolation of nucleic acid
- PCR
- Purification of the PCR products
- Robotic deposition of the DNA to create a microarray
- Hybridization of a fluorescent probe to the microarray
- Detection of fluorescence by a laser-based scanning microscope
- Analysis of the fluorescent image

Due to this high level of complexity, some investigators focus on relative quantitation between two samples. Absolute quantitation has remained a desired but unrealized goal for some investigators. Having said this, extremely useful data have been generated in numerous laboratories, and the ultimate goal is to elevate experimental sensitivity and true quantitation to the highest level.

ARRAYING TECHNOLOGIES

We briefly review currently available methods used to create microarrays via microfluidic delivery, and we reference other chapters for more detailed descriptions. Table 1 outlines a number of conceptually different methods of making microarrays.

Table 1. Microarray Technologies

Replicator Pins (Solid Pin)
Each pin picks up one sample at a time from one stationary location and transports it to one location on one chip.

Collecting, Holding and Depositing Devices (Quill and Split Pins)
One device performs all the spotting functions. Each device picks up a volume of fluid from one well, holds that fluid within it, and transports it to each of multiple locations where it stops. That same device is moved down to deposit a fraction of that volume at each location.

Pin and Ring Technology
Two different devices are used to make spots. One device takes a volume of fluid from one well and holds that volume while it is transported to each of multiple locations where it stops. A separate device takes a fraction of that transported volume and deposits it at each location.

Capillary Transport and Deposition
A capillary is permanently connected between the source of fluid. The other end of the capillary is moved to each location where a sample is deposited by contact.

Jetting
One device takes a volume of fluid from one well, holds it while it is transported to each of multiple locations where it does not stop. At each desired location a suitably timed high speed actuation takes place that projects a fraction of the transported volume.

Syringe

Since capillary tubes have been well developed for a variety of other applications, they were an obvious choice for early microarray manufacturing and have been used successfully by several groups. Graves et al. (7) present a summary of this work. Actual reported deposition volumes exceed 4 mL, or about 2 orders of magnitude greater than current techniques. As far as the authors are aware, there are no longer commercially available micro-syringe systems at this time.

Solid Pin "Replicator"

Replicating tools were initially designed to duplicate the contents of one microtiter plate into another, and subsequently this was extended to replicates, or arrays, using membranes and subsequently slides, as substrates. A typical configuration for the replicating head is an array of solid pins, generally in an 8 × 12 format, spaced at 9-mm centers that are compatible with 96- and 384-well plates. The pins are dipped into the wells, lifted, translated over the target, lowered to touch and transfer, and then repositioned over the wells to repeat the process. For a 96-well plate, the replicator configuration may have no more than 12 pins to fit the typical 25 × 76-mm size of glass slides. This printing device is dipped into a

96-well microplate, then lifted, and transferred onto one microscope slide. The process is repeated for the next slide, and so on.

After the spotting is repeated on all the slides on the platter, the next set of wells is addressed, and so on until the last well of the last microplate has been sampled. In their original application, solid pins were rounded at the tip and did not need to be located with great accuracy as they are quite small compared with the diameter of the microwells. More recently, solid pins with a flat tip have been developed that have a precise diameter. Replicators are best used to make low-density arrays because the printing process is inherently slow. When used in a highly stringent process such as the delivery of subnanoliter volumes of samples on a large number of microscope slides, several problems are encountered:

- The transfer time from the microplate to the slides can be a few seconds.
- The travel distance to reach the slides is variable, and therefore evaporation causes the proximal slides to have larger spots than the distal slides, although each spot on each slide should have approximately the same number of molecules.
- Large surface-to-volume ratio of nanoliter drops of fluid places conflicting demands on the humidity and temperature control. To minimize evaporation and drying of the sample, high humidity is absolutely necessary. However, the disadvantage of high humidity with respect to delivery is that it may prevent the sample from drying sufficiently on the substrate, thereby causing sample migration and spreading.

This technology is very well suited for ± studies with low-density arrays configured on either membranes or slides. All other methods have a sample "reservoir" at the point of delivery and therefore do not require that the printing head return to the microplate between printed slides. The standard GMS instrument loads sample into the ring from a standard microplate located on the bed of the printing instrument, although in one GMS design the entire plate is transported to the delivery location.

Quill and Split Pin

Quill-based arrayers, as described first by Schena et al. (14), and later variations on the same theme, withdraw a small volume of fluid within the depositing device from a microwell plate by capillary action. A robot then moves the head with quills to the desired location for dispensing. The quill carries the sample to all spotting locations, where only a small fraction from that larger fluid volume is deposited. The forces acting on the fluid held in the tweezer must be overcome for the fluid to be released. Accelerating and then decelerating by impacting the quill on the glass slide by tapping achieves this. When the tip of the quill hits the glass slide, the meniscus is extended beyond the tip and transferred onto the substrate. As the spotting process proceeds, the mass of the volume remaining in the quill decreases, thereby decreasing the volume deposited. Although carrying a large volume of fluid minimizes spotting variability, quill-based deposition over a very large number of slides may change gradually. Because tapping on the surface is required for fluid transfer, the choice of material is critical to ensure a practical life of small tipped quills. Strong surface tension forces prevent the fluid from escaping. The opening at the gap of the quill is reported to be between 10 and 100 μm, with the larger gaps used early in the development of the technology.

The quill represents an ingenious design adaptation and has found successful use in early research applications. There are some shortcomings to the quill design, however, one of which is that clogging of the gap can be a problem. The small points required to produce small spots are susceptible to clogging from dust, particulates, evaporated buffer crystals, or other contaminants. The tapping on the surface required for printing with quills tends to cause deformation of the tip that may change the spot diameter, or in severe cases cause the quill to close. Therefore, replacement cost is an important factor in the use of quill technology. An excellent description of many quill and split pin designs can be found at the Brown Lab Web site (http://cmgm.stanford.edu/pbrown/).

Jetting and Piezoelectric Pumps

Jetting is a technology that finds its roots in electrocardiography dating back to the late 1930s. Its success in other fields expanded after technological advances made it possible to produce ink with very predictable properties that was free of particulates at the micron level. Piezo pumps offer the capability of controllable and extremely fast jetting rates (up to many kilohertz) and repeatable volume deposition. Piezo pumps offer the benefits of consistency of drop size for a given homogenous fluid, speed, and (potentially) low cost. Most are unidirectional pumps that need to be directly connected, by flexible capillary tubing, to a source of sample supply or wash solution. The capillary and jet orifices must be of sufficient inner diameter so that long molecules are not sheared. The void volume of fluid contained in the capillary may be 100 to 500 μL and not recoverable. Bidirectional pumps suffer from low suction capability.

Piezo-based jetting technology has been explored for an additional application, the potential to synthesize defined sequences in situ. Currently, additional development needs to be performed for jetting to be highly competitive for biochip fabrication. Some of the problems that need to be overcome include the fact that high velocity of the fluid through the jetting orifice may cause shearing of DNAs. Small "satellite" droplets can be formed outside the central spot, delivery heads can become clogged by particulates and air bubbles that can interrupt fluid liquid flow, and some jetting technologies have relatively large void volumes. Although ink-jet technology offers a very rapid method of delivery and shows great potential, the entire process has not yet achieved its expected promise, as evidenced by the relatively small number of commercial instruments available at this time.

Pin and Ring

The PAR arraying method was initially described by Rose (13) and represents an alternative to the methods described above. It is designed to capture the benefits of microarray manufacture, while overcoming certain shortcomings of previously employed microfluidic designs. The basic goal was to create microarrays whose primary requirement was consistency of spotted material, so that truly quantitative data could be obtained from array-based experiments. It also meant designing a system that would have few moving parts and would therefore be reliable by design, flexible in operation, relatively simple to manufacture, easy to clean, and compatible with samples that contain particulates such as microorganisms or even cells.

The basic mechanism of sample pickup is shown schematically in Figure 3. A single ring and pin are centered over the source well poised to acquire the sample (Figure 3A). The pin is kept stationary, while the ring assembly moves down independently into the well of a microtiter plate (Figure 3B). When the ring is immersed within the solution, it fills with fluid, and this fluid, held in place by surface tension and the surface activity of the inner portion of the ring, forms a meniscus after the ring is raised (Figure 3C). The inner surface of the ring is treated to increase the "holding power" of the contained liquid. Both the materials used to construct the ring and the surface properties of the inner face of the ring are major factors that contribute to the amount of material withdrawn from the well. In Figure 3D the filled ring is ready to move to the sites of deposition. It should be noted that the ring has an uptake reservoir sufficient to print 600 to

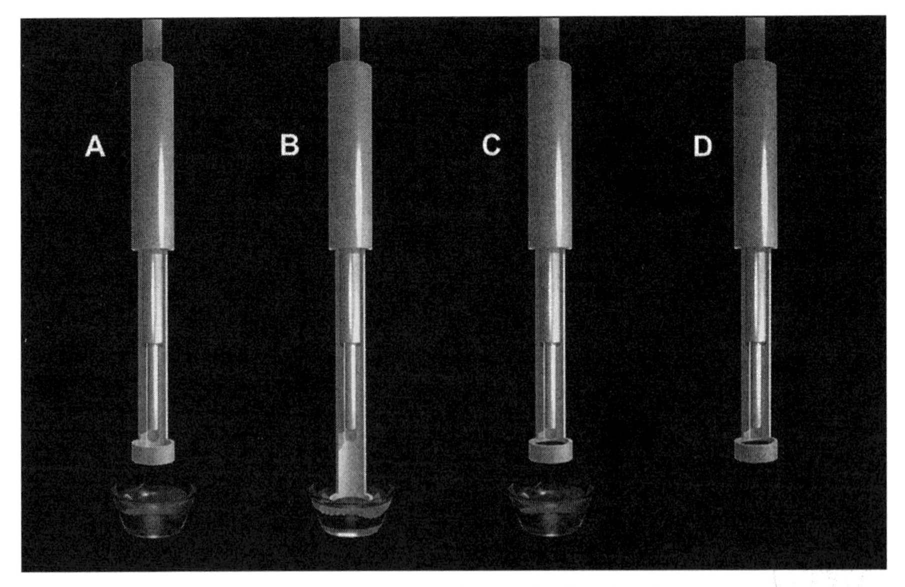

Figure 3. Schematic of sample uptake with the Pin and Ring technology (A–D).

1000 spots, depending on both the diameter of the pin and the printing substrate. Rings can be selected to hold 0.5 to 3.0 μL.

Figure 4 illustrates the actual printing process. The pin-ring assembly is positioned over the target, the ring maintains its Z-height, and the pin is lowered through the meniscus (Figure 4A). When the pin has protruded far enough through the meniscus, the punctured portion of the meniscus snaps back into the ring and leaves a pendant drop of the sample liquid at the end of the solid pin. When the meniscus of liquid on the end of the pin makes initial contact with the substrate, the combination of surface tension and gravity transfers the droplet onto the substrate (Figure 4B). The droplet is released when the pin is lifted away from the substrate (Figure 4C), and the process is ready to be repeated when the pin is moved back through and above the meniscus of fluid in the ring (Figure 4D). Of course, the printing cycle can be repeated at the same location for delivering additional sample, shifted slightly for printing duplicate spots, or moved to another slide.

It should be remembered that the pin and the ring perform different and independent functions. In this respect, PAR differs from the other devices that carry the sample supply within the delivery device. The pin and the ring can be optimized for the two different functions and may be modified to allow a larger sample uptake. Simple substitution of pins with different diameters allows the user to choose many different spot sizes. It also offers a number of additional capabilities and benefits. The ring withdraws a relatively small sample from the microwell plate and transports it to the delivery location. The ring is configured to minimize fluid exposure and thus evaporative sample loss in dry printing environments. In addition, it is an open structure with no occluded regions, affording easy cleaning, which reduces the level of cross-contamination between samples even with very short wash cycles. The open design of the ring makes it

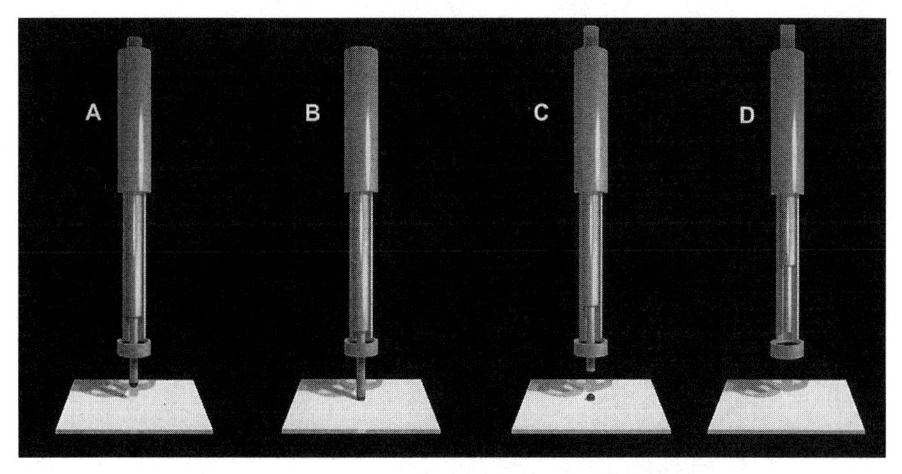

Figure 4. Schematic of the printing process with the Pin and Ring technology (A–D).

resistant to materials that cause failure in traditional spotting methods: dust, particulate matter, high-viscosity fluids, debris, buffers or salts, and other material. The ring geometry can accommodate a wide variety of volumes and fluid viscosity because the ring is not subjected to any shock or acceleration during spotting, and the ratio of ring to pin diameter is large. Figure 5 is an illustration of four pin and rings assembled into a single spotting head. Other configurations are easily accomplished.

Consistency

The pin performs the spotting function only. It has a sequence that is repeated in an identical manner for every cycle and at each slide location, so the same process is used to print all the spots (Figure 6). As the pin passes through the ring and captures a volume of fluid on its tip, it has a transit of about 50 milliseconds, thereby minimizing any evaporative loss of the aliquot deposited. The final result is positional accuracy that allows highly consistent and reproducible microarrays (Table 2). Results

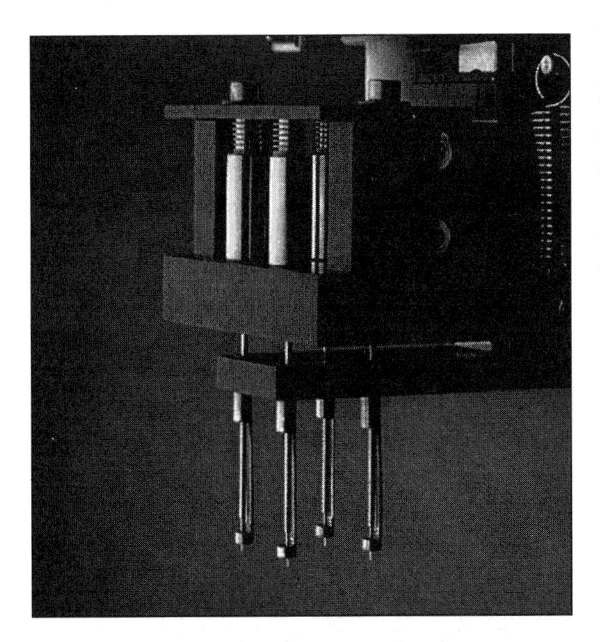

Figure 5. Photograph of the printing apparatus of the Pin and Ring technology.

Figure 6. Microarray printed with the Pin and Ring technology. Plant DNA printed as 150-μm spots on polylysine-coated slides stained with eosin. (Data courtesy of Ray Samaha, EOS Biotechnology.) (**See color plate A4.**)

Table 2. Pin and Ring Printing Precision* (Shown as Percent SD)

Instrument	Quadrant 1	Quadrant 2	Quadrant 3	Quadrant 4	Total number
1	2.3	2.2	2.4	2.1	1600
2	4.8	4.7	4.0	4.8	1600
3	5.1	3.3	4.7	5.2	1600
4	7.1	7.1	6.6	5.5	1600
5	5	6.5	6.0	6.6	1600

*Shown are printing precision data for five different Pin and Ring Instruments (Genetic MicroSystems) with a four-pin delivery head printing four separate subarrays (quadrants 1–4). Percent standard deviations of spot diameters for each 20 × 20 quadrant are shown. A total of 4 × 20 × 20 (1600) spots was analyzed for each instrument.

from the delivery with a single pin were analyzed as a quadrant Q (Table 2). Spots resulting from the production of a single pin in a 20 × 20-spot array were measured and the percent standard deviations of average spot diameters produced by individual pins calculated. The data show that individual pins deposit material precisely and with remarkable consistency. Pin-to-pin performance measurements obtained with five different instruments (randomly selected from early production runs) over a period of several months showed a very small level of variation. The volume of the fluid to be deposited by a PAR system is influenced by its viscosity, however. The size of the resulting spot is a function of the pin diameter, pin material, viscosity of the fluid, and dynamics of the interaction between the fluid and the surface, for example, aqueous solutions placed on hydrophobic or hydrophilic surfaces will spread to different degrees before evaporating.

Flexible Substrates

Another important feature of the PAR architecture is that it is capable of spotting on soft substrates such as agar, gels, or delicate membranes including nitrocellulose or nylon. Figure 7 is an example of DNA spotted on a nitrocellulose membrane resting on a wet, denaturing piece of filter paper. Particularly noteworthy is the fact that the spots are consistent and that there is neither deformation nor penetration of the fragile membrane. Successful spotting has also been accomplished on polyacrylamide gel and agar (data not shown). Figure 8 shows a Southern blot (15) microarray printed on nitrocellulose attached to a glass slide.

The principle advantage of spotting onto membrane substrates is that the binding chemistry between DNA and nitrocellulose has been well characterized

(15) and is highly effective. Simply placing a sample onto a microarray substrate is more straightforward than attaching the sample molecules in a manner that is resistant to the rigors of hybridization and washing. For fluorescent detection schemes, however, the intrinsic fluorescence of many membranes poses a problem that is yet to be completely overcome, although early experiments are yielding promising results (Figure 9).

CONCLUSIONS

Microarrays prepared via microfluidic delivery have broad applicability in life science research and development. The manufacture of microarrays has generated a large variety of different processes, and no one method dominates the field, partly because at least two different markets exist: the research market, in which versatility and repeatability may be more important than throughput, and the microarray production market, in which repeatability, unit cost, and throughput dominate as requirements. In addition, each market can be split according to assay applications.

The solid pin "replicator" can be successfully used to make microarrays for applications requiring qualitative answers, and when not much need for a complex and high-density array configuration exists. Examples include certain genotyping, mutation detection, and mapping experiments. When high throughput and unit cost are the dominant requirement, jetting may become the preferred technology. More development is required, however, to develop reliable and nonde-

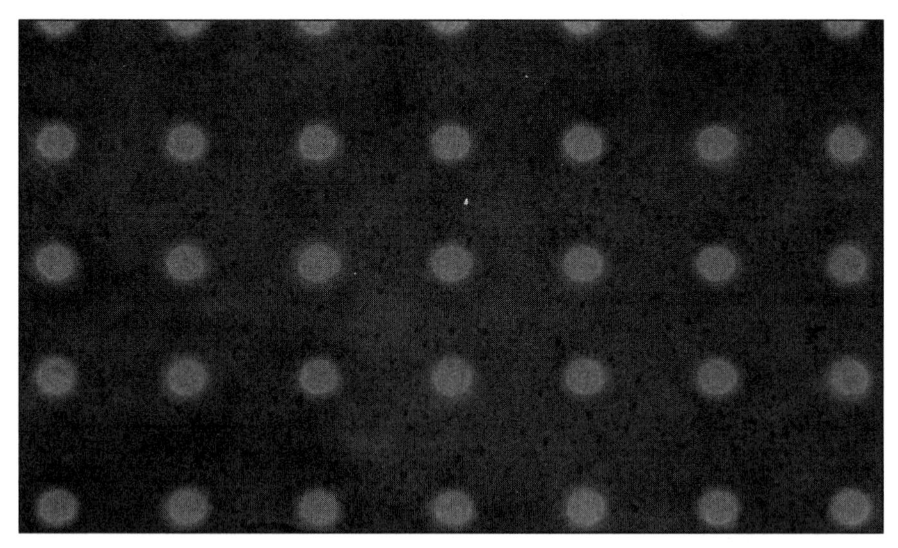

Figure 7. Pin and Ring printing on nitrocellulose. Shown is a transmission light micrograph of DNA spotted directly onto a nitrocellulose membrane. (Data courtesy of Dr. Bertrand Jordan.) **(See color plate A4.)**

structive fluid handling processes utilizing this technology.

For applications in which quantitative analysis is required, such as gene expression monitoring, consistency of spotting is a prerequisite, and the PAR technology should be considered. It provides the most consistent spotting of any microarray delivery technology developed to date. It is also adaptable to a broad range of sample types delivered to a broad choice of substrates.

MICROARRAY SCANNING SYSTEMS

Microarray manufacture is but a first step in the microarray experimental cycle. Additional downstream requirements include hybridization with labeled targets and data acquisition. Data acquisition is the role of the imaging system and, preferably, the imaging system is capable of viewing the entire array in a single

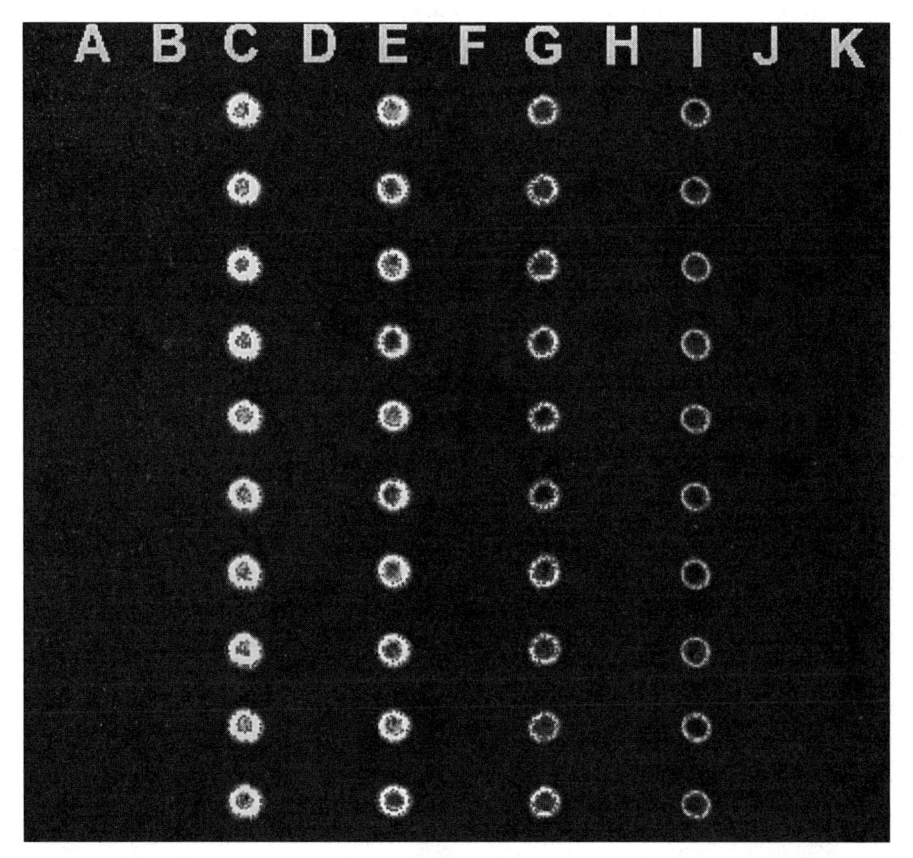

Figure 8. Southern blot microarrays made with the Pin and Ring. (A–K) A dilution series of oligonucleotides interspersed with a nonhomologous sequence was printed onto a nitrocellulose-coated glass microscope slide, and then hybridized with a Cy5-labeled target. No sample carryover is detected. (**See color plate A5.**)

Table 3. Microarray Scanning Systems

Optical design	Manufacturer
XYZ stage/microscope objective	GSI Lumonics
Preobjective scanning	Hewlett Packard, Amersham Pharmacia Biotech/ Molecular Dynamics
Charge-coupled device (CCD) camera	API, Genomic Solutions
Flying objective scanning	GMS, Axon, Virtek

image. Three basic technologies exploit conventional confocal fluorescent microscopes to meet the needs of microarray imaging (Table 3).

All the noncharged coupled device (CCD) instruments are epifluorescent pseudo-confocal laser scanning microscopes. They are pseudo-confocal systems because they have two optical paths. One path is the excitation (or laser) path that defines the pixel size, typically between 3 and 10 μm. It has a relatively low NA, around 0.1, and consequently a greater depth of field. The other path is the emission [or photomultiplier tube (PMT)] path, which is designed to maximize energy collection and therefore has as high a numerical aperture as possible. This also has the tradeoff of a correspondingly small depth of field.

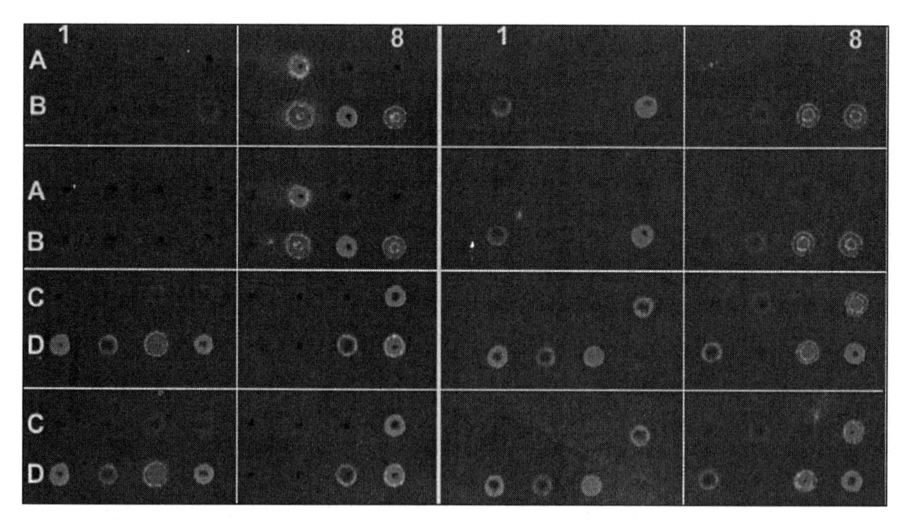

Figure 9. Oligonucleotide arrays made with the Pin and Ring. Shown is a composite image of oligonucleotide microarrays printed on nitrocellulose and hybridized with PCR products from two different individuals (separated by vertical yellow line). Each array contains 16 unique probes with duplicate rows of spots indicated by capital letters. Specificity, repeatability, and signal to background definition are well suited for single-nucleotide polymorphism (SNP) and expression analyses. **(See color plate A6.)**

The dramatic benefit of confocal microscopy, namely, rejection of out of focus background signal, is not practical for biochip array scanning. The depth of field of a true confocal microscope that uses the same NA for both lenses is $d = 2\lambda/NA^2$. For a 0.7-NA lens, this yields a depth of field of approximately 2 μm in the visible wavelengths. This small depth of field is used to create "optical sections" of biological material and has proved invaluable in revealing the three-dimensional structures of tissues and other samples. However, the practical considerations for microarray reading demand a large depth of field, which is required to accommodate a lens, stage, and slide tolerances and imperfections. A minimum of 30 μm is typical for commercial instruments.

A number of performance tradeoffs are required in the design of microarray scanning instruments. A useful approximate figure of merit (FM) has been derived to assist in the evaluation of an instrument's design. The laser wavelength used and the absorption coefficient of the chosen probe have to be considered in this comparative evaluation:

$$\text{FM} = \text{laser power (mW)} \times \text{relative absorption coefficient (nondimensional)} \times 1/\text{scan rate} \times \text{objective collection efficiency } (\text{NA}^2)$$

All three technologies have been developed to meet the required performances and will be reviewed here.

System Performance

Alexay et al. (1) have tabulated the criteria for analysis of fluorescent samples, and their data are summarized in Table 4. They judiciously point out that uniform performance over the entire slide is "critical for precise results." Consistent data at all locations within the field of view, and for all wavelengths, is a critical requirement. The handling of the substrate is an integral part of this design requirement.

Fluorescently labeled biological material can be located on either the front or back of the microscope slide for successful imaging. Because the flatness and thickness of the glass substrate are frequently not uniform at the micron level, it is desirable for the instrument to have as great a depth of field as possible. The method of substrate support is also important. In addition to stationary location, it should not deform or tilt the substrate, as field flatness is critical for uniform intensity. Also, the mechanism that transfers the slide into the instrument should be amenable to interfacing with an automatic loading station for unattended operation. The resolution of these design requirements greatly influences the choice of optical/mechanical architecture.

Optical System Design

The combined requirements of low light sensitivity, large dynamic range, and low cost led to a single-detector design, eliminating multielement solid-state de-

Table 4. Analysis of Fluorescent Samples*

Specification	Value	Note
Scan area	4.2 mm diagonal	20 × 65 mm is typical for most microarrays
Resolution	3.5 μm	10 μm is sufficient for 100-μm spots
Numerical aperature, pixel definition	0.4	
Intensity uniformity (*x*,*y*)	95%	Required for preobjective scanning
Spatial uniformity (*x*,*y*)	98%	
Dynamic range	>5 orders	
Polychromatic range	500–750 nm	550 and 670 nm for Cy3 and Cy5, respectively
Scan line speed	80 Hz	2 min/slide/channel acceptable
Repeatability static	1%	
Thru-focus sensitivity	1% signal change over 20 μm	
Field flatness variation	± 10 μm	Necessary for preobjective scanning only
Sample thickness	Negligible	
Working distance	>3 mm	With no interference, 0.5 mm is acceptable

*Data from Alexay et al. (1).

tectors (both linear and area CCDs). Below 600 nm, the best cooled single-element solid-state detector is an order of magnitude less sensitive (signal-to-noise ratio) than PMTs. The critical significance of repeatability and field uniformity is best achieved when every slide location is addressed in an identical and preferably telecentric manner for both excitation and collection.

Preobjective Scanning

Conventional laser scanning epifluorescent microscopes (Figure 10) have been commercially available for the past 25 years. Zeiss, Nikon, Bio-Rad, and Olympus are common names in that field. Because fluorescence is a low-energy phenomenon, a high numerical aperture is preferable. This yields a very small field of view, well under 1 mm^2, so that multiple images are required to cover the large field of view contained in microarrays. By limiting the spectral range to 1 or 2 wavelengths and accepting a lower NA, it is possible to design an instrument with a larger field of view (up to 10 mm). The Avalanche™ from Amersham Pharmacia Biotech/Molecular Dynamics is a scanner of this construction. The Avalanche objective is a nine-element design with a 10-mm field of view. It has a comparatively low NA, approximately 0.25.

To make the best use of this remarkable lens, the microscope slide is translated under the objective and the beam scanned over the microscope slide. To address an area larger than 10 mm wide, the slide is translated sideways and scanned to capture a second swath. The two swaths are later stitched together by the instrument's computer. The major benefit of a 0.25 NA is a large depth of field, approximately 16 µm. This instrument is also closer to a true confocal microscope than the other technologies described. However, this does not reduce its sensitivity to background signal, as the depth of field is kept high to accommodate imperfections in the field-flattening performance of the objective lens. Alexay allocates 20 µm to accommodate for errors in the lens' field flatness.

Classic Scanning Systems Design

The Minsky (10) design (Figure 11) offers a cost benefit, as it is designed around a commercial microscope objective with six elements; the tradeoffs are limitations in scan speed and reliability. The working clearance of high-power objectives is quite small, and the slide needs to be held reliably to avoid cata-

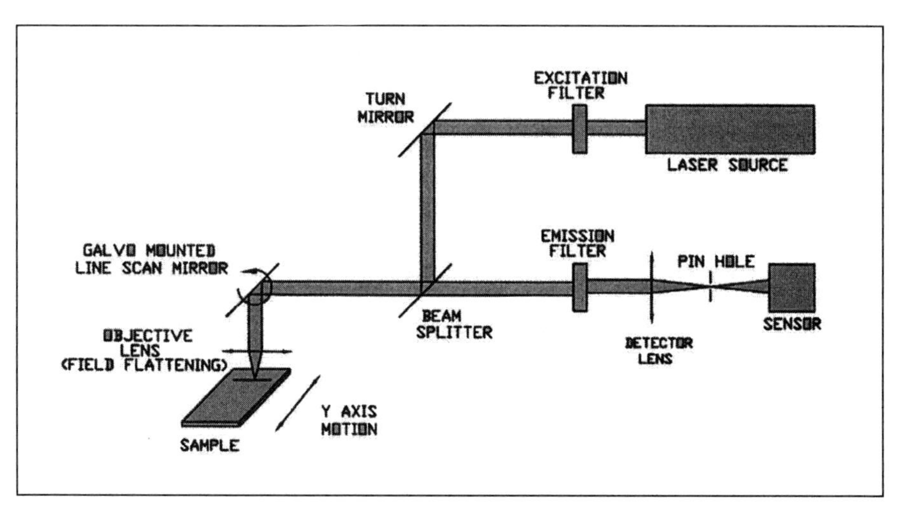

Figure 10. Preobjective line scan architecture for microarray detection.

strophic interference in a dynamic environment. The slide anchor mechanism must also be light enough to permit a reasonable scan rate without excessive vibration. Finally, high acceleration forces may affect the integrity of the microarray if an antifade or other liquid material is used with a coverslip. The Smith and Brown Labs at Stanford have designed and implemented a functional Minsky scanner (http://cmgm.stanford.edu/pbrown/). It is an effective, low-cost design; the spot size is defined by the laser beam diameter, and the slide stage moves mechanically side to side and simultaneously along its long axis. A variation of the design is commercially available from GSI Lumonics as the ScanArray® series of instruments. In the GSI Lumonics instruments, the slide is held in place by a spring-loaded device that forces the slide into contact with the top surface to a registration plane. That entire mechanism is then mounted on a stage driven by a galvanometer in the relatively fast X-axis and slower stepper motor driving the Y-axis (11). This design requires less than 3 minutes to scan an 18 × 18-mm field of a slide at 2 wavelengths scanning at 10-μm resolution.

Flying Objective Scanning Microscope

The flying objective architecture avoids the expense and difficult requirements posed in designing a classic large, flat field of view lens by replacing it with a relatively simple objective lens that is moved across the sample area. In other words, instead of moving the slide and its holding device in the fast axis, the beam of light is moved instead. This approach requires good mechanical alignment throughout the entire range of motion and carries the additional benefit of a high NA objective lens without compromising large depth of field. It also offers a

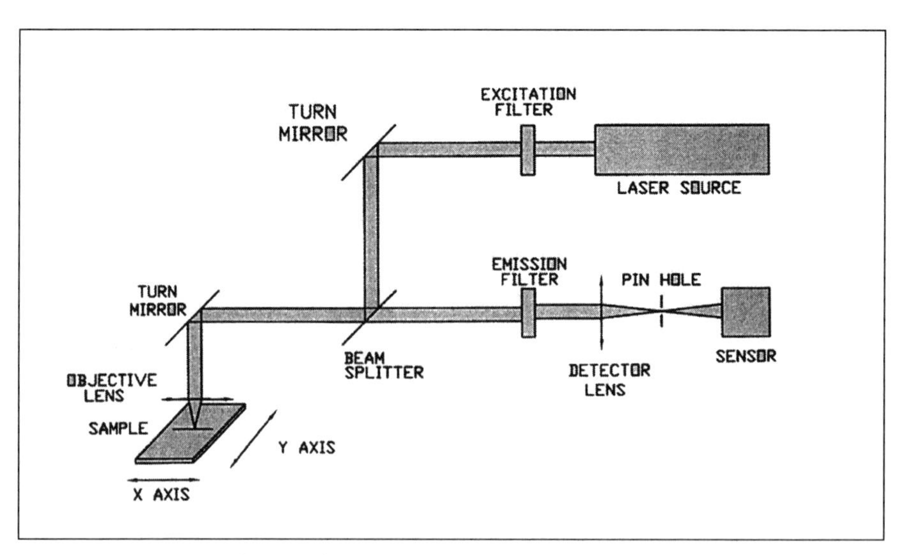

Figure 11. Moving stage architecture for microarray detection.

high degree of measurement uniformity across the field of view. All pixels are acquired in an identical manner so that field flatness is not a problem. We believe that this is the most desirable choice in scan architectures for microarray applications in which flexibility, speed of acquisition, energy collection efficiency, and cost are the primary design goals. There are two basic designs currently available: one oscillates an objective lens mounted on a linear rail, and the other oscillates an objective lens mounted at the end of a rotating arm.

Rectilinear flying objective microscope. Hueton and Van Gelder (8) describe the application of this concept. Briefly, and by way of specific commercial examples, the scanning system offered by Axon Instruments uses a voice coil to move a lens in a linear motion, while the Virtek instrument oscillates a lens using a scanner in a construction similar to that described by Montagu and Pelsue (11).

In both examples, the path length of the collimated beam of excitation and emission light changes between the moving lens and the stationary part of the optical path as the beam is scanned over the target (Figure 12). The properties of the optical system must remain unchanged as the length of the optical path changes to image the full 25-mm width of the microscope slide (with even greater demands made when scanning across larger formats). Such a requirement demands both an exceptionally stable and rigid structure as well as one that will be robust for extended use. The complexity of building a reliable, low-cost, low-inertia, stable high-speed linear scanning system is rewarded by the direct acquisition of a cartesian image. Data may be acquired in both directions of the lens' motion, thus doubling the effective speed of the system.

Rotary flying objective microscope. An alternative method to creating a rapidly scanning beam of light is the rotary architecture (GMS patent applied

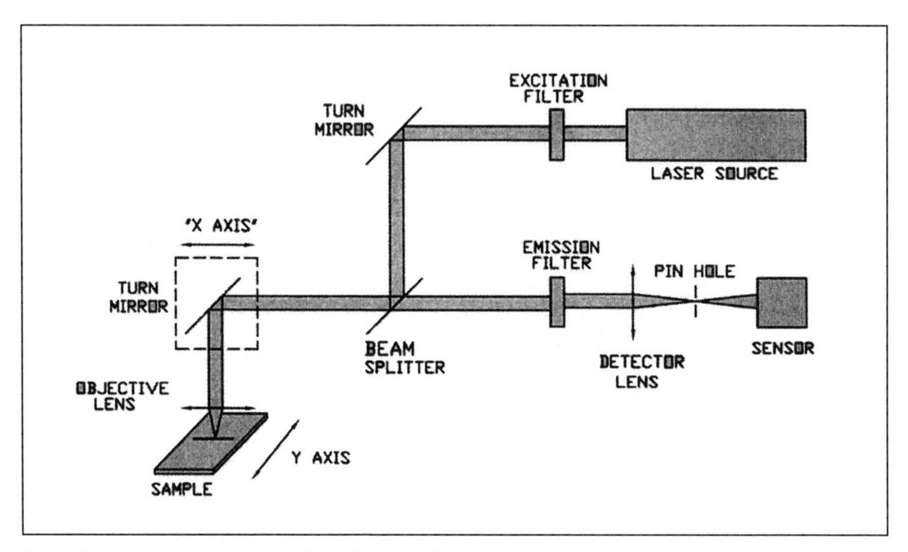

Figure 12. Rectilinear flying objective architecture for microarray detection.

for) that offers both speed and a constant optical path length. A periscope couples the scanning objective to the stationary elements of the microscope and thus keeps the light path length constant as it oscillates to scan the light beam (Figure 13). As the arc covering the width of the slide is in polar coordinates, the resultant image is instantly converted to cartesian coordinates by the instrument's computer so the image is then directly correlated with the microarray on the substrate. A micro-, aspheric, one-element lens is carried on a counterbalanced arm that extends perpendicular to the shaft of an oscillating motor and that is fitted with a position transducer (a galvanometer). The arm also carries mirrors that receive laser light on the axis of rotation via a periscope and send it out the length of the arm and down through the lens (Figure 14).

The field of view has the advantage of always being on axis. Consequently, this optical design eliminates all the common sources of lateral or chromatic aberrations found in preobjective line scan systems or linear translators. Because the moving lens weighs only a fraction of a gram, the load on the galvanometer is modest. The scan rate is 60 Hz with a 70% duty cycle. The oscillating optical subsystem is coupled to the stationary beams through a "periscope" on an axis with the armature of the scanner. This does not require critical alignment because the light beam at this stage is collimated, and the motions are periodic.

The slide handling is a critical aspect of all microscopes, especially when autoloading is used. The industry standard has defined slides and stages such that slides are supported on the back surface and brought into focus. A number of automatic focusing algorithms have been standardized and are well accepted. The GMS 418 Array Scanner offers a simple and novel slide-handling concept, integrated with the flying objective design. The oscillating mechanism maintains the objective lens in a plane normal to the axis of rotation. If the slide is held in a second plane, the two planes intersect to form a line. The design holds that line normal to the axis of rotation and normal to the line of symmetry of the scan

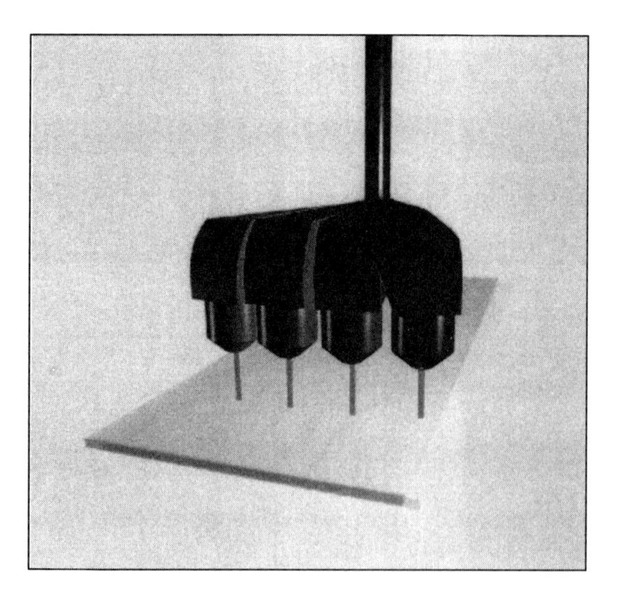

Figure 13. Rotary architecture for microarray detection. A stroboscopic depiction of the rotating arm moving over the entire width of a slide is shown. The light beam is shown as discrete rays, but it scans a continuous path from side to side. **(See color plate A6.)**

motion. Under these conditions, it is possible to tilt the plane of the slide to focus the slide and keep it in focus as it translates under the objective. The supporting plate may be moved by a stepper motor (with manual override) for autofocus achieved by measuring the high-frequency components of the light reflected by the sample.

Software and Scanning Coordinate Conversion

The user interface on the GMS 418 Array Scanner is a flexible and simple Windows® Graphical User Interface (GUI). The GUI provides the operator an easy means for controlling the stage position and movements for inserting and scanning of slides, as well as selecting excitation sources, power levels, and PMT gain (i.e., anode voltage). The control software automatically selects the appropriate emission filter for a given excitation wavelength.

Once the desired scanning parameters have been set, a controller D/A channel outputs the sawtooth drive signal to the galvanometer. A pixel clock, located in the microscope, uses the position feedback from the galvanometer to produce the needed pixel and line sync information to drive the A/D channels and control the motion of the stage between scan lines. This ensures that each image is in direct pixel registration with other images taken under the same positional parameters; therefore, two images taken at different wavelengths are in precise register. Fluorescence and laser power values are acquired for each pixel, and normalization of fluorescence data to fluctuations of laser excitation power is performed by the systems DSP. PMT and laser power signals are processed through matched, sophisticated high-order, phase-corrected analog filters before being digitized in order to provide enhanced signal-to-noise performance.

The sensitivity, dynamic range, and speed of the system combine to provide very high performance under a wide variety of conditions. Figure 15 illustrates the

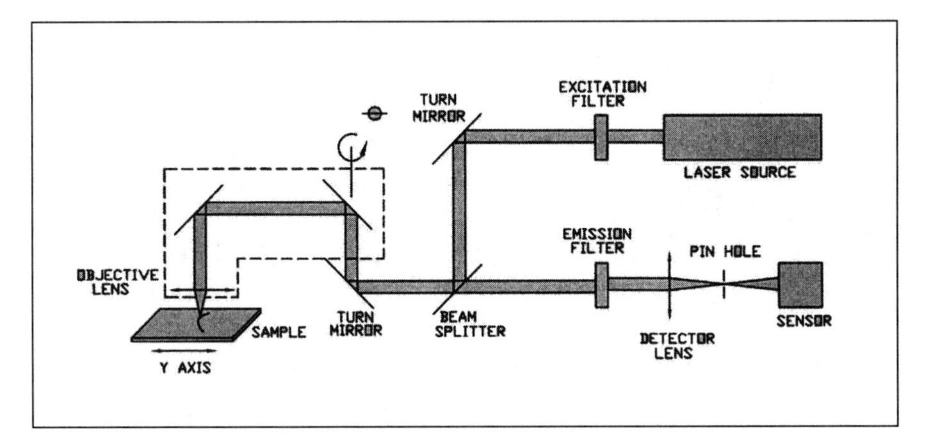

Figure 14. Rotary flying objective architecture for microarray detection.

GMS 418 Sensitivity

$FL/\mu m^2$:	0.28	0.85	2.8	8.5	28	LOD (S/N > 3:1)
Nitrocellulose:						8.5
Poly-L-Lysine:						0.85
Amino-silane:						0.85

Cy3-end-labeled oligosspotted onto various surfaces; values are fluors/μm^2;
all samples diluted in 250 ng/µL Lambda DNA; SSC blanks spotted between each sample; 75% laser power, 75% PMT gain used for all scans;
stock concentration of 0.25mM = 0.25 nmole/uL ~ 150,000,000 molecules/100pL;
spot diameter 150 µm; spot area 17,663µm2; spot volume 33pL

Figure 15. Sensitivity of the GMS 418 scanner. (See color plate A7.)

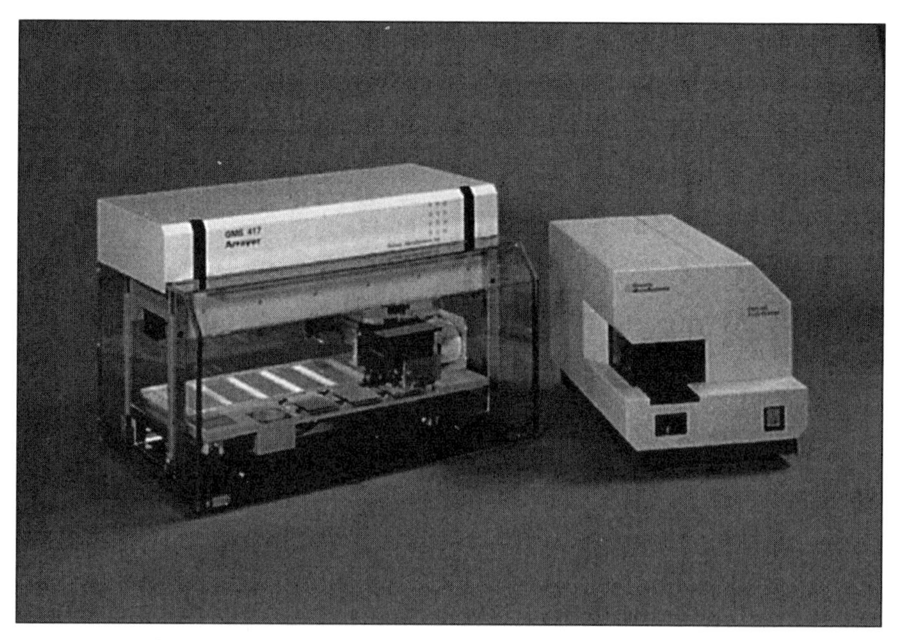

Figure 16. Complete microarray system from Genetic MicroSystems. Shown is a photograph of the GMS 417 arrayer and GMS 418 scanner, with sensitivity data for the latter.

sensitivity of the system using glass slides pretreated with polylysine and aminosilane, as well as nitrocellulose membranes bonded to slides. The scanning speed is 60 Hz, and at 10-μm pixel size covers an 18 × 18-mm area in 1 minute at two wavelengths. Averaging or integrating a series of images can be used to minimize random noise.

CONCLUSIONS

The array-making technology described here provides a novel means for very rapid microarray manufacture, offering consistent spot size, a range of spot diameters, and flexibility in terms of array configuration. The unique PAR design is intrinsically simple and allows microfluidic delivery of an extremely wide variety of samples from nucleic acids to proteins, and even delivery of microbes and cell suspensions. The design is inherently rugged and performs reliably over extended periods of operation. The oscillating periscope also fulfills the goals of simplicity, rugged design, and high performance. Together, the arrayer and scanner provide an integrated system (Figure 16) for manufacturing and imaging microarrays and give particularly good access to the unlimited power of microarrays by individual investigators.

ACKNOWLEDGMENTS

The authors wish to thank the numerous contributors to the design and development of these instruments, in particular, Peter Honkanen, Nate Weiner, Peter Flowers, Tim Woolaver, Frank Pagliughi, and Jim Overbeck. We also thank our many collaborators in academia and industry and are particularly grateful to Vivian Cheung (Children's Hospital of Philadelphia), Buddy Brownstein (Washington University School of Medicine), Tom Freeman (Sanger Center), and Ray Samaha (EOS Biotechnology, currently at Mendel Biotechnology).

REFERENCES

1. **Alexay, C., R. Kain, D. Hanzel, and R. Johnson.** 1996. Fluorescence scanner employing a macroscanning objective. SPIE (The International Society of Optical Engineering) **2705/63**.
2. **BioRobotics Technical Information T-1833A.** 1996. Beckman Instrument Company.
3. **Brown, P.O. and D. Shalon.** 1998. Methods of fabricating microarrays of biological substances. U.S. Patent 5,807,522.
4. **Cheung, V.G., M. Morley, F. Aguilar, A. Massimi, R. Kucherlapati, and G. Childs.** (1999). Making and reading microarrays. Nat. Genet. (Suppl.) *21*:15-19.
5. **Fodor, S.A., J.L. Read, M.C. Pirrung, L. Stryer, A. Tsai Lu, and D. Solas.** 1991. Light-directed, spatially addressable parallel chemical synthesis. Science *251*:767-773.
6. **Fodor, S.A., R.P. Rava, X.C. Huang, A.C. Pease, C.P. Holmes, and C.L. Adams.** 1993. Multiplexed biochemical assays with biological chips. Nature *364*:555-556.

7. **Graves, D.J., H.J. Su, S.E. McKenzie, and P. Fortina.** 1998. System for preparing microhybridization arrays on glass slides. Anal. Chem. *70*:5085-5092.
8. **Hueton, I. and E. Van Gelder.** 1995. High-speed fluorescence scanner. U.S. Patent 5,459,325.
9. **Lemieux, B., A. Aharoni, and M. Schena.** 1998. Overview of DNA chip technology. Mol. Breeding *4*:277-289.
10. **Minsky, M.** 1961. Microscopy apparatus. U.S. Patent 3,013,467.
11. **Montagu, J.I. and K.A. Pelsue.** 1985. Positioner for optical element. U.S. Patent 4,525,030.
12. **Roach, D.J. and R.F. Johnston.** 1998. High-speed liquid deposition device for biological molecule array formation. U.S. Patent 5,770,151.
13. **Rose, S.D.** 1998. Application of a novel microarraying system in genomics research and drug discovery. JALA *3*:53-56.
14. **Schena, M., D. Shalon, R.W. Davis, and P.O. Brown.** 1995. Quantitative monitoring of gene expression patterns with a complementary DNA microarray. Science *270*:467-470.
15. **Southern, E.M.** 1975. Detection of specific sequences among DNA fragments separated by gel electrophoresis. J. Mol. Biol. *98*:503-517.
16. **Southern, E., K. Mir, and M. Shchepinov.** 1999. Molecular interactions on microarrays. Nat. Genet. (Suppl.) *21*:5-9.

4 A Systems Approach to Fabricating and Analyzing DNA Microarrays

Jennifer Worley, Kate Bechtol, Sharron Penn, David Roach, David Hanzel, Mary Trounstine, and David Barker
Molecular Dynamics, Inc., Sunnyvale, CA, USA

INTRODUCTION

The explosion of biological information catalyzed by the Human Genome Project has ushered in a new era in our understanding of how biological systems work. It is humbling to realize that less than 10% of human genes, and less than half of the genes of *any* free-living organism, can be assigned a clear function. It is exhilarating to know, however, that the rapid pace of gene discovery in both human and other organisms will give biologists the tools needed not only to discover gene function but also to gain an understanding of biology at higher levels, the levels of metabolic and cellular systems. The complete mapping and sequencing of human and other genomes is an accomplishment of enormous importance, both for the wealth of information gained and because this information provides the reagents that can be used to begin to unravel the operation of genes in functional networks. These reagents are the genes themselves, which, when arrayed by the thousands, can be used as a massively parallel assay of the expression of individual genes within the complex mixture of RNA expressed (21). Brown and Botstein (4) describe the promise of DNA microarrays with their review of early work in yeast, in which the complete genome sequence is known.

Gene expression analysis in eukaryotic cells poses a particular challenge to microarray technology. It is likely that there are up to 30 000 distinct species of mRNA in a human cell (3). Of these, 99% are rare, occurring at a frequency of less than 1 copy per 20 000 mRNA molecules (25). To make matters worse, half of the mass of mRNA in a cell is the product of only about 300 genes, leaving the technology with the problem of sorting out some 29 700 different needles in

Microarray Biochip Technology
Edited by Mark Schena

the haystack. Microarray technology, then, should be sensitive enough to detect the rare messages and should have a large enough dynamic range to quantify the abundant messages as well.

At Molecular Dynamics, we gained early experience with the technical problems of microarrays by building the first commercial GeneChip® scanner for Affymetrix (9). The oligonucleotide array system of Affymetrix is sophisticated and powerful but does not allow custom fabrication of arrays in the user's laboratory. Consequently, we focused on developing a complementary technology that is more adaptable to individual needs. This technology is based on spotting microarrays of cDNAs or polymerase chain reaction (PCR) products representing expressed genes. Spotted arrays are created by depositing a small volume of DNA in solution onto the array surface using a pen or pin deposition tool. The DNA samples can be human genes obtained from commercial vendors or can be amplified from personal cDNA libraries from any organism without the need for sequence information. Once the DNA samples are prepared, small batches of arrays can be spotted and evaluated, and the array design can be changed simply by reformatting the sample plates.

Although parallel expression analysis using spotted arrays is beginning to be put to practical use for scientific discovery, the technique is complex and not reliable unless instrumentation, chemistry, and software are integrated. Since the initial publication by Schena and colleagues (20), the demand for spotted microarray technology has greatly outstripped its availability. As the use of these arrays grows rapidly, it becomes increasingly important to understand the fundamental science behind the technology. The goal of most microarray applications is to be able to see a change in a biological system. To accomplish this, the variability inherent in the technology must be kept to a minimum. Sources of technical variation include the array surface chemistry, DNA attachment and availability for hybridization, cDNA sample labeling, and hybridization conditions to achieve quantitation and specificity. In addition, the instrumentation and fluorescence chemistry used to make and analyze the arrays must be thoroughly characterized to show that they do not compromise the biological results. In this chapter, we describe the integrated microarray system developed at Molecular Dynamics and Amersham Pharmacia Biotech. Figure 1 illustrates the many steps, instruments, and chemistries involved in a complete system and indicates the ones we address here.

INSTRUMENTATION

Spotter

Our microarray spotter design goal was to allow users to create easily customized microarrays from their own clone libraries in their own laboratories. Spotted array manufacture uses robotic systems to create moderate-density arrays of samples on a specially prepared surface. The quality, density, and reproducibility of the arrays depend on the design of the system. Spotting tool design and surface characteristics affect the spot size and density of the array. The design of the total system must include precise positioning, minimum carryover between samples, and uniform sample deposition to create reproducible microarrays.

In the past 3 years, our spotting technology has evolved through three generations. The Gen III Spotter deposits subnanoliter volumes of DNA samples in duplicate from standard 384-well microplates onto coated glass slides. Spotting is fully automatic once samples and slides have been loaded into the instrument. The system will deposit nearly 10 000 spots onto 36 slides in approximately 4 hours of unattended operation.

The Gen III Spotter has been designed for throughput, reproducibility, and ease of use. It consists of a microplate hotel for 13 384-well plates, a slide-holding tray, an array of 12 deposition pens, an automated pen wash station, and motorized positioning stages to control pen, slide, and microplate placement precisely. The spotting is performed in a humidified enclosure (Figure 2).

System operation is controlled by an integrated software package that has been designed for convenient user interaction. The user selects from a graphical menu to set up the spotting session, selecting spotting and wash parameters. During operation, spotting progress is continually displayed, allowing the user to monitor the course of the spotting session easily.

Motorized translation stages position microplates, slides, and deposition pens precisely during the spotting process. Early-version systems used lead screw-

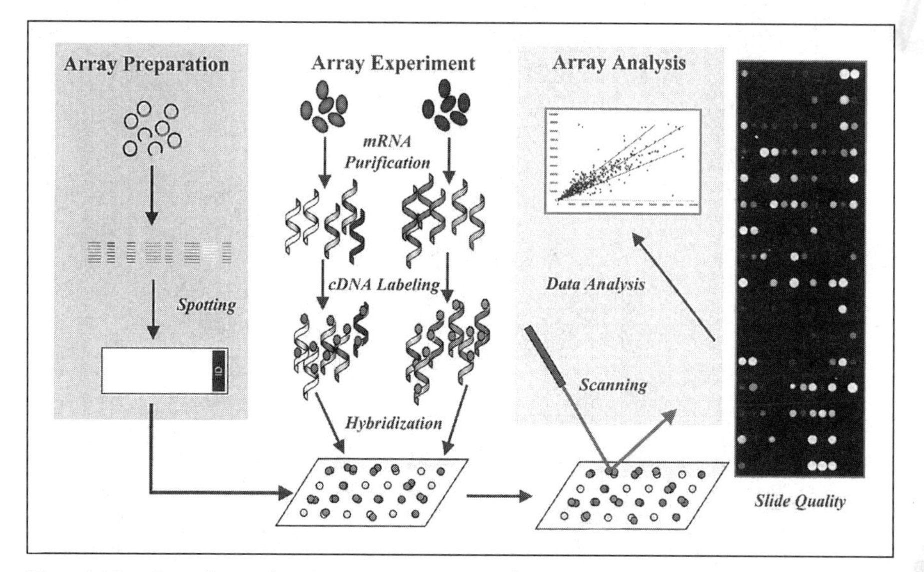

Figure 1. Flow chart of two-color microarray experiments. The major phases of array preparation, differential gene-expression experiment, and analysis are listed, with the topics discussed in this chapter labeled in italics. **(See color plate A7.)**

driven stages; the Gen III system uses linear motors to achieve significantly faster speeds. The position of the stages is monitored via encoder feedback to ensure precise, reproducible spot positioning.

The spotter chamber is enclosed and humidity controlled to minimize dust and environmental influences on spotting. Transparent doors use antistatic plastic panels to reduce dust problems further. The doors slide open to provide convenient user access while minimizing air turbulence and exchange with outside air. During a spotting session, however, the doors are secured with safety interlocks to prevent possible user injury from rapidly moving parts.

At the beginning of a spotting session, standard 384-well microplates loaded with purified spotting samples are placed in the microplate hotel. Specially coated slides are placed onto the spotting tray and held in position with vacuum. The pen array (Figure 3), consisting of 12 pens, is dipped into the first set of sample wells in the first microplate and then touched down on each of the slides to be spotted. After these samples have been deposited, the pens are cleaned in the pen wash station, and the process is repeated for each set of wells in each microplate in the hotel, until spotting is complete.

There are a number of design approaches for microarray spotting tools, including pins, tweezers, and capillary tube formats. The capillary pen design we

Figure 2. Spotter compartment of the Molecular Dynamics Generation III Array Spotter. The compartment includes a microplate hotel at the left, slide-holding tray in the foreground, spotting pens near the back of the tray, and wash station to the left of the tray.

have developed has a number of advantages. The mass of the pen is very small, and, as a result, there is no observed tip deformation during use that could change spotting characteristics, eliminating the need to "resharpen" the pens periodically. The pen consumes only a very small volume of sample (about 200 nL) and is easy to clean. The small diameter of the pen permits sampling from wells of even higher density microplates as plate standards change and sample throughput increases. The capillary action-based design is inherently simple, leading to robust performance.

The pen is fabricated from a short stainless steel capillary with a slot through one wall to act as a vent and to facilitate efficient cleaning (19). The spotting end has been formed to ensure reliable, repeatable fluid transfer to the slide by capillary action. During fabrication, the pens are cleaned and coated to achieve the desired surface characteristics. The pens are individually spring loaded, which ensures proper contact with the slides during sample deposition and prevents deformation of the pens during the deposition process. The pens are matched in a pen array, and the system is tested before shipment to ensure spot-to-spot reproducibility within a 15% standard deviation. The spotting volume varies between pen sets but is approximately 0.8 to 1 nL. The pens are thoroughly cleaned in the pen wash station after spotting each set of samples to prevent carryover. After they are lowered into the wash manifold, a series of cleaning solutions is forced

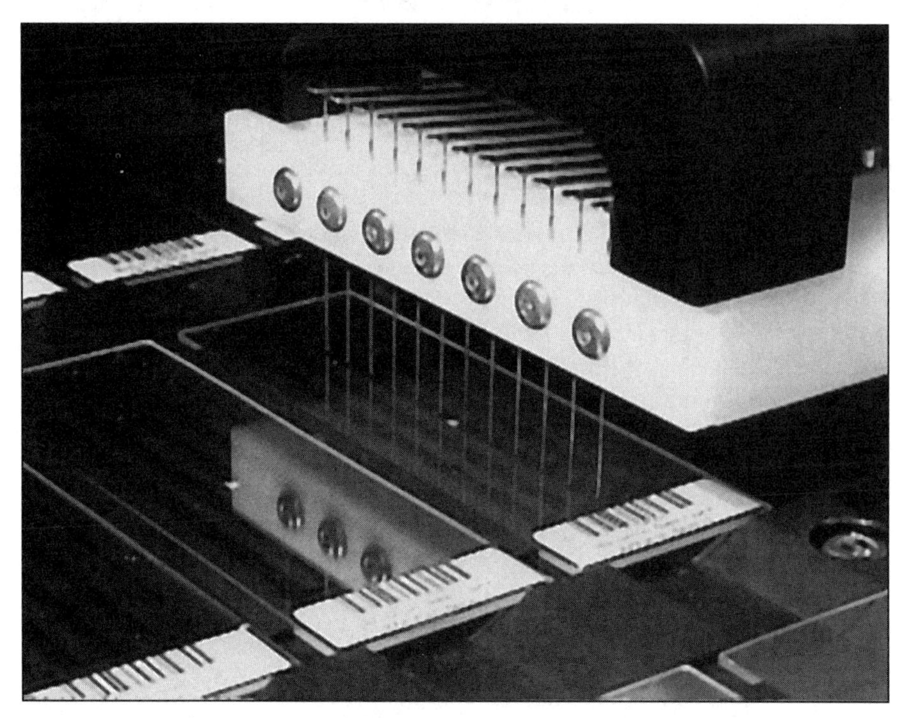

Figure 3. An array of 12 pens positioned to deposit DNA on a precoated, bar-coded slide.

both through and around each pen. They are then dried with nitrogen before the next samples are loaded. The time and sequence of desired wash solutions are user-selectable options, allowing easy conversion between spotting chemistries. Under normal conditions, carryover is reduced to less than 1 part in 10 000 by a 20-second wash cycle.

Scanner

The introduction of gene expression microarrays necessitated the development of novel imaging technology, since the size of microarray elements (50 to 200 μm) falls in between the size of gel bands or dots on membrane arrays and objects normally imaged by a microscope. In addition to moderately high resolution, the system must image a fairly large area (in our case, 2 × 6 cm) in a short time. It must distinguish at least two fluorescent spectra to accommodate differential gene expression experiments using two fluorescent dyes. The wide range of message abundance levels requires an instrument with a low fluorescent detection threshold to allow quantification of rare messages and a large linear dynamic range to measure abundant ones simultaneously. The entire area of the microarray must be scanned uniformly to ensure reproducibility. All the systems designed for this application can be categorized as either *staring* or *scanning*.

Staring systems illuminate and detect a large field of view at one time. These systems employ an arrayed-element detector, usually a charge-coupled device (CCD) and often stitch together many overlapping images to cover the entire slide. Optical filters placed in front of the detector are used to separate fluorescent emission spectra. The major technical challenges to using these systems for quantitative applications are uniformity of illumination and variability of the individual elements of the detectors.

Scanning systems use lasers to illuminate and detect each pixel sequentially. The use of lasers makes it easy to design these systems to provide optimal excitation power, which is a photon flux that causes 50% of the fluorescent molecules to be in the excited state (23). To detect multiple fluors, more than one excitation laser is often used. Optical emission filters separate the emission spectra in multilabel arrays, and a photomultiplier tube (PMT) converts the signal to a digital image.

Microarray imaging requires optics with sufficiently high resolution to ensure that there are enough pixels per arrayed spot for reliable quantitative analysis. Microscanners usually use a standard microscope objective to achieve the necessary resolution. These lenses have high light collection efficiency (measured by NA) but have a limited and nonuniform field of view. Using a standard 4× objective, for example, the signal collected at a point 75% of the distance out to the edge of the field is <50% of the signal collected at the center of the field (Figure 4A). Calibration can compensate for much of this nonuniformity, but such corrections reduce the dynamic range of the system. As a result, most scanning systems move the lens or sample, so that only the center of the lens is used for excitation and detection. One method scans by moving the lens across the sample, whereas the other uses a stationary microscope objective and moves the sample on the *x*- and *y*-axis (20). Both approaches require rapid motion of either lens or sample stage to keep scan times manageable and maintain throughput capacity of the microarray system as a whole.

Many microscanners use confocal optics to improve image quality. A confocal imaging system uses a pinhole placed in a focal plane of the optics system to pass light emitted from the slide surface selectively while rejecting stray light. Confocal imaging reduces background compared with a non-confocal system. However, a confocal scanner must maintain a highly stable system alignment to ensure optimum focus and resolution for the life of the scanner. Moving parts must be designed so that any wear is negligible and will not compromise the performance of the system over time.

Our approach was to design an optical system optimized for microarray imaging. Microarrays, unlike most microscopic samples, are not directly viewed by a human operator. Eliminating direct viewing led to the development of a unique objective lens that allows high-resolution, confocal scanning over a wide scan field (1,10–12). Whereas a typical 4× microscope objective can image a field about 3 mm wide (of which less than 1 mm width offers good collection efficiency and low distortion), in contrast, our lens allows a galvo-driven mirror to scan the laser beam across a 10-mm-wide scan field and collect the resulting fluorescence. Thus the lens remains stationary over the slide while the laser beam scans across the field (Figure 5). The slide is advanced in the *y*-axis past the scanning laser beam. This design eliminates the need for rapidly moving lenses or

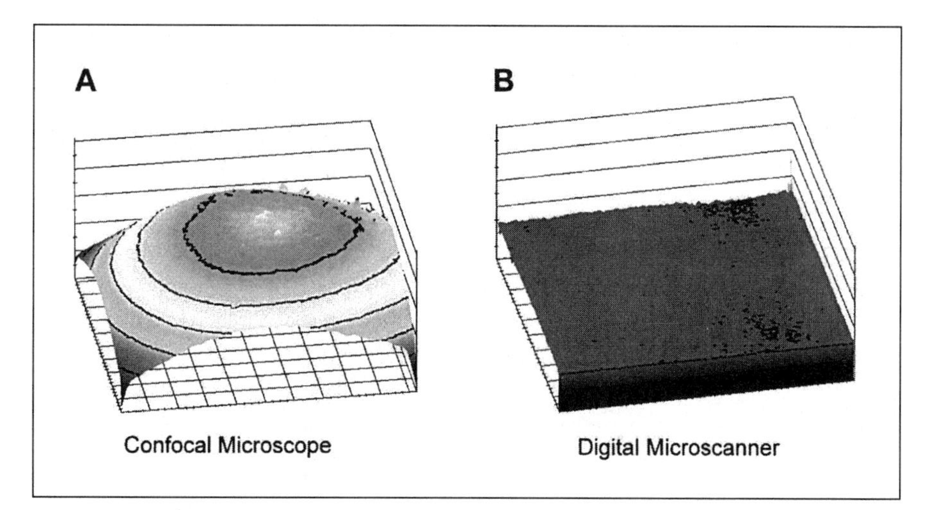

Figure 4. Comparison of field uniformity. (A) Confocal microscope. (B) Molecular Dynamics Microarray Scanner. A thin uniform fluorescent substrate was scanned in both a confocal microscope (4× objective) and the microarray scanner. In both figures the *x*- and *y*-axes map the field of view, and intensity in arbitrary units is plotted on the *z*-axis. [Reprinted with permission from D.K. Hanzel et al. (8)]. **(See color plate A8.)**

complex stages and delivers an inherently more uniform response, which we have measured to be ±4% over the entire 10 × 60-mm scan field (Figure 4B). The instrument uses multiple lasers, confocal optics, and this unique wide-field objective to collect a quantitative microscopic image.

Quantitative characterization of the scanning system. The functional characterization and optimization of the scanner is essential for accurate quantitation of microarray data. Our microscanner design was tested for linear dynamic range, detection threshold, and field uniformity (Figure 4) to evaluate its quantitative performance.

The dynamic range of an array scanning system can be defined in at least three ways: *(i)* the electronic dynamic range of the scanner, *(ii)* the chemical dynamic range of the fluorescent dyes used, and *(iii)* the biological dynamic range of the system under study. Our design objective was to be certain that the biological data were not limited by electronics or fluorescent chemistry. The electronics of the system have five orders of linear dynamic range. To determine the linear dynamic range of the fluorescent dyes and their detection threshold, dilution series of Cy3 or Cy5 dye solutions were spotted on treated microscope slides and scanned at several different PMT voltages. Increasing the voltage of the PMT is equivalent to increasing the gain of an amplifier. The total fluorescence of each

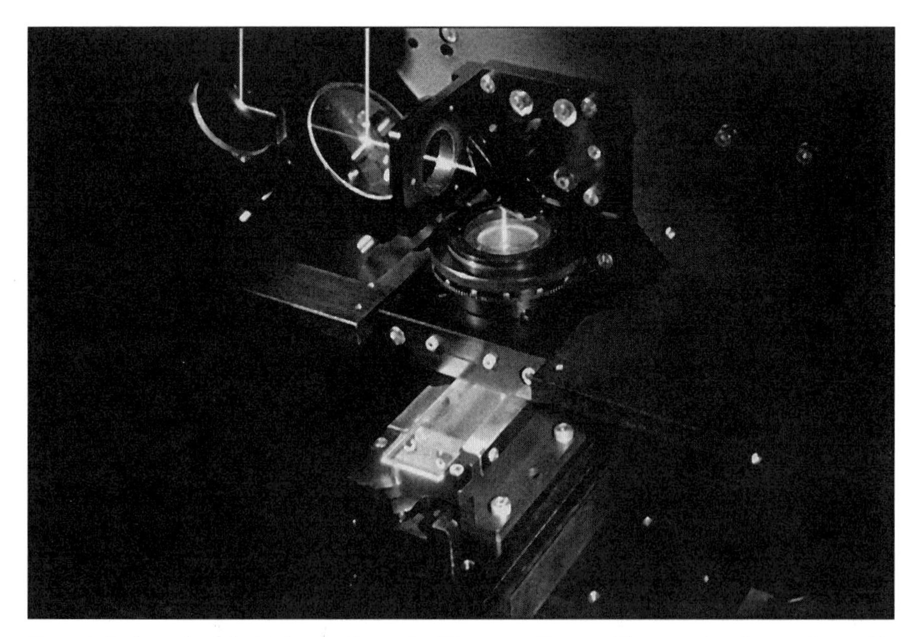

Figure 5. Light path of the Molecular Dynamics Microarray Scanner. The green laser beam reflects off a dichroic beam splitter and is scanned by a galvo-controlled mirror through the wide-field objective lens. Fluorescent light emitted from the surface of the slide is collected by the lens and "descanned" by the mirror. The fluorescent light (indicated by the orange line) passes through the dichroic beam splitter and is reflected by a mirror toward the confocal pinhole and the PMT detector. **(See color plate A8.)**

spot was measured and was divided by the amount of dye in the spot to calculate the detection efficiency. We define the linear dynamic range of the system as the range of sample concentrations that has the same detection efficiency. For our scanner, a PMT gain of 600 V provides the largest linear dynamic range for Cy3, approximately four orders of magnitude (Figure 6).

The limit of detection (LOD) of a system is limited by the background signal, and by the noise, or variation, of the background. We define LOD as the amount of fluorescent dye for which the signal-to-noise ratio (SNR) is 3. The SNR is defined as:

SNR = (signal – background)/(standard deviation of background)

As signals approach the background, quantitative accuracy diminishes. Although features may be visible below this point, signal quantitation is less reliable. Clearly, the LOD depends critically on the background signal from the microarray glass and coating. Using clean, coated microscope slides on our scanner, the LOD is about 0.2 amol of Cy3 per spot (3 fluor molecules/μm^2) at a PMT voltage of 600 V (Figure 7) and about 0.4 amol of Cy5 per spot (6 fluor molecules/μm^2; data not shown). Experiments with other substrates having less background show that the scanner itself can achieve an order of magnitude lower LOD, offering opportunity for significant improvement in the future.

Tissue microscanning. As an aside to our main topic, we note that the microarray scanner produces quantitative, digital images that resemble low-magnification microscope images. Because the microscanner collects an image of an entire microscope slide at a moderately high spatial resolution (5–10 μm), entire tissue sections can be viewed and analyzed (8). The uniformity of signal response, which far exceeds that of a microscope objective, allows reliable quantitation of the digital image. Finally, the extended linear dynamic range of the microscanner allows quantitation of very bright and very dim objects in the same image. We believe that digital microscanning combined with in situ hybridization will enable high-throughput analysis of *spatial* gene expression in tissue samples.

SLIDES

Microarray Slides and Coating Chemistry

We use microscope slides made from low-fluorescence glass as substrates for cDNA microarrays. While suitable slides are easily obtainable, the glass must be scrupulously cleaned and coated with a chemical that can bind DNA to the surface. The quality of the slide coating is central to the quality of the data ultimately obtainable from the slide. Because poor slide coatings can lead to poor DNA retention, the coating must be uniform with no bare patches. In addition, the coating must be able to withstand conditions such as boiling, baking, and soaking in warm detergent-containing liquids for extended periods of time. For detection of microarray data, the coating must be nonfluorescent, and for confocal scanning instruments such as the Molecular Dynamics microarray scanner, the substrate glass must be flat.

When coating glass microscope slides, pretreatment of the slides to remove organic compounds that can be deposited during their manufacture appears to be critical. In our experience, hot nitric acid works very well for this purpose. At Molecular Dynamics, we have developed a vapor-phase method for coating the cleaned slides with 3-aminopropyltrimethoxysilane. This coating method is preferable to solvent-based methods because it is expensive to dispose of solvents, and they can contribute to fluorescence background if they are not of superior (and hence expensive) quality. In addition, we believe that vapor-phase coatings are more likely to be deposited as a monolayer on the slide surface. A monolayer is preferable because it should be more reproducible than attempting to deposit a given number of layers routinely. After silane deposition, we wash the slides in a cascade bath with deionized water. This removes any silane that is not covalently attached to the glass, and the water catalyzes unreacted methoxy groups to cross-link to neighboring silane moieties on the slide.

Characterization of the slide coating can be difficult. There are, however, surface analysis tools that can provide useful results. One such technique that we have investigated is electron spectroscopy for chemical analysis (ESCA) (18). In

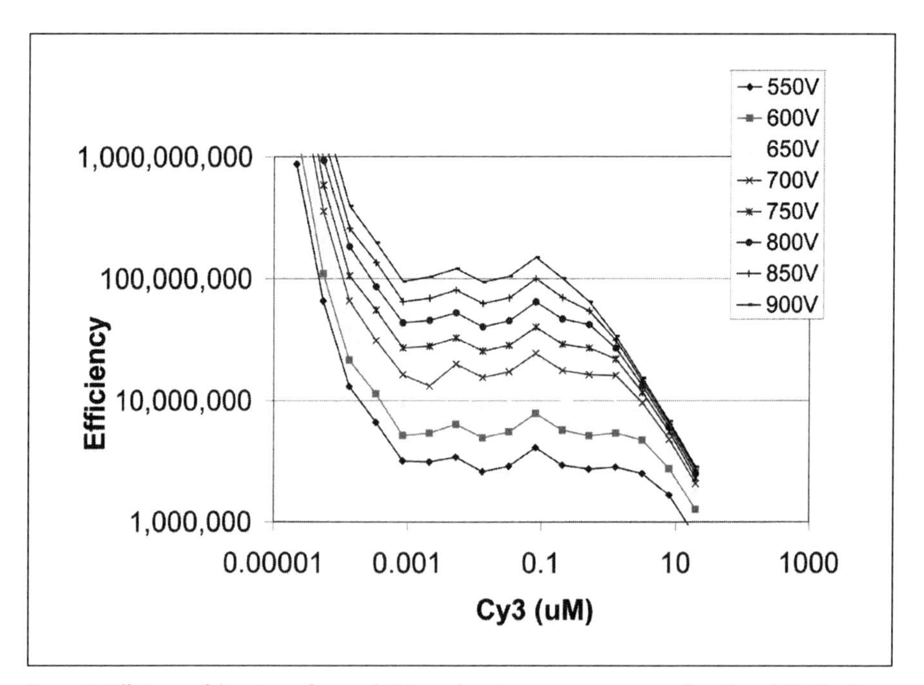

Figure 6. Efficiency of detection of spotted Cy3 on the microarray scanner as a function of PMT voltage. The flat regions of an efficiency plot reveal the linear dynamic range of the system. (**See color plate A9.**)

ESCA, the surface is bombarded with photons in the form of X-rays. Electrons are emitted from the surface with an energy characteristic of their atomic source. Silicon, carbon, and oxygen are easily detected but are not very useful since these elements are found in the underlying glass as well as in the coating. However, the presence of nitrogen can be very informative, and preliminary data collected on Molecular Dynamics coated slides suggest that the density of amines can be quantified by ESCA. This method is useful for research purposes, but its cost is prohibitive for routine quality control.

Another surface analysis tool that has proven very useful is spectroscopic ellipsometry (22). Spectroscopic ellipsometry is an optical technique for determining the properties of thin films on surfaces. Spectroscopic ellipsometry measures the change in polarization of light reflected from a surface and provides a determination of film thicknesses from ten to several thousand angstroms. Figure 8A shows a theoretical plot of coated and uncoated glass substrates, as measured by spectroscopic ellipsometry. The curves are simulated theoretical Ψ (psi) and Δ (delta) spectra (22) of an uncoated glass sample (substrate) and of two simulated thin film samples of 10 and 20 Å. For an ideal uncoated slide, psi has a minimum at zero and delta goes through an inflection from a low value at or near zero to a very high value. Adding layers onto the glass surface results in a less pronounced minimum for psi, which increases from zero. The slope for the transi-

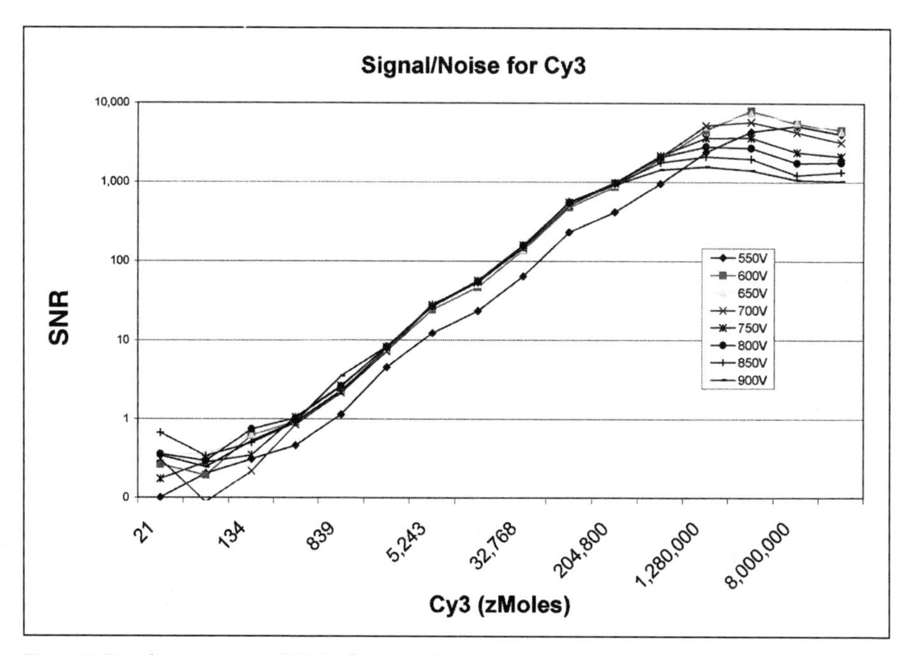

Figure 7. Signal-to-noise ratio (SNR) of spotted Cy3 on a prototype microarray scanner. Scans were conducted at a series of PMT voltages. These plots are used to determine the limit of detection (LOD) of dye fluorescence and the acceptable values for PMT voltage. 10^3 zmole = 1 amole. (**See color plate A9.**)

tion from low to high in delta becomes shallower and continues to decrease with increasing layer thickness. For measuring very thin coatings, the slope in delta is used, because the slope corresponds to the thickness of the coating on the glass.

The usefulness of spectroscopic ellipsometry was determined by deliberately producing slides that had a gradient of silane across the slide. Figure 8B shows spectroscopic ellipsometry measurements taken at points across the slide in the direction of increasing silane thickness (measurements 1 through 4). As can be seen, the slope in delta decreases as measurements are taken across the slide, indicating an increase in coating thickness. Slides produced using the vapor-phase monolayer protocol were shown to have a uniform coating that was estimated, within the errors of the method, to be equivalent to a monolayer. Spectroscopic ellipsometry proved to be an excellent technique for process characterization, but was too costly and time-consuming for routine quality control (QC) of slides. At this time, for routine QC, we perform a hybridization experiment on selected slides to ensure uniform quality.

Microarray DNA Attachment Chemistry

For cDNA or PCR product arrays, we have found that noncovalent attachment is suitable. This is because the fragments are long (on the order of several hundred base pairs) and, despite covalent end attachment, the DNA also interacts electrostatically along the backbone of the attached fragment. We took advantage of this phenomenon to develop a noncovalent method of binding, which can be less expensive and more versatile than covalent attachment because existing clone libraries can be used without expensive resynthesis and functionalization. Previous workers (13,14) had used sodium thiocyanate (NaSCN) to bind DNA to glass beads, and we adopted this approach for amino-silanized slides. We found that, although spotting in NaSCN onto bare glass does result in some binding of DNA, binding is vastly improved by spotting onto silanized glass.

The protocol we have developed is described in detail elsewhere (M. Trounstine et al., manuscript in preparation). Briefly, it involves spotting PCR fragments in a solution of 3 to 5 M NaSCN. After spotting, the slides are heated at 80°C for 2 hours to dehydrate the spots. Prior to hybridization, the slides are washed in isopropanol for 10 minutes, followed by boiling water for 5 minutes. These steps remove any DNA that is not bound tightly to the glass and help reduce the background because loosely attached DNA may redistribute during hybridization, causing uneven backgrounds and smeared spots. We have found that contaminants such as detergents and carbohydrates (including glycerol) in the spotting solution have an adverse effect on binding DNA to the surface.

To determine how much of the spotted DNA is retained throughout a hybridization experiment in our system, we spotted Cy3-labeled PCR products and then measured the fluorescent intensity after heating and also immediately following mock hybridization. We took care to minimize the impact of environ-

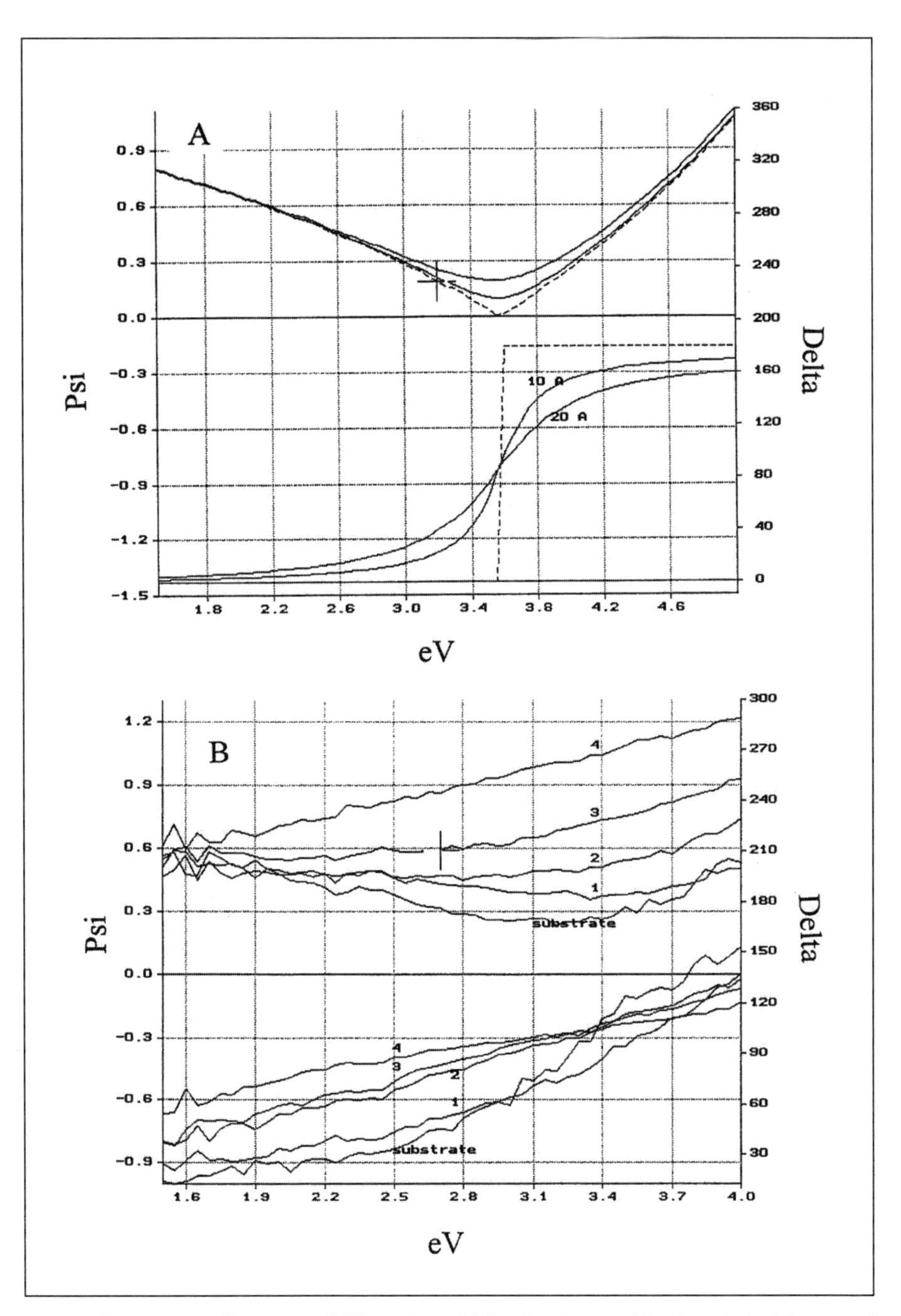

Figure 8. Spectroscopic ellipsometry of slide coatings. (A) Simulated curves of the theoretical psi (lower set of curves) and delta (upper set of curves) spectra of a pure, uncoated glass sample (dashed lines) and of two simulated thin film samples of 10 and 20 Å. (B) Measurements taken at four positions (1–4) across a slide produced to have a gradient of silane across the slide. Substrate refers to an uncoated control slide. The slope in delta decreases as measurements are taken across the slide, indicating an increase in coating thickness.

mental factors such as dryness, quenching, or photo bleaching on the fluorescent response. For DNA concentrations of 100 ng/μL (corresponding to 100 pg/spot for a 1-nL deposition), the retention is between 20% and 30% for fragments that range in size from 250 to 750 bp (S. Penn et al., manuscript in preparation). We have found that this binding chemistry is reproducible and easy to use. Spotted slides can be stored desiccated for many months, unprocessed, without deterioration of performance. In addition, the PCR products in the source microplates can be reused for several months if frozen at -80°C between uses (M. Trounstine et al., manuscript in preparation).

DYES AND PROBES

Probe Preparation and Quality

We prepare probe populations from RNA by first-strand cDNA synthesis using dye-labeled nucleotides. Factors that we have found important for deriving high-quality, quantitative results in the microarray system include *(i)* starting with clean RNA, free of contaminants such as polysaccharides, protein, and DNA; and *(ii)* removal of unincorporated nucleotides and small fragments (<200 bases) following cDNA synthesis. This latter step proved to be critical for reducing the background signal at hybridization.

A good diagnostic for labeled probe preparations that are likely to provide high-quality hybridization results is the specific activity of dye incorporation, as determined by the absorbance at 260 and 550 (Cy3) or 650 nm (Cy5), using published extinction coefficients (17). Very high dye incorporation (specific activities of >1 dye molecule/20 nucleotides) results in a decreased hybridization signal compared with probe with lower dye incorporation. This is consistent with the results of Randolph and Waggoner (17), who showed that high dye incorporation significantly destabilized the hybridization duplex, as evidenced by a requirement for reduced stringency. Very low specific activity (<1 dye/100 nucleotides) gives unacceptably low hybridization signals (M. Trounstine et al., manuscript in preparation). As with many dyes, careful handling of both the labeled nucleotides and probes is important, especially for Cy5, which is more easily photobleached and more susceptible to freeze-thaw degradation than is Cy3.

Probe Matching in Dual-Label Experiments

Both dye characteristics and scanner properties affect the detection of the probes in dual-label experiments. Thus, whereas Cy3 has a lower quantum yield (0.15) than Cy5 (0.28), this is balanced by the fact that the solid-state green laser used to excite Cy3 is more powerful than the red HeNe laser used to excite Cy5 in the Molecular Dynamics Microarray Scanner. In addition, the PMT used to collect the fluorescent signal is more sensitive in the green (Cy3) part of the spectrum than in the red (Cy5) (8).

The Cy3 and Cy5 signals can be balanced during data collection by using a higher PMT-voltage setting during scans of the dye with the weaker signal. Although the SNR is not improved by increasing the PMT setting (Figure 7), the balancing of signals from the two dyes allows the software to pick up a greater proportion of the spots with low signal in the less intense color. This can result in more spots, providing usable data for later analysis.

HYBRIDIZATION

Is the Arrayed DNA in Excess?

For accurate quantitation of gene induction, it is essential that the arrayed target DNA be present in excess relative to the labeled probe cDNA. We mentioned earlier that, in our system, we have 20% to 30% retention of 250 to 750 bp bed DNA following boiling, washing, and mock hybridization. Assuming 20% retention, this corresponds to approximately 50 pg of double-stranded DNA (0.2 fmole) present in each spot for probe binding. We found, in a study of 14 slides of arrayed *Escherichia coli* genes (Analysis of Gene Expression in a Bacterial Model System, below), that the most intense protein-coding genes in the study were *tufA*, a protein chain elongation factor, and *ompA*, an outer membrane protein. The most intense binding obtained from these had 8.5 pg of probe hybridizing to the DNA spots. This is well below the 50 pg of DNA calculated to be remaining on the substrate and corresponds to a hybridization occupancy of 17% (assuming 20% retention of the arrayed DNA) and 11% occupancy (assuming 30% retention). We conclude from these results that in our system, the arrayed DNA is in excess for all the protein-coding genes in our *E. coli* system (M. Trounstine et al., manuscript in preparation).

We have confirmed this observation using a single 500-bp PCR product of the *AmpR* gene spotted at eight widely separated sites per slide. The slides were hybridized with Cy3-labeled cDNA (made from in vitro synthesized RNA) at concentrations ranging from 0.004 to 4 ng/μL, with a vast excess of salmon sperm DNA as carrier. At the lower probe concentrations, up to 0.1 ng/μL (corresponding to the range expected for individual *E. coli* mRNAs), the hybridization was linear with respect to concentration. However, at higher probe concentrations, the curve no longer showed a linear response. These results indicate that it may be dangerous simply to assume that the spotted, or substrate, DNA is in excess (5). To ensure accurate quantitative results for highly expressed genes, it is important to determine the amount of spotted DNA actually retained following the hybridization step and whether the hybridization signal is within the range of linear response for the system being used.

The results described above suggest that binding DNA to silanized glass slides is a very different process from spotting DNA on porous filters or membranes.

The binding capacity of membranes is quite large, so the assumption that all spotted DNA is retained is probably valid. On the other hand, glass arrays have the significant advantage of low fluorescence background and a flat surface, enabling high-density, two-color, confocal fluorescence imaging. Furthermore, the availability for hybridization of DNA bound to glass appears to be greater than on membranes. Typically, membrane-based arrays have approximately 4% of the membrane-bound DNA available for hybridization (24). Our data suggest that the availability of DNA on glass arrays is significantly greater than on membranes.

Single- Versus Dual-Labeled Probe

It has been widely assumed that using a mixture of two labeled probes provides the same results as using single cDNA samples. To test this assumption directly, we measured expression of 46 *E. coli* genes in dual- and single-label experiments and compared the results. We found that the relative expression of the genes was similar with both labeling schemes over a wide range of expression levels (M. Trounstine et al., manuscript in preparation). Thus, combining two labeled probe populations does not affect the quantitation of expression results. However, as noted above, it is important to confirm that the arrayed DNA is in excess for each system being studied.

Probe Depletion During Hybridization

Based on an estimate of the diffusion coefficient of a 50-mer, we calculate that its movement due to diffusion over an 18-hour hybridization is <1 mm. Larger probe molecules would diffuse even more slowly. Thus, in the absence of mixing, each arrayed spot is in effect sampling from its immediate or nearly immediate environment. This suggests that it may be important to have replicate spots well separated on a slide, as adjacent replicates may be depleting probe from the same microenvironment of the hybridization solution.

Hybridization Solution and Temperature

Membrane hybridizations are typically carried out at 65°C; however, this temperature has proved to be too stringent for some high-density glass-based array experiments, at least in part because the presence of the fluorescent tags destabilizes the DNA duplex (17). To optimize hybridization conditions, we first tested a series of hybridizations consistent with methods used in the literature, using 5× standard saline citrate, 0.1% sodium dodecyl sulfate at various temperatures. At 60°C, we detected signal above background from only 86 genes of 400 genes tested. At 50°C, we detected 209 genes, the gene expression patterns were correct, and the negative controls were negative. It should be noted, however, that the details of the labeling and hybridization conditions will affect the optimal hybridization temperature.

On changing to a hybridization buffer containing formamide (16), we saw improved results. The hybridization reaction rate is slower in formamide (2), and the fluorescent signals are reduced compared with more traditional salt-based hybridization. However, the results are much more reproducible both across the slide and between slides. Under these buffer conditions, the best hybridization results (consistent gene expression and no detection of the negative controls) were achieved at 42°C. Although the signal was higher when we decreased the hybridization temperature to 36° or 30°C, we began to see spurious signal in the negative controls.

ANALYSIS OF GENE EXPRESSION IN A BACTERIAL MODEL SYSTEM

A Simple Method for Evaluating Slide Quality

We have developed an analysis system using duplicate spots to assess both slide quality as a whole and reproducibility of induction results for each gene. We describe the system here using the simple and well-characterized prokaryotic model system of heat shock in *E. coli.* The arrayed DNA includes spots for both mRNAs and rRNAs. To synthesize the labeled probe cDNA, we used random primers with total cellular RNA (depleted of small RNAs <200 bases). Although the *E. coli* mRNA lacks polyadenylation and cannot be easily purified away from rRNA, we found that we were easily able to analyze messages over a wide range of expression levels without interference from either labeled rRNA-specific probe or adjacent rRNA spots. This is in contrast to the interference that can be seen from intense adjacent spots using radiolabels or in non-confocal detection of fluorescence.

Our study used 14 replicate slides, with each slide including 200 *E. coli* open reading frames, most of which are represented by both 250- and 500-bp PCR products. Each PCR product was represented by two widely separated spots (a spot pair) on each slide. Labeled probes were prepared from four different sets of RNA, each set derived from a control (37°C growth) and an experimental (42°C growth for 3 minutes) culture. For each slide, RNA from one set was labeled using one of two labeling choices (Cy3 for controls with Cy5 for heat shock, or the reverse). This approach is intended to be a practical evaluation of reproducibility and reliability under real-life conditions, which include a broad range of variation sources, such as biological sample variability, differences in efficiency and recovery in individual labeling runs, dye stability, and hybridization conditions and stringency.

Accepting a Slide

After hybridization, we find it is useful to evaluate the slide as a whole and eliminate unreliable slides on a basis more stringent than gross failure. This is not an approach that has been reported by other workers, but is an important step when dealing with single experiments or small numbers of slides, when the biological significance of the results must be determined by the internal consistency of one or two experiments rather than by agreement among multiple repeats or smooth changes during a time series. This is particularly relevant for small and valuable samples, such as biopsies, which by their very nature are not available for repeat experiments.

We evaluate each slide statistically for overall slide quality using two measures: *(i)* the average variation of replicates in the spot pairs (defined as the standard deviation of spot values/mean), and *(ii)* the R^2 value for the least-squares line drawn through the scatter plot of duplicate spots. Based on these statistical evaluations, a clear difference emerges between slides in which the replicates agree well and those that have a large number of poor replicates. For example, among the 14 slides in our study, 13 had an average variation of replicates that ranged from 8% to 22%, whereas one had an average variation of replicates of 34% and was rejected. It is important to note that although inspection of the scanned images of all 14 slides showed no visible differences between slides, the heat-shock induction results from the rejected slide proved to be of poor quality.

Accepting Induction Replicates

We evaluate spot pairs by calculating the change in gene expression (induction or I = heat-shock fluorescence/control fluorescence) for each spot separately and then determining the closeness of the two induction values. We generally accept a spot pair if the two induction ratios are within 50% of each other. This approach determines that the experimental/control ratios are consistent without requiring that the chemical substrate and arrayed DNAs be deposited identically in both spots, thus removing these sources of error from the results. Other causes of poor spot-pair replicates include high or variable background, low average fluorescent signal on the slide, or speckles of bright fluorescence in one or both spots.

Expression of Heat-Shock and Non-Heat-Shock Genes

A scatter plot of the experimental versus control fluorescent signal for each accepted spot on a good slide (Figure 9) shows that the vast majority of spots, over a wide range of gene expression levels, falls along a straight line. The non-heat-shock genes are expected on average to maintain constant expression levels relative to each other under the growth conditions used. A small group of genes (the heat-shock genes) shows significantly increased relative expression at 42°C compared with 37°C.

Because it is not always possible to obtain equal signals from the pool of reference (non-heat-shock) genes due to factors such as unequal input of dyes and different dye stability, we find it useful to normalize the fluorescent signal for each probe on a slide to the median signal of the probe on that slide. This type of normalization takes into account overall differences between the probes (6,7,15,16). After normalization and calculation of the induction ratio for each spot, we average the induction ratios for each accepted spot pair to obtain the final results.

Our heat-shock expression results will be discussed in detail elsewhere (M. Trounstine et al., manuscript in preparation). However, it is worth noting that our microarray results consistently showed *ibpB*, *ibpA*, and *dnaK* among the most highly induced genes, with *groES*, *groEL*, and *htpG* as the next most highly induced group. Both *lon* and *dnaJ* were not significantly induced. These results are consistent with the results of Chuang et al. (6) using membrane dot blots of clones from the Kohara ordered DNA library of *E. coli K12*. The microarray results have the advantage that each arrayed spot represents a single open reading frame, whereas the earlier studies of necessity used large clones, which contain several genes.

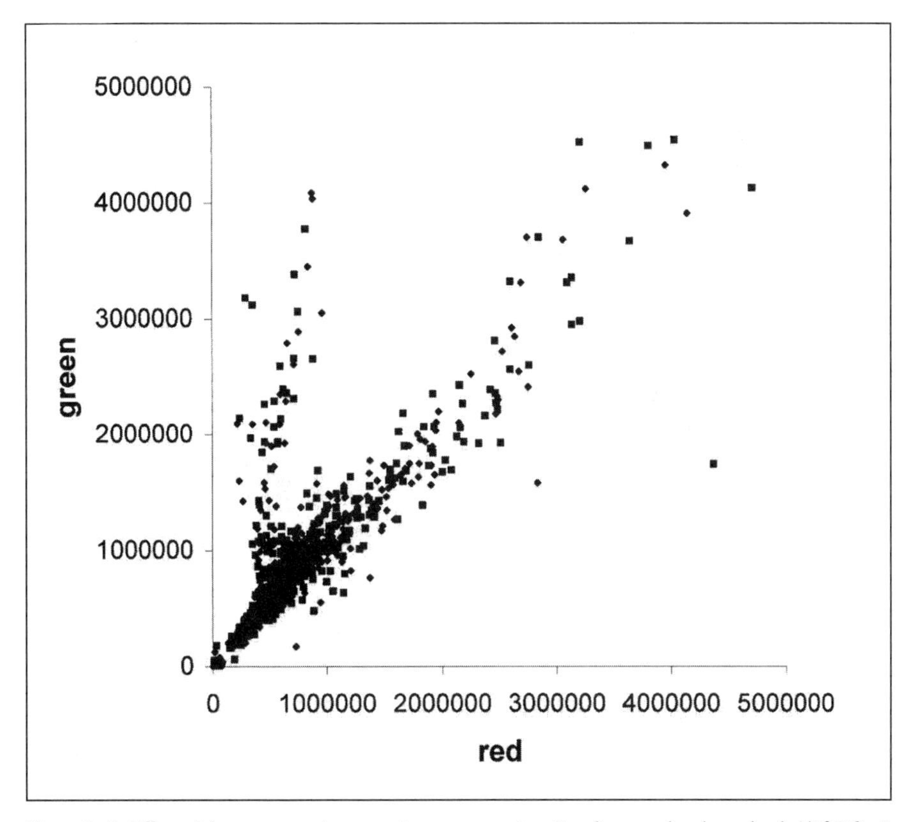

Figure 9. A differential gene expression experiment comparing *E. coli* exposed to heat shock (42°C for 3 minutes, green) and control (37°C, red) growth conditions. The fluorescent signal from the experimental condition is plotted versus the signal from control condition for each spot in all accepted spot pairs. Each of 200 genes is represented by as many as four data points using two different arrayed DNA lengths and replicates for each PCR product. Spot pairs with induction ratios within 50% of each other were accepted. Although not visible from this linear plot, the dynamic range of accepted values is about three orders of magnitude.

CONCLUSIONS

In this chapter, we have characterized our microarray system and described some of the factors that are important for obtaining reproducible, quantitative results from gene expression experiments. We described the advantages of the spotter and pen designs. For reading the microarray results, we developed novel wide-field optics that produce a quantitative, flat-field fluorescence image.

We describe the slide cleaning and coating methods that have resulted in uniform slide preparation. For noncovalent attachment of the target cDNA, we developed an NaSCN-based attachment protocol that produces 20% to 30% retention of the spotted DNA throughout the washes and subsequent hybridization process. We optimized RNA isolation, synthesis of first-strand cDNA labeled probes, and matching of probes for dual-label experiments. Hybridization conditions were optimized for buffer and temperature.

For our model bacterial system, we demonstrated that the substrate DNA is in excess of the probe and found that the availability of the arrayed DNA for hybridization appears to be greater than on membranes. We also demonstrated that we obtain similar results whether we use two single-label experiments or a single dual-label experiment. We described a method to evaluate data using replicate spots to assess slide data quality as a whole and to confirm the reproducibility of induction results for each gene.

The integrated system described here provides the basis for obtaining reproducible and reliable microarray data. In this rapidly moving field, we and others will continue to make improvements in automation, sensitivity, quantitation, and analysis. Our complete system, as well as the scanner, software, intellectual property licenses, and technical support, is available through commercial programs developed by the Molecular Dynamics Microarray Business Segment. Additional information is available electronically at http://www.mdyn.com.

ACKNOWLEDGMENTS

We would like to thank Dr. Lars Markwort and Dr. Michael Kristo for elipsometry and ESCA data. For a critical reading of the manuscript, we thank Dr. Jeanne Loring. We owe the instrumentation and methods described to a long list of talented scientists and engineers who have contributed their dedicated efforts over the past 4 years. In particular, we thank Dr. Richard Johnston for early leadership during the creation of prototype systems, Bob Loder and Ron Galwey for making them work, Robert Kain for inventing the array scanner lens and for leading the ongoing spotter and scanner engineering teams, Siobhan Pickett and David Jenkins for developing and verifying methods, and all members of the Molecular Dynamics Advanced Research Team for their contributions and encouragement.

REFERENCES

1. **Alexay, C., R. Kain, D. Hanzel, and R. Johnston.** 1996. Fluorescence scanner employing a macro scanning objective. SPIE Proceedings, Fluorescence Detection IV. *2705*:63-72.
2. **Bertera, A.L., M.W. Cunningham, M.R. Evans, and D.W. Harris.** 1990. Filter hybridization and radiolabeling of nucleic acids, p. 99-133. *In* P.J. Greenaway (Ed.), Advances in Gene Technology, vol. 1. JAI Press Ltd., London.
3. **Bishop, J.O., J.G. Morton, M. Rosbash, and M. Richardson.** 1974. Three abundance classes in HeLa cell messenger RNA. Nature *250*:199-204.
4. **Brown, P.O. and D. Botstein.** 1999. Exploring the new world of the genome with DNA microarrays. Nat. Genet. *21*(1 Suppl):33-37.
5. **Cheung, V.G., M. Morley, F. Aguilar, A. Massimi, R. Kucherlapati, and G. Childs.** 1999. Making and reading microarrays. Nat. Genet. *21*(1 Suppl):15-19.
6. **Chuang, S.E., D.L. Daniels, and F.R. Blattner.** 1993. Global regulation of gene expression in Escherichia coli. J. Bacteriol. *175*:2026-2036.
7. **Granjeaud, S., C. Nguyen, D. Rocha, R. Luton, and B.R. Jordan.** 1996. From hybridization image to numerical values: a practical, high throughput quantification system for high density filter hybridizations. Genet. Anal. *12*:151-162.
8. **Hanzel, D.K., J.Q. Trojanowski, R.F. Johnston, and J.F. Loring.** 1999. High-throughput quantitative histological analysis of Alzheimer's disease pathology using a confocal digital microscanner. Nat. Biotechnol. *17*:53-57.
9. **Hueton, I. and E. VanGelder.** 1995. High-speed fluorescence. U.S. patent 5,459,325.
10. **Kain, R.** 1998. Micro-imaging system. U.S. patent 5,754,291.
11. **Kain, R.** 1997. Fluorescence imaging system. U.S. patent 5,672,880.
12. **Kain, R. and C. Alexay.** 1997. Fluorescence imaging system compatible with macro and micro scanning objectives. U.S. patent 5,646,411.
13. **Maitra, R. and A.R. Thakur.** 1994. Multiple fragment ligation on glass surface: a novel approach. Indian J. Biochem. Biophys. *31*:97-99.
14. **Maitra, R. and A.R. Thakur.** 1992. Silanization of DNA bound baked glass permits enhanced polymerization by DNA polymerase. Curr. Sci. *62*:586-588.
15. **Nguyen, C., D. Rocha, S. Granjeaud, M. Baldit, K. Bernard, P. Naquet, and B.R. Jordan.** 1995. Differential gene expression in the murine thymus assayed by quantitative hybridization of arrayed cDNA clones. Genomics *29*:207-216.
16. **Piétu, G., O. Alibert, V. Guichard, B. Lamy, F. Bois, E. Leroy, R. Mariage-Sampson, R. Houlgatte, P. Soularue, and C. Auffray.** 1996. Novel gene transcripts preferentially expressed in human muscles revealed by quantitative hybridization of a high density cDNa array. Genome Res. *6*:492-503.
17. **Randolph, J.B. and A.S. Waggoner.** 1997. Stability, specificity and fluorescence brightness of multiply-labeled fluorescent DNA probes. Nucleic Acids Res. *25*:2923-2929.
18. **Ratner, B. and D. Castner.** 1997. Electron spectroscopy for chemical analysis, p. 43-98. *In* J.C. Vickerman (Ed.), Surface Analysis: The Principal Techniques. John Wiley & Sons, New York.
19. **Roach, D.J. and R.F. Johnston.** 1998. High-speed liquid deposition device for biological molecule array formation. U.S. patent 5,770,151.
20. **Schena, M., D. Shalon, R.W. Davis, and P.O. Brown.** 1995. Quantitative monitoring of gene expression patterns with a complementary DNA microarray. Science *270*:467-470.
21. **Southern, E., K. Mir, and M. Shchepinov.** 1999. Molecular interactions on microarrays. Nat. Genet. *21*(1 Suppl):5-9.
22. **Thompkins, H.G.** 1993. A User's Guide to Ellipsometry. Academic Press, New York.
23. **Tsien, R.Y. and A. Waggoner.** 1995. Fluorophores for confocal microscopy: photophysics and photochemistry, p. 267-279. *In* J.B. Pawley (Ed.), Handbook of Biological Confocal Microscopy, 2nd ed. Plenum Press, New York.
24. **Vernier, P., R. Mastrippolito, C. Helin, M. Bendali, J. Mallet, and H. Tricoire.** 1996. Radioimager quantification of oligonucleotide hybridization with DNA immobilized on transfer membrane: application to the identification of related sequences. Anal. Biochem. *235*:11-19.
25. **Wan, J.S., S.J. Sharp, G.M. Poirier, P.C. Wagaman, J. Chambers, J. Pyati, Y.L. Hom, J.E. Galindo, A. Huvar, P.A. Peterson, et al.** 1996. Cloning differentially expressed mRNAs. Nat. Biotechnol. *14*:1685-1691.

5 The Flow-Thru Chip™: A Three-Dimensional Biochip Platform

Adam Steel, Matt Torres, John Hartwell, Yong-Yi Yu, Nan Ting, Glenn Hoke, and Hongjun Yang
Gene Logic, Inc., Gaithersburg, MD, USA

INTRODUCTION

The field of microarray technology, whether the arrays are designed for DNA genosensors, combinatorial synthesis, or "molecular tongues or noses," has become increasingly more active during the last decade of this millennium (24). The concept of fabricating miniaturized molecular assay systems that are capable of performing thousands of analyses simultaneously has its roots in the electronics and semiconductor chip industry. The arrangement of transistors, resistors, capacitors, etc. that make up discrete electronic components on a semiconductor chip are analogous to an array of molecular recognition elements on a "biochip." The "state" of each element on the biochip is determined by the abundance of the complementary analyte (and/or interferents) in a sample mixture. Molecular recognition events are translated into an analytical signal via a "signal transduction" mechanism (e.g., fluorescence or chemiluminescence), which is measured by some detection device. Biochip recognition elements and analytes can be, as examples, small molecules, nucleic acids (DNA or RNA), proteins (ligand receptors, antibodies, and enzymes), and cells. The promised dividends for investing in biochip development are increased information content, higher throughput, and smaller sample and reagent requirements (8).

Microarray development is truly a multidisciplinary field, not only because of the variety of applications including diagnostics, drug screening, forensics, and plant biology, but also because of the variety of sciences and technologies involved in the development process. Biochip development routinely draws on such diverse fields as material science, physical chemistry, biochemistry, nucleic acid chemistry, molecular biology, genetics, toxicology, electrical and mechanical

Microarray Biochip Technology
Edited by Mark Schena

engineering, optics, image and data analysis, database management, and automation. As biochip research continues to expand, it is very likely that more fields will be tapped for contributions and that more and more research fields and industries will incorporate and benefit from the technology.

The most common and advanced examples of biochips are the DNA chips. Several factors have resulted in nucleic acids being the primary target for initial biochip analysis. Of the nucleic acids, DNA is fairly convenient to handle, whereas RNA is less so, owing to ribonuclease suspectibility. A large body of widely accepted protocols for handling nucleic acids already exists (11). DNA is stable under standard laboratory conditions and interacts favorably with surfaces, unlike proteins that tend to denature on surfaces. DNA is also inherently self-complementary. By immobilizing one strand of the DNA duplex on the microarray surface, the complementary strand can be captured almost exclusively, under appropriate conditions. From a significance standpoint, nucleic acids represent important biological markers. Common nucleic acid-based determinations that can be performed on chips include resequencing, differential gene expression, diagnostics, and genotyping. Thus, development of convenient microarray technology for quantitative determination of nucleic acids is of high value to the scientific, medical, diagnostic, biotechnology, and pharmaceutical communities.

There are two major components to every DNA chip system, the chip and the sample. In this chapter we will concentrate on issues related to the chip and comment sparingly on the sample, although many design factors pertain to both components. DNA chips are typically fabricated on glass or silicon substrates. Probes are assembled in an array on substrates by two different approaches: in situ synthesis (19,22) and postsynthesis covalent attachment (17). It should be noted that we refer to the surface-immobilized nucleic acid segment as the probe and the solution-borne nucleic acid as the target. Nearly all the methods used thus far to produce DNA arrays, whether mechanical or lithographic, produce arrays on a two-dimensional solid substrate. With careful consideration of this two-dimensional geometry, one can determine two major obstacles that limit microarray performance. First, as the spot density increases, spot size decreases, translating to a smaller number of recognition elements per spot. The sensitivity limit at spots of decreasing dimensions may become limited because of the dependence of DNA binding on the concentration of the immobilized probe (1,4,26). Also, if probe molecules are too densely packed on the microarray surface, hybridization of target is inhibited by steric interference. The upper limit for detection is proportional to the number of potential binding sites in the spot: the more binding sites, the larger the number of targets that can be captured. Hence, reducing spot size on a two-dimensional surface may ultimately limit the breadth of the dynamic range of the measurement. Second, mass transport of analyte to the immobilized recognition element on the surface represents the rate-limiting step in all but the slowest molecular interactions. The diffusion coefficient for DNA molecules is small compared with that of small molecules and decreases with target length so that the capture rate is slow unless significant convective currents are introduced to enhance mass transport (27). Thus, the hybridization rate is slow, thereby limiting the number of chip assays that can be conducted in a given work period.

The potential limitations of a two-dimensional geometry have prompted thinking on alternative platforms for DNA chips that would overcome the obstacles without introducing new significant challenges for fabrication and signal detection. One approach to enhancing chip performance is to increase the surface area by using a three-dimensional (instead of a two-dimensional) substrate. The impetus for adding depth to the chip is to increase the surface area available for binding capture probes (Figure 1). The third dimension has been implemented on chips in the form of a gel pad and, as we shall discuss at length in this chapter, in the form of the Flow-Thru Chip™ (9,20).

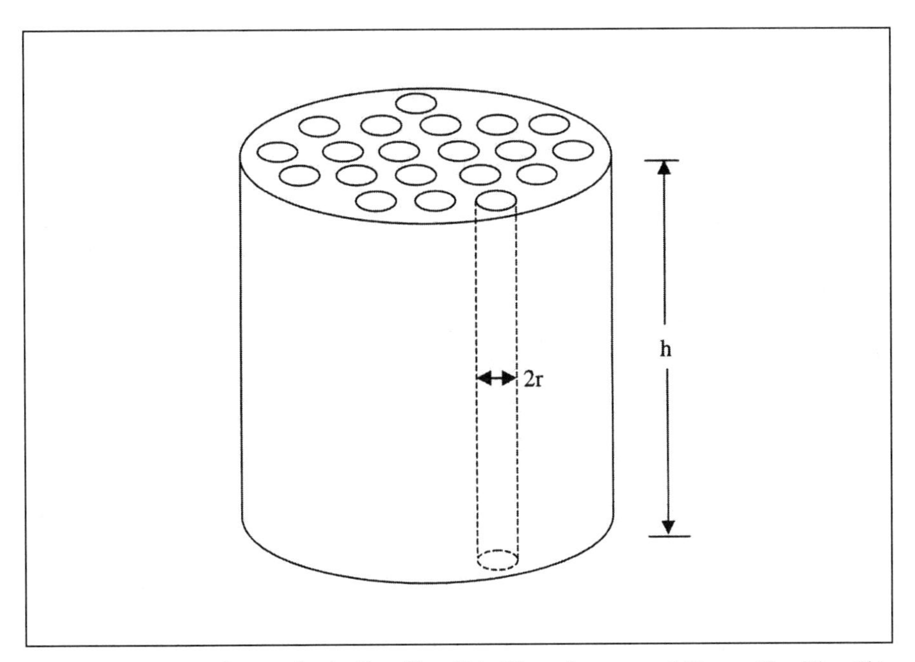

Figure 1. Increase in surface area for the Flow-Thru Chip. The surface area available on a Flow-Thru Chip (S_{FTC}) is greater than that for a planar chip(S_{PC}). The increase afforded by a Flow-Thru Chip compared to a planar chip is given by:

$$\frac{S_{FTC}}{S_{PC}} = 1 + k\left(\frac{2h}{r}\right)$$

In the equation, k is the open area fraction, h is the chip thickness, and r is the radius of the individual microchannels in the chip. For example, for k = 0.5, h = 500 µm, and r = 5 µm, the surface area increases by 101-fold on a Flow-Thru Chip relative to a planar chip.

THE FLOW-THRU CHIP CONCEPT

Gene Logic has developed an advanced sensor platform in which molecular interactions occur within the three-dimensional volumes of ordered, porous substrates rather than at a two-dimensional surface. We have termed the technology the Flow-Thru Chip. The Flow-Thru (FT) concept is illustrated in Figure 2. In the example of an FT chip shown, a uniformly porous substrate is used in place of a flat, impenetrable material. The pores, or microchannels, connect the upper and lower faces of the chip in such a manner that fluid can flow through the chip. Probes are immobilized on the walls of the microchannels. In the case of DNA analysis, the probe molecules are single-stranded nucleic acid, although any capture element could in principle be immobilized on microchannels. In the example discussed here, a single probe type is immobilized within a spot that contains several discrete microchannels. Spots containing individual probe sequences are dispensed over the surface of the microchannel-containing substrate to create a microarray. The FT concept is akin to membrane separation technology, but for DNA probe arrays target is retained only at chip sites that contain complementary probes.

We note that the organized nature of the three-dimensional channels on an FT substrate is significantly different from the three-dimensional properties of gel-pad technology (9,20). Both types of substrate increase significantly the capacity for probe and target binding by providing additional "surface." In the case of gel-pad technology, however, the "surface" is a polymeric network structure. The mobility of DNA within the polymeric gel is much different than in solution, resulting in nonuniform hybridization within the gel pads unless extended binding times are allowed (16). In contrast, target solution passes through the microchannels in the Flow-Thru Chip such that mobility of the target is unimpeded, and hybridization is therefore uniform across each microchannel feature.

Flow-Thru Chip technology is currently in the early stages of product development. Even at this early stage, however, a number of theoretical advantages of the three-dimensional geometry over conventional two-dimensional surface have been borne out by experiment. This chapter describes the FT concept and demonstrates the advantages of this technology that have been realized to date.

Advantages of Flow-Thru Chip Geometry

The theoretical advantages of developing microarrays on organized porous materials include: *(i)* improved responsiveness and dynamic range due to the increased surface area compared with a flat surface geometry, *(ii)* reduced assay times due to enhanced mass transport within the channels, *(iii)* more uniform probe deposition and higher array densities due to improved wetting properties of microporous materials, and *(iv)* smaller sample and reagent volume requirements due to the reduction in the reaction volume. The FT substrate geometry has been recognized by the United States Patent office (3). We will discuss each of these advantages in turn and then provide the data that demonstrate how certain of the advantages have been realized in Flow-Thru Chip development to date.

Increased responsiveness and dynamic range. Responsiveness and dynamic range are attributes that are commonly associated with detector performance, but these parameters can also be defined for a microarray assay. In terms of an assay, responsiveness is given by the slope of the derived signal versus analyte concentration curve and dynamic range by the analytical signal range over which the response curve is linear. The dynamic range is usually defined at the lower end by the detection limit and at the upper end by an analyte concentration for which the signal deviates from the linear portion of the curve by some defined amount (e.g., 10%). We will consider detection schemes and detector properties later in the chapter; at this point, we consider the influence of using an FT substrate on assay responsiveness and dynamic range in comparison with a flat substrate.

The amount and distribution of probes in a spot are determining factors for assay responsiveness and dynamic range. The larger the amount of probe immobilized within a spot, the greater the responsiveness to target, the more signal per unit concentration, and the higher the binding capacity for target providing that the concentrations are not so high as to prevent hybridization by steric hindrance. The amount of probe that can be immobilized in a given region is a function of the substrate surface area. The distribution of probes within the spot is

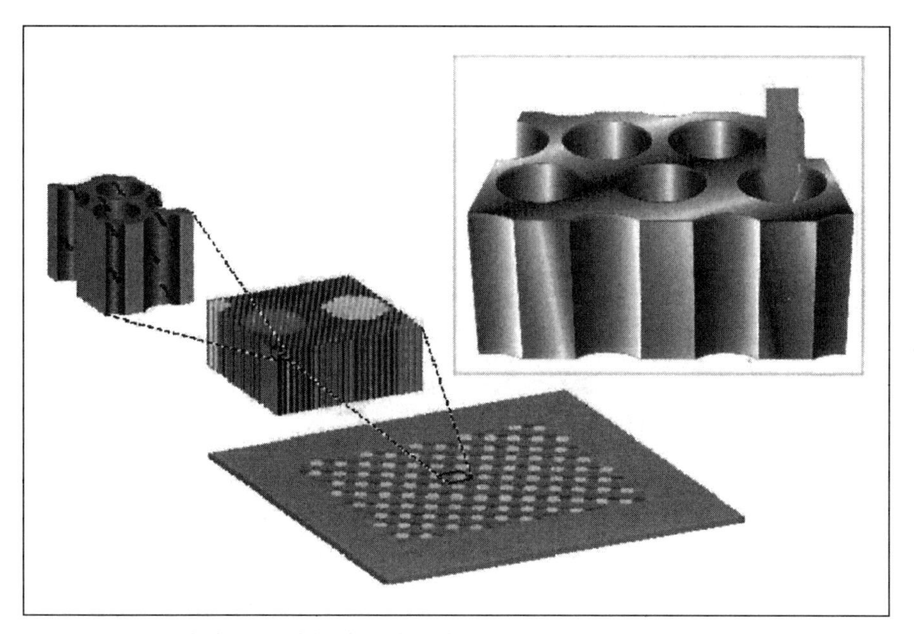

Figure 2. Conceptual schematic of the Flow-Thru Chip. The chip is composed of an ordered array of microscopic channels that transverse the thickness of the substrate. Capture elements (DNA probes) are deposited on the chip in spots. Each spot incorporates several individual microchannels. **(See color plate A10.)**

important because the ability of target to bind to probe is a function of the probe surface density. The optimal probe density for hybridization is a function of probe length; shorter probes can be immobilized at higher densities than longer probes because steric crowding is a function of probe length. When probe density is too high, target access is reduced, and when probe density is too low, assay responsiveness and capacity is diminished.

The FT geometry provides enhancements in responsiveness and dynamic range by providing more surface area for probe immobilization per spot diameter. The additional surface area is gained by adding "depth" to the substrate rather than expanding the lateral dimensions. This is important because microarray signal is eventually detected by reading the emission of photons with a two-dimensional detection device. Assuming that signal can be collected from a three-dimensional substrate with good efficiency, the FT concept is amenable to a microarray format. As we show later, a three-dimensional FT substrate allows a high probe density while localizing the signal to a small region of the detector so that signal-to-noise characteristics are also maximized. The lower limit of detection is, in principle, a single fluorescent molecule but in practice is a function of detector sensitivity. The upper end of the dynamic range corresponds to the maximum number of probe-target pairs that can be formed for the given probe density on the microarray surface. Due to the increased surface area provided by the FT geometry, the breadth of the dynamic range should be greater than a flat substrate. The increase in responsiveness and dynamic range should be proportional to the ratio of the surface areas of the two substrate geometries.

Enhanced reaction rates. A limitation of flat surface geometries is that mass transport of target molecules to the surface is rather inefficient. Mass transport of target molecules to the chip surface is inefficient because the reaction volume, the region in space in which the target exists, is quite large. Physically confining the probe and target to as small a volume as possible enhances the rate of hybridization (7). To enhance the mass transport in flat substrate systems, convective fluid currents are often introduced by shaking, bubbling, and the like. However, a stagnant boundary layer of fluid on the order of 100 μm thick remains (2). Transport of target molecules that are large and slow-moving over such a relatively long distance prior to encountering the probe is the rate-limiting step in the hybridization reaction. Hence, the hybridization step with flat chips may require 6 to 20 hours to complete (14,23,28).

In the current FT geometry, the target solution flows through microchannels with a diameter that is significantly smaller than the boundary layer thickness. Inside the microchannels, targets are in close proximity to the probes immobilized on the walls. Bringing the reactants close together results in fast mass transport due to diffusion. The residence time of target molecules in the microchannels is controlled by the bulk fluid flow rate through the chip. The bulk flow rate can be adjusted such that the residence time in the microchannels is long relative to the time needed for target to migrate the short distance to the substrate wall via diffusional transport. The enhanced mass transport of target to reactive site increases the rate of hybridization by speeding up the rate-limiting step. Hence, assays run on a Flow-Thru Chip are faster than assays that use a two-dimensional substrate.

Uniform probe distribution and increased spot density. The wetting behavior of microporous substrates results in more homogeneous probe distributions than on flat substrates. The microchannels are capillaries; thus, liquid is drawn into the void volume of the chip by capillary action. Two distinct advantages of the FT substrate result from this property: *(i)* probe immobilization is facilitated due to slower evaporation of small (nanoliter) droplets of probe-deposition solution, and *(ii)* higher density arrays are possible because the same volume of liquid has a smaller footprint on the FT substrate than on a flat substrate due to reduced spreading on the surface. With regard to the former advantage, rapid evaporation of the probe-deposition solution may result in inconsistent immobilization of the probes on the surface when chemical cross-linking is used. Dehydration may prevent the reaction of the cross-linking reagent with the surface. Rapid evaporation of delivered fluid also may cause nonuniform spot shapes that can present significant challenges for image analysis.

Smaller sample requirements. In the Flow-Thru Chip, most of the probe is immobilized on the inner channel walls of the substrate; hence, most of the hybridization events occur within the very small void volume of the channels so that an accordingly small volume of sample can be used for the hybridization reaction. In examples of the Flow-Thru Chip system described in this chapter, in which mechanical pumping of liquids in a macroscopic sample loop is performed, this advantage is not realized. Future implementations of the technology will employ flow cells engineered to use smaller sample volumes by incorporating micromechanical fluid pumping or electrophoretic pumping. Theoretically, the total volume required for a hybridization is an amount large enough to fill the channels in the chip.

Technical challenges. The advantages of the Flow-Thru Chip geometry are not gained without a price; in fact, two significant technical challenges are presented by the substrate geometry, namely, plumbing and detection. To realize the enhancement in hybridization rate brought about by confining the probe and target to the small volume of the channels, it is necessary to ensure that the carrier liquid passes through the chip microchannels. In addition, fluid flow must be uniform across the entire chip to ensure signal uniformity. Uneven fluid flow across the array could lead to hybridization differences that are independent of differences in target concentration. By using a microporous substrate rather than a nonporous substrate, the hybridization signal is generated throughout the thickness of the chip rather than in a single plane. Hence, detection systems developed for two-dimensional geometries (confocal imaging) are not appropriate for the Flow-Thru Chip, rather, detection principles that differ considerably from confocal scanners are more appropriate for three-dimensional chip geometries. Fortunately, these issues are questions of engineering rather than of fundamental research, and approaches have been designed to overcome or accommodate them.

Fluidics. Fluid delivery to the chip is a primary concern for implementation of the FT geometry. A number of factors have been considered in the design of the fluid interface to the chip. First, the diameter of the channels in the FT substrate is generally small, on the order of 10 μm, such that fluid flows through the substrate only when a pressure gradient is present across the chip. The fluid flow through the microchannels is laminar because the diameter is so small (Reynold's number <<2300). Because a pressure gradient is necessary for fluid flow, a seal on the edges of the chip is important. Improper sealing results in a fluid leakage around the edge of the chip instead of through the microchannels. For reproducible chip performance, the fluid flow must be uniform across the entire face of the chip. Nonuniform flow results in nonhomogeneous array performance due to differences in local fluid distribution. A means to redistribute the target solution is necessary for detection of low-abundance targets. Once a target enters a channel, it cannot migrate laterally into other channels until it has exited the microchannel. Hence, recirculation of the target mixture through the chip is necessary to ensure that a particular target has the opportunity to bind with the complementary probe. Finally, the fluid interface should permit visual access to the chip for imaging by the detection device.

All these aspects were considered in the design of the Gene Logic Flow-Thru Chip cartridge. A picture of a cartridge is given in Figure 3. The prototype cartridge is manufactured from anodized aluminum with a single inlet and outlet that are connected to an external fluid delivery system via a pin and septum. A compression-fit window is located on the top of the cartridge above the chip. Sealing is achieved by use of viton gaskets. Viton gasket material was selected because the hardness is low, the chemical resistance and biofouling properties are good, and fluorescence emission is low in the wavelength ranges that are used for signal detection. The fluid enters the cell from underneath the chip and flows up through the chip and then out through an exit port above the chip. Uniform fluid distribution has been accomplished by machining the base of the cartridge with a proprietary contour that normalizes the pressure gradient across the face of the chip. Details of uniform fluid distribution measurements are included below. Recirculation of the target material is accomplished by making the flow cell part of an isolation loop that contains a peristaltic pump for continuous fluid pumping around the loop. Alternative pumping methods (e.g., incorporation of the pump and loop into the chip cartridge and electrophoretic pumping) are currently being investigated. The fluid handling system has been engineered to the point at which reproducible results are generated in a facile manner.

Detection in three dimensions. We made an assumption above that "signal can be collected from a thick substrate with good efficiency." This assumption is not trivial because conventional imaging optics are designed to provide a thin focal plane, and the signal in the Flow-Thru Chip is distributed throughout the thickness of the chip. Most of commercial microarray readers use confocal scanning optics (5,8). The motivation for developing scanning microarray readers is very much tied to the nature of the substrate. Most microarrays are currently produced on microscope slide glass where the binding reaction and signal generation occurs within a single plane. In the three-dimensional geometry of the Flow-Thru Chip, the binding reaction and signal generation occur throughout the thickness of the chip. Unlike imaging other optically thick samples, light can be collected more efficiently from the FT substrate. Even though the signal originates from throughout the thickness of the chip, photons are guided up through the chip because the microchannels act as a fiberoptic bundle. From an imaging standpoint, the FT and flat substrate geometries differ significantly but both are amenable to imaging.

Selecting an imaging mode often involves tradeoffs between two levels of resolution—lateral and depth of field. Conceptually and mechanically, static imaging is simpler to implement than confocal scanning. Static imaging gives good lateral resolution when the area of interest is the same size or smaller than the area of the imaging element, which is often a charge-coupled device (CCD) chip. For larger areas such as 1 × 3-in microscope slides substrates, only expensive, custom-built CCDs are large enough to image the entire microarray in a single detection step. For microscope slide formats, confocal scanning is probably a more economical solution. Depth of field becomes more important for optically thick samples like the Flow-Thru Chip. CCD imaging uses conventional optics, which have a relatively large depth of field, and the thickness of the optical field that impinges on

Figure 3. A Flow-Thru Chip mounted in a cartridge interfaced to the fluid delivery system. The chip is visible in the center of the cartridge. The cartridge interfaces to the fluid delivery system via two pins that are below the cartridge.

the detection device is on the order of several to hundreds of micrometers. Collecting light from a thick optical section when the signal is generated from a much thinner plane creates undesirable artifacts such a diffuse halos around the spots. By incorporating confocal optics, crisp, high-resolution images can be obtained. Confocal optics necessitate the use of higher power illumination sources (i.e., lasers) and more sensitive detectors [i.e., photomultiplier tubes (PMTs)]. An image of the microarray is constructed from discrete spot measurements by scanning over the entire area of interest; hence, a tradeoff for being able to image a larger field of view is increased time for image acquisition.

Selecting the imaging mode for the Flow-Thru Chip requires analysis of the same tradeoffs in resolution—lateral and depth of field—as other microarray systems, but differences in geometry impose a different weight to the selection criteria. In designing the Flow-Thru Chip system, we are not constrained to use any particular set of physical dimensions so that a chip size that matches a detection system has been selected. Confocal techniques are neither compatible nor advantageous with the FT geometry. In terms of compatibility, opaque walls can make it difficult or impossible to focus within the interior of the substrate. More importantly, the confocal concept of acquiring signal from a very thin optical slice conflicts with the three-dimensional geometry of the FT chip. For the FT chip, the detection method must integrate the signal generated across the entire thickness of the chip, while still providing adequate lateral resolution to distinguish individual spots. It is critical that photons be collected with good efficiency from all points in the chip. Low-magnification microscope objectives have large depths of field in accordance with the small NA of the objective. This effect is readily realized in the difficulty in producing an in-focus or crisp image using a low-power objective. More detailed discussion is presented below. Bear in mind that the chip thickness can be selected to match the capabilities and limitations of the imaging system.

The size and configuration of the Flow-Thru Chip suggests the use of CCD imaging rather than confocal scanning. Designs for an FT chip reader incorporate a low-power objective with matched Flow-Thru Chip and CCD dimensions for good lateral resolution. The illumination source is optimized to provide uniform, collimated illumination to the entire chip area. Detection is accomplished using a sensitive, large-format CCD camera that images the entire chip at once. A prototype custom imaging system has been built with these characteristics that acquires a chip image in less than 10 seconds.

The imaging described in this chapter was conducted using a Nikon Eclipse E800 microscope (Tokyo, Japan) that has been fitted with a Nikon PCM-2000 confocal accessory and a Hamamatsu C4742-95 Orca monochrome camera (Hamamatsu City, Japan) for image capture. The Orca camera has a 1280 × 1024-pixel CCD chip with 6.7 × 6.7-μm pixels, 3 binning modes, and a 12-bit A/D converter. Image capture and analysis were performed using SimplePCI©, a software application that provides image capture, enhancement, and analysis in an integrated package (Compix, Cranberry Township, PA, USA). The microscope is equipped with 1, 2, 4, 10, and 40× infinity-corrected objectives, and NAs of 0.04, 0.06, 0.20, 0.45, and 0.95, respectively. Higher NAs result in greater lateral and depth resolution. The imaged area is 8.5 × 6.8 mm for the 1× objective and 0.22 × 0.17 mm for the 40× objective.

FLOW-THRU MATERIALS AND PROPERTIES

Substrate Properties and Materials

Advantageous properties of substrates for the Flow-Thru Chip share many attributes with substrates for traditional microarrays: ease of manufacture and processing, mechanical strength, compatibility with detection systems, and affordability. In general, microarrays are fabricated using substrates that have a high density of attached molecules, low nonspecific biomolecule adsorption, a long shelf life, and good mechanical strength (21). The substrate material should allow efficient immobilization of probes either directly or through an intermediate surface coating and should have sufficient tensile strength and chemical resistance to withstand diverse hybridization conditions. For these reasons, fused silica and quartz make good microarray substrates.

In addition to these characteristics, FT substrates contain a high density of noninterconnected, monodispersed microchannels that connect the upper and lower faces of the substrate. The diameter and packing density of the microchannels can be varied to match the requirements of specific applications. The surface area enhancement observed with FT chips is a function of the microchannel radius, the open area ratio, and the chip thickness (Figure 1). Table 1 shows how the surface area enhancement varies with a series of common parameters. It is clear that the surface area enhancement increases for thicker chips with smaller, more densely packed microchannels (Table 1). However, the physical dimensions that provide the largest surface area in theory may not be workable in practice. In particular, the pressure required to pass liquid through the chip increases dramatically as the microchannel diameter is reduced. Also, reliable quantitation of signals from chips of increasing thickness becomes quite difficult due to the optical properties of the microchannel matrix. In the case of fluorescence detection, excitation light must be able to penetrate the chip substrate and excite the fluorescent reporter groups, and emission light must be able to escape from within the chip, that is, the chip should behave much like a fiberoptic bundle. Optical methods such as fluorescence are the most compatible with an optically conductive substrate with minimal autofluorescence. Ideally, minimal attenuation of the impinging illumination should occur due to scattering or adsorption by the substrate.

Three potential substrate materials that display desirable attributes for the Flow-Thru Chip include glass capillary arrays, electrochemically etched porous silicon, and metal oxide filters. Template-assisted porous polymer formation also

Table 1. Increased Surface Area for Flow-Thru Substrates*

Open Area Ratio	Channel Radius (µm)	Chip Thickness (µm)		
		100	500	1000
0.1	1	21	101	201
0.1	5	5	21	41
0.1	20	2	6	11
0.5	1	101	501	1001
0.5	5	21	101	201
0.5	20	6	26	51
0.75	1	151	751	1501
0.75	5	31	151	301
0.75	20	9	39	76

*Increase in surface area for flow-thru substrate compared with a flat substrate for a variety of microchannel radii, open area ratios, and chip thickness values.

promises to provide materials for use as FT substrates (29). These materials are discussed in greater detail below.

Glass capillary arrays. Glass has several advantages for use as a microarray substrate. First, a large body of research has been devoted to immobilization of biomolecules onto glass substrates, and robust attachment strategies have already been developed (15,21). Second, glass has good tensile strength and chemical resistance. Finally, glass with low autofluorescence is available so that the substrate does not contribute appreciably to background levels. Glass in the form of an FT substrate is quite conveniently produced for a number of very different applications (e.g., as intensifiers in night-vision goggles). The material is known as microchannel plate glass or glass capillary arrays.

Flow-Thru Chips can be made from glass capillary arrays that are commercially available from a number of vendors. Each chip made from a capillary array consists of many parallel glass capillary tubes fused together in a uniform and mechanically rigid matrix. Standard microchannel sizes range from 2 to >50 µm, with a variation of <6% within a single chip. Arrays are manufactured in a variety of open area ratios, although open areas >60% greatly diminish the structural integrity of the capillary array. An image of a glass capillary array with 10-µm channels at an open area ratio of 50% under 100× magnification is given in Figure 4. The hexagonal packing of the capillaries is a result of the manufacturing procedure, as is the hexagonal superstructure. Alternate packing arrangements are available by custom order. The capillary glass is available in a range of dimensions, although the standard thickness is 0.5 mm. The optical properties of glass capillary arrays are quite good in terms of light transmission; however, background levels are typically higher than for microscope slide glass due either to scattering or to higher fluorescence from the glass itself. The capillary arrays reported here were made from a different composition glass than typical microscope slide glass. The capillary array glass has a high lead oxide content, 74% by weight. The glass capillary arrays that have been used here are not polished. Cross-talk of transmitted light on capillary arrays could limit spotting density if spots appear broader than their physical dimensions. More detailed information concerning several aspects of the capillary array glass for chip production appear following description of other potential Flow-Thru Chip substrates.

Macroporous silicon. Silicon has many potential advantages as a substrate for microarrays, not the least of which is that it is one of the most highly studied materials known. The compatibility of silicon with a host of analytical techniques for surface analysis has resulted in tremendous knowledge of materials and methods to control its surface properties. Silicon presents a pristine, well-ordered surface in comparison with glass, which is amorphous and quite heterogeneous. The formation of pore arrays with high aspect ratios by electrochemical etching was reported in 1990 (12). The pores are arrayed in patterns that are predetermined by an initial photolithography step. However, a limited number of pore patterns

Figure 4. Brightfield image of glass capillary array Flow-Thru Chip under 100× magnification. The channels are 10 μm in diameter and are packed at a 50% open area ratio. The dark region is a 40-nL aliquot of a probe solution. The hexagonal packing of the channels, a result of the method used to produce the capillary array, controls the spot shape by capillary wetting of the channels.

are accessible to the electrochemical etching process. The product of pattern and pore diameter produces a constant porosity material during efficient etching (13). Pore diameters can be produced in the range of 0.3 to 20 μm. Aspect ratios of up to 250 have been produced, meaning that the process easily produces 10-μm pores across a 300-μm-thick silicon wafer (aspect ratio = 300 μm/10 μm = 30). In addition, the pore diameter can be varied over the length of the pore by up to a factor of approximately 3, resulting in tapers of up to 30°. The potential advantages of tapered channels may prove to be extremely useful for fluid handling and enhancing detection from within the chip.

Potential disadvantages of using silicon as a chip substrate may result from the optical and electronic properties. Although the extreme opaqueness of the interchannel bulk silicon dramatically reduces optical cross-talk between channels, silicon adsorbs light more efficiently than glass. The emission from a clean silicon surface is attenuated because the electronic structure of the bulk silicon quenches the fluorescence by providing an alternative mode of energy transfer. When a fluorophore is deposited on a clean silicon surface (only native oxide present), the fluorophore emission intensity is only about 10% of the brightness from a spot of the same solution on silicon that has a 250-nm thermally grown oxide. The intensities on the thermally grown oxide and glass are similar, but the signal-to-background ratio is higher on silicon due to the stronger adsorption of scattered photons. Growing a thermal oxide on top of bulk silicon produces a silica (silicon oxide) that can be treated similarly to glass, which is a mixture of silica and other metal oxides.

The current limitation to using macroporous silicon is economic rather than scientific. Currently, macroporous silicon substrates are much more expensive than similarly shaped glass, largely because the silicon material is produced only in small batches for research purposes by large semiconductor companies. Mass production of electrochemically etched silicon should result in economies of scale, making the macroporous silicon cost competitive with glass capillary arrays and other FT chip substrates.

Metal oxides. Membrane filters made from metal oxides can have geometries suitable for use in Flow-Thru Chips. Commercially available membrane filters of this type are typically made of aluminum oxide that has been etched to produce a precise, nondeformable honeycomb pore structure. Filters are available in pore sizes of 0.2, 0.1, and 0.02 μm with narrow pore size distributions. Channel diameters <1μm in diameter require application of a larger pressure gradient to move fluids through the membrane. The small pore sizes also indicate that a large surface-area enhancement should be possible with metal oxide membrane filter substrates. The optical properties of the metal oxides indicate that interchannel signal cross-talk may also be a potential limitation of the material.

Physical Properties of the Flow-Thru Chip Substrate

Due to the availability of straightforward methods for probe attachment, glass capillary arrays were selected as the initial substrate for Flow-Thru Chip development. Glass capillary arrays were purchased from Galileo Electro-Optics Corporation (Sturbridge, MA, USA). The material is supplied in 48 × 48 × 0.5-mm thick plates. In the configuration currently used, plates are divided into 16 12 × 12-mm chips for further processing. The channels are 10 μm in diameter and packed at a 50% open area ratio. Thus, for a given plate face cross-sectional area, the capillary array provides 100× the surface area of a flat substrate (Table 1). A number of parameters for this particular substrate geometry are given in Table 2.

The optical properties of the glass capillary arrays for Flow-Thru Chip use have been studied in some detail. The detection limit for tetramethylrhodamine fluorescent dye on the capillary glass is 1 attomol, or 600 000 molecules, in a 175-μm diameter spot of dye. At this level, the detection limit is determined by background fluorescence of the substrate rather than the sensitivity of the microscope and camera. Two sources of background have been identified for this chip material, scattering and autofluorescence. Scattering from the rough face of the glass between the microchannels can be minimized by using polished substrates. The substrate material is a high lead content glass so that autofluorescence effects will be best minimized by optimizing the excitation and emission filters.

To take full advantage of the enhancements gained by adding thickness to the chip, the detection method must be able to detect signal from the entire thickness of the chip. In imaging terminology, this means that a large depth of field is required in the measurement device. The effects of depth of focus and depth of field are illustrated in Figure 5. The data depict the ratio of the intensity of emission from a fluorescent spot as the focal plane is raised from 250 μm above the base of a 500-μm-thick chip to 500 μm above the substrate surface. The image is only crisp and in focus for those exposures taken when the focal plane is at or very near the top face of the chip. Although image quality degrades with all objectives, the intensity of the signal is not effected for the lower magnification lenses until the focal plane has been pulled back further from the substrate surface. For the 2× microscope objective, which views slightly more than one-fourth of the active chip area, the depth of field encompasses the entire chip thickness.

Now that we have described the desired attributes of a Flow-Thru Chip substrate, we shall move on to describe how the Flow-Thru Chip has performed. Provided below are experimental results of a prototypical implementation of the Flow-Thru Chip.

EXAMPLE OF BIOSENSING: DNA SENSOR

Elements of DNA Measurement System

The natural course of events in using a biochip begins with two parallel tracks, chip preparation and sample preparation, that converge at the point of running

Table 2. Physical Dimensions*

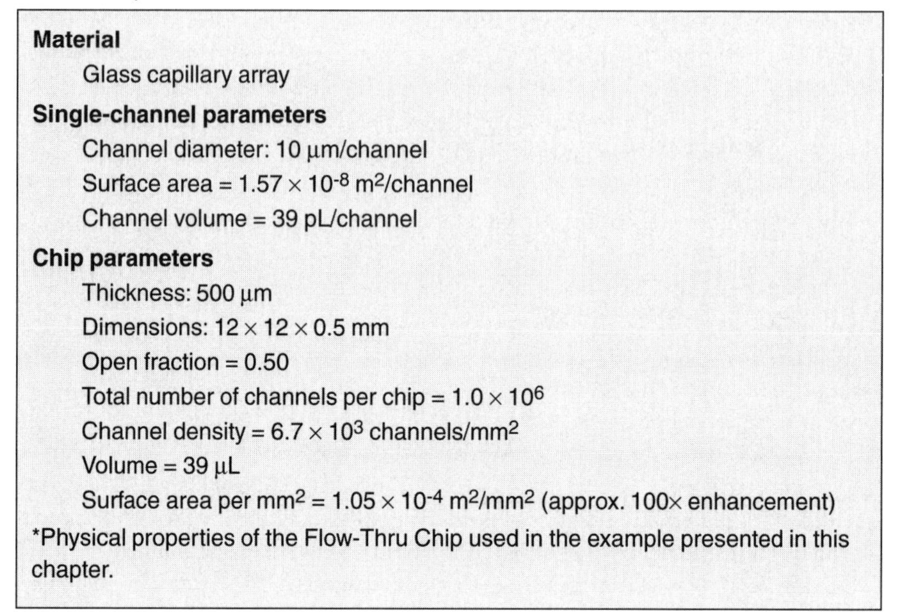

Material

Glass capillary array

Single-channel parameters

Channel diameter: 10 μm/channel

Surface area = 1.57×10^{-8} m^2/channel

Channel volume = 39 pL/channel

Chip parameters

Thickness: 500 μm

Dimensions: $12 \times 12 \times 0.5$ mm

Open fraction = 0.50

Total number of channels per chip = 1.0×10^6

Channel density = 6.7×10^3 channels/mm^2

Volume = 39 μL

Surface area per mm^2 = 1.05×10^{-4} m^2/mm^2 (approx. 100× enhancement)

*Physical properties of the Flow-Thru Chip used in the example presented in this chapter.

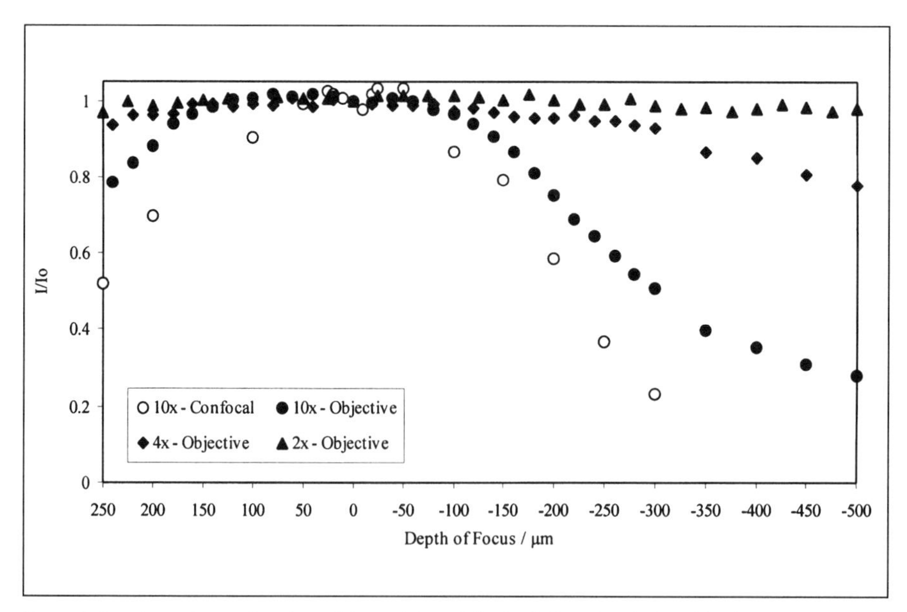

Figure 5. Spot intensity as a function of the position of the focal plane. Focal plane (depth zero) is at the surface of the chip. Positive values indicate focusing into the chip and negative values focusing above the surface of the chip. Spot intensities are given as a ratio to the intensity measured at a focal plane depth of zero. The data are shown for three microscope objectives. Confocal and normal imaging are compared for the 10× objective.

the chip assay. Once the sample is introduced to the chip, the hybridization signal can be monitored in situ when a label is incorporated in the target or ex situ at the conclusion of the assay. After image acquisition, data processing is performed, and analysis of the resulting hybridization pattern is conducted. The chip itself is just one of several steps in a microarray experiment.

The chip preparation tract begins with the raw material that is subsequently processed to produce a ready to run device. The first step in our present processing scheme is to clean and activate the substrate surface. The activated surface is modified with an ultrathin film that incorporates specific binding sites for attachment of probe. The microarray is produced, or spotted, by deposition of probe bearing solutions in a predetermined pattern. Finally, the chip is treated with a blocking agent to prevent nonspecific binding of target molecules. The method of treatment in each of these steps is described in some detail below.

The sample preparation tract is a entire technology area that in the end controls the viability of any microarray assay. Numerous schemes for target selection, amplification, and labeling are available. The advantages and disadvantages of each must be weighed in relation to the desired information output of the biochip assay. Because the focus of this chapter is to demonstrate the advantages of the three-dimensional Flow-Thru Chip geometry, detailed sample preparation methods will not be disclosed. For the work that is included herein, targets containing a single fluorescent tag have been synthesized using standard phosphoramidite chemistry. We note that target sample mixtures prepared from cell cultures have been successfully assayed using a Flow-Thru Chip with sensitivity comparable to that of synthetic target mixtures.

Chip Preparation

A primary objective in chip preparation is to immobilize probe molecules permanently in a functional conformation such that the appropriate target can bind to the probe and nonspecific binding of other components in the target mixture is reduced to a level that provides useful signal-to-background ratios. Biomolecule immobilization can be accomplished by various means including adsorption, entrapment, and covalent attachment. Covalent attachment is the preferred method for "permanent" immobilization; however, great care must be exercised when developing a covalent bonding scheme to ensure that the activity of the probe is not adversely effected. Since few organic functional groups will react directly with an activated silica surface, an intermediate layer is advisable for biomolecule immobilization. Functionalized organosilanes have been used extensively as such an intermediate layer on glass and silicon substrates (15,21), and this is the approach that has been taken in production of the Flow-Thru Chip as described in this chapter.

Cleaning. Effective cleaning of the substrate prior to silanization is extremely important to reproducible chip production. The cleaning process should both remove contaminates (e.g., oils, dirt, and detergents) and generate reactive hydroxyl groups on the silica surface (21). A uniform and dense film of reactive hydroxyl, or silanol (Si-OH) groups is critical for effective immobilization of biomolecules. If the substrate surface is not properly activated, the silane film will not be uniform, leading to inconsistent biomolecule immobilization and poor microarray performance. A number of protocols have appeared in the literature. Common to many methods are degreasing and acid activation steps. Degreasing is commonly accomplished with an organic solvent (e.g., methanol) or by strong oxidizing conditions (e.g., basic hydrogen peroxide or sulfochromic acid). Acid activation is typically performed using mineral acids such as hydrochloric, nitric, or sulfuric acid.

Of particular concern with the FT geometry is the effectiveness of the cleaning (and subsequent silanization) step along the interior of the channels. Under static conditions, the channels fill by capillary action, but, once they are filled, the only exchange between fluid inside and outside of the channels occurs by diffusion. In the hybridization step, the chip is placed inside a custom-built flow cell where fluid is pumped through the chip. Such a geometry for the chip cleaning and silanization steps is not economical, so another means to enhance fluid distribution throughout the chip is needed. Sufficient fluid exchange has been realized by performing the degreasing and activation steps with the solutions immersed in an ultrasonic bath. The large surface area afforded by the capillary array is ideal for bubble nucleation and growth caused by ultrasonic excitation. Drying the capillary glass presents an additional concern at this stage because evaporation from the microchannels is slow. An efficient means of drying chips is first to de-wet the capillaries by placing the wetted chip on filter paper and then dry the chip in an oven at 80°C for at least 1 hour.

Silanization. Silanes are routinely used to modify surfaces both to reduce nonspecific adsorption and to provide organic functional groups suitable for covalent attachment of biomolecules (21). Many solid supports, including glass and silicon, can be modified to contain surface hydroxyls that react with methoxy/ethoxy residues of a silane. Silanes are commercially available that contain an ever-increasing number of reactive functional groups suitable for biomolecular binding, either directly or via a cross-linker. Factors in determining which silane to employ include stability of the silane functional group, choice of reactive group on the biomolecule (amine, thiol, carbohydrate, etc.) and the type of cross-linker (homo-bifunctional or hetero-bifunctional). Substrates are silanized either by immersing the substrate into a silane-containing solution or by vapor deposition via refluxing the silane. Limited data are available to suggest which method provides the superior silane film for probe immobilization and nonspecific adsorption inhibition, owing to the difficulty in analyzing these surfaces by standard analytical techniques. Solution phase silanization has been used for all chip preparations described herein, primarily because it is easier to produce a moderate number of chips in this manner than by gas phase deposition.

Silane films containing amine, epoxide, and mercaptan functional groups have been tested for use on the Flow-Thru Chip. The amine- and mercaptan-containing silanes produced more uniform and reproducible films than the epoxide. It should be noted that certain silanes degrade over time with improper storage. It is recommended that organosilanes be stored in a dry, low-light environment. The stability of the silanes varies according to the organic functional groups that have been incorporated, and we have found that the epoxysilane group degrades most rapidly among those listed, which is the most likely cause of poor probe immobilization in that case. The selection between the amine- and mercaptan-containing silane was made based on the available cross-linkers and modified amidites for probe synthesis. A hetero-bifunctional cross-linker was selected over homo-bifunctional linkers to avoid linking reactive groups of the same type (surface-surface, probe-probe linkages as opposed to the desired surface-probe linkage). Using a hetero-bifunctional cross-linker requires that the probe have a different chemistry than the surface. Several extensive reviews of bioconjugate techniques are available, such as a paper by Greg Hermanson of Pierce Chemical Company (10). Gamma-maleimidobutyryloxy-succinimide (GMBS) was selected to bind the mercaptan (maleimide) to a primary amine (succinimide).

The relative stability of the amine and mercaptan groups formed the basis of our decision to modify the surface with the mercaptosilane and the probes with an amine-bearing group. Mercaptans are oxidized slowly on exposure to air, so that the silanized chips must be used as soon as possible to yield maximal probe immobilization. Amine-modified oligonucleotides are quite stable in buffer at 4°C.

Chips were immersed in a 2% solution of mercaptopropyltrimethoxysilane (MPTS; Aldrich, St. Louis, MO, USA) in dry toluene (Fisher Scientific, Pittsburgh, PA, USA) for 1 hour. Excess silane solution was removed by blotting the chips on a piece of filter paper, followed by 10-minute serial sonications in dry toluene, 1:1 toluene/absolute ethanol, and absolute ethanol. Silanized chips were baked at 80°C for 16 hours to ensure complete hydrolysis of the methoxy residues.

The number of mercaptan groups immobilized on the FT substrate was determined using a thiol quantitation kit from Molecular Probes (Eugene, OR, USA). The available mercaptan surface coverage is 2.6×10^{13} groups/cm^2 prior to the baking step. Baking the chips results in a decrease in available mercaptan surface coverage to 2.0×10^{13} groups/cm^2, presumably due to oxidation of the mercaptan to the sulfinate or sulfonate. Chips can be stored in a dry, low-light environment for several days without much loss of mercaptan activity. However, chips are spotted typically within hours of removal from the baking oven.

Probe immobilization. It is important to consider the effect of immobilization on biomolecule activity when designing immobilization strategies. When using a cross-linker, there are two reaction steps that must be completed sequentially. The order of the reactions is typically inconsequential for solution phase reactions. However, when one of the reactions occurs on a surface, the reaction order becomes more critical due to local concentration and cross-reactivity effects. For example, markedly lower immobilization efficiencies are observed on an MPTS-modified surface that is first treated with the cross-linker GMBS and then spotted with amine-bearing probes than when GMBS and probe are mixed prior to deposition. Clearly, the preferred method would be to activate the entire chip and spot only the probe, as has been reported in the literature (18), but the resulting coverage is lower than by the alternative method.

Probes were synthesized (Research Genetics, Huntsville, AL, USA) with a primary amine at the 3′ end of the sequence. Probes were reacted with the heterobifunctional cross-linker GMBS (Pierce, Rockford, IL, USA) in 1× saline sodium citrate buffer (Sigma, St. Louis, MO, USA). The reaction mixture was 60 μM probe and 600 μM GMBS to give a linker-to-probe ratio of 10:1. The reaction was allowed to proceed for 1 hour at room temperature prior to spotting. A 10-nL aliquot of the reaction mixture of probe and GMBS was spotted onto the silanized chips using a Packard BioChip™ piezoelectric spotter (Packard, Meriden, CT, USA). For the size of a 10-nL spot, the amount of probe deposited is 1.4×10^{13} probes/cm^2, which is less than the number of potential mercaptan binding sites of 2.0×10^{13} thiol groups/cm^2. Binding efficiencies are typically in the 30% to 60% range, so the final probe density in each spot is on the order of 6×10^{12} probes/cm^2. Literature values for probe surface densities on silanized glass surfaces range from 10^{12} to 10^{13} probes/cm^2 (1,9,18). This range of probe surface densities has been reported to be optimal for DNA hybridization (23,25). Spotted chips were allowed to dry in a dehumidified, low-light box for 24 hours prior to blocking. Spotted chips have been stored for as long as 1 month prior to use without observing significant loss of performance.

A piezoelectric spotter has several advantages for use with the Flow-Thru Chip. The single most important aspect is that piezo-driven devices are noncontact devices, that is, the liquid delivery device never touches the substrate. Although the capillary glass does have some structural rigidity, it is still a 50% open area ratio glass. Contact spotting (e.g., pin methods) tends to scratch the surface or, even worse, fracture the chip. The Packard technology allows reproducible delivery of nanoliter-scale volumes at machine resolution (i.e., 10 to 30 μm). Present production rates are admittedly slower than other methods. A batch of 32 chips, each containing 384 distinct spots, requires as long as 12 hours to complete. At this stage of development, large-scale production is not an overriding concern, whereas production of reproducible, unblemished chips is a necessity.

Blocking nonspecific adsorption. Nonspecific adsorption of target can create a high level of background signal, which, in turn, reduces the sensitivity of the assay. The importance of treating chips with a blocking reagent prior to hybridization is shown in Figure 6, which contains images from the hybridization of fluorescently labeled target to a small array. Figure 6A corresponds to an unblocked chip and Figure 6B to a chip that was treated with 250 μL of 5× Denhardt's blocking reagent (Sigma) for 15 minutes. Hybridization times were 2 hours for both chip assays. On the blocked chip, three spots are clearly visible. The missing spot in the upper left corner corresponds to a negative control. On the unblocked chip, all spots are visible, but they appear in an inverted fashion. In this case, the probes themselves act as a blocking agent to block nonspecific adsorption. Little signal is observed in these spots because the target is distributed over the entire surface area of the chip, and less is available for binding to probes.

The blocking procedure for the Flow-Thru Chip involves soaking a chip in 250 μL of a proprietary blocking solution for 15 minutes, blotting dry, and baking at 80°C for 1 hour. Blocked chips are stored in a dehumidified, low-light box until use.

Realizing the Advantages of the Flow-Thru Chip Geometry

The FT substrate geometry offers several potential advantages over flat substrates for chip applications. Among these are: *(i)* increased responsiveness and dynamic range; *(ii)* enhanced reaction rates; *(iii)* uniform probe deposition and higher array densities; and *(iv)* smaller sample volume requirements.

Above, we addressed how the FT geometry produces each of these advantages in theory. Now we turn to data demonstrating that these advantages can be realized. The first and second proposed advantages can be shown by comparing hybridization events for identical arrays on a Flow-Thru Chip and on a traditional flat glass substrate. In the experiment described here, both the reaction rate and detection sensitivity were measured using real-time fluorescence. The third

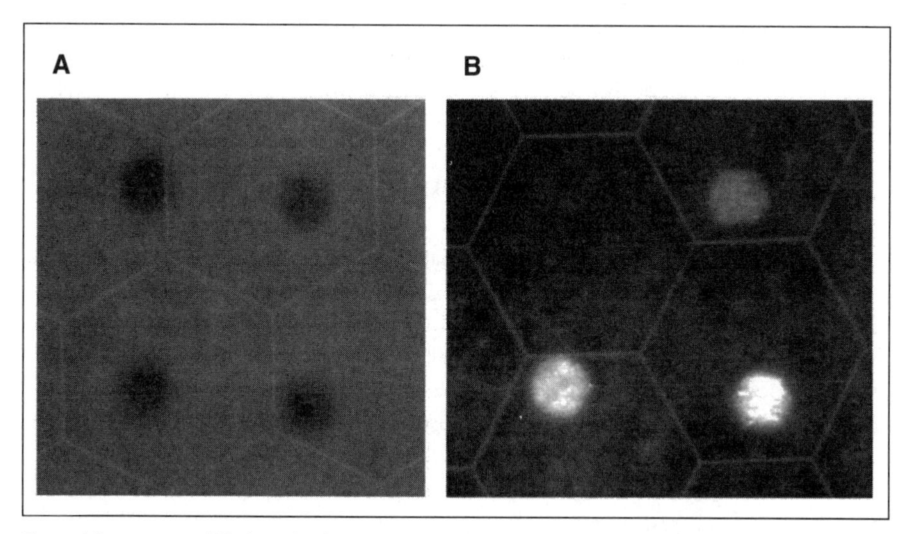

Figure 6. Importance of blocking the chip prior to hybridization. Hybridization of a 50-nM target solution to an unblocked (A) or blocked (B) chip. The images are on different contrast scales for display purposes. In A, the target binds nonspecifically to the entire substrate, except in the regions where probe was deposited. On the blocked chip (B), labeled target binds predominantly within the spots where complementary probe has been deposited.

advantage is demonstrated by imaging fluorescently labeled probes deposited on FT and flat substrates. The images show that the probe is more uniformly distributed on the FT substrate.

Experimental description. The probe sequences used for all experiments can be found in Table 3. The sequences correspond to human cytoplasmic β-actin (GenBank accession # X00351), c-*myc* proto-oncogene (EGAD # HT3717), and interleukin 2 (GenBank accession # U25676). A primary amine was incorporated into the sequence via an amino-modified dT (Glen Research, Sterling, VA, USA). Target sequences are listed in Table 3 as well. The targets were synthesized with a 5′ Texas Red phosphoramidite for fluorescence detection (Synthetic Genetics, San Diego, CA, USA). Hybridizations were performed at room temperature in 5× saline-sodium phosphate-EDTA buffer SSPE (Sigma).

FT chips were prepared as described above. For comparison, two-dimensional microarrays were prepared by cutting microscope slides to fit into the Gene Logic fluid cartridge. The flat glass chips were treated and spotted the same as FT chips. The flow rate in the system is 200 µL/min. In this manner, both microarrays received the same flux of target solution under fluid flow conditions. The chips were placed in the cartridge such that fluid flowed through the microscopic channels of the Flow-Thru Chip but flowed over the surface of the flat glass chip.

In a standard hybridization experiment, a chip was mounted in the reusable chip cartridge. The cartridge was connected to the fluid delivery system and filled with hybridization buffer. The chip was washed with hybridization buffer for 15 minutes prior to injection of the target. When the target was injected, the fluid delivery system was switched from pass-through mode to recirculation mode. The target sample loop holds 87 µL, and the recirculation loop, including the volume of the flow cell, is approximately 1 mL. With fluorescently labeled targets, hybridization was monitored in real time by mounting the chip cartridge on a microscope. For targets that use an indirect labeling scheme (e.g., biotin-streptavidin staining, chemiluminescence) a posthybridization staining step was required. For a concentrated sample in which the target concentration was >10 nM, signal was observed within minutes. More dilute samples require longer hybridization times before the signal elevates above the background level.

Sensitivity and speed. Two valuable advantages gained by using a three-dimensional substrate rather than a traditional flat substrate—sensitivity and speed—are observed in the comparison of the hybridization of 50-nM fluorescently labeled target GL-T1 to identical arrays on FT and flat substrates. Hybridization was monitored in situ to demonstrate the sensitivity and reaction rate enhancements afforded by the three-dimensional substrate geometry. Figure 7 contains the results of the real-time measurements. The graph shows the signal-to-background fluorescence ratio from the perfect match probe GL-P1 spot on the FT and flat substrates. The signal increase over the background level is quite dramatic for the Flow-Thru Chip.

After 10 hours of target recirculation, when both samples had reached a satu-

Table 3. Sequences of Probes and Targets*

Sequence Name	Sequence	Length
GL-P1	5′-CCC-AGG-GAG-ACC-AAA-AGC-3′	18
GL-P2	5′-ACC-ATT-TTA-GAG-CCC-CTA-3′	18
GL-P3	5′-AAG-ATG-GTA-AGC-ATA-AAA-3′	18
GL-P4	5′-CTA-TAG-TGA-GTC-GTA-TTT-3′	18
GL-T1*	5′-GGC-TTT-TGG-TCT-CCC-TGG-GAG-TGG-GTG-GAG-GCA-GCC-AGG-GCT-TAC-CTG-TAC-ACT-GAC-TTG-AGA-CC-3′	65
GL-T1*	5′-CCC-TAG-GGG-CTC-TAA-AAT-GGT-TTA-CCT-TAT-TTA-TCC-CAA-AAA-TAT-TTA-TTA-3′	51
GL-T1*	5′-ACT-TTT-TTA-TGC-TTA-CCA-TCT-TTT-TTT-TTT-CTT-TAA-CAG-ATT-TGT-ATT-TAA-3′	51

*All targets were synthesized with a single Texas Red fluorescent dye at the 5′ end of the sequence.

rated intensity value, the signal-to-background ratio was 970 for the Flow-Thru Chip data and only 22 for the flat glass. One should bear in mind that these measurements were taken in situ, with the fluorescently labeled target in the flow cell, resulting in significant background levels and seemingly low signal-to-background ratios for the flat glass substrate. The relative ratios are not affected by making the measurements in the presence of the labeled target. The FT intensity is 44 times that of the flat glass intensity at the saturated level. The rate of hybridization is also different. Although it is difficult to determine from the figure as presented, the rate at which the intensity increases is faster on the Flow-Thru Chip. The relative rate of hybridization on the two substrates is given by the ratio of the t_ms, the time it takes for the signal to reach one-half of the saturated value. Hybridization on the Flow-Thru Chip occurs at approximately 5.6 times the rate of that on a flat glass substrate.

The relative performance of the microarrays on the two substrates is well understood in terms of the differences in geometry. The theoretical surface area enhancement is 100× for the glass capillary array substrate used in this example. The observed enhancement was 44× more signal on the FT chip than on the flat glass chip. At this stage in development, we consider such enhancement quite considerable. The theoretical enhancement may not be realized though, due to differences in probe immobilization levels on the two substrates or because the current detection configuration is not capable of collecting photons from the entire thickness of the chip.

Theoretical treatment of the relative rates of hybridization on the two substrates is complicated due to the nature of the fluid mechanics in both cases. By confining target molecules within the microchannels, mass transport distances are greatly reduced in comparison with the diffusion distances at flat substrates (2,6). Bringing the probe and target close together dramatically increases the rate of reaction. If mass transport limitations and spatial confinement alone are considered, then the rate of hybridization in the FT substrate should be approximately 100× faster than at the flat substrate. The relative rate in the mass transfer limited case at both substrates is given by the ratio of the square of the diffusion layer thicknesses in each case: 5 μm on the FT and an assumed 50 μm on the flat substrate. The diffusion layer thickness on a flat substrate was assumed to be the same as the thickness of the stagnant layer. The observed rate of enhancement was nearly a factor of 6 on the Flow-Thru Chip. When corrected for the relative spot diameters, the observed rate of enhancement increases to approximately 10-fold faster on the FT chip. Two factors may limit the increases in rate that can be attained. In the FT geometry the target cannot migrate laterally once it enters the chip, so the only means for target redistribution is to complete the circuit in the fluid handling recirculation loop. Because lateral redistribution is possible on the flat substrate (6), the effective rate will be somewhat increased inside a closed circuit like that used here. Also, the hybridization reaction itself may be the rate-

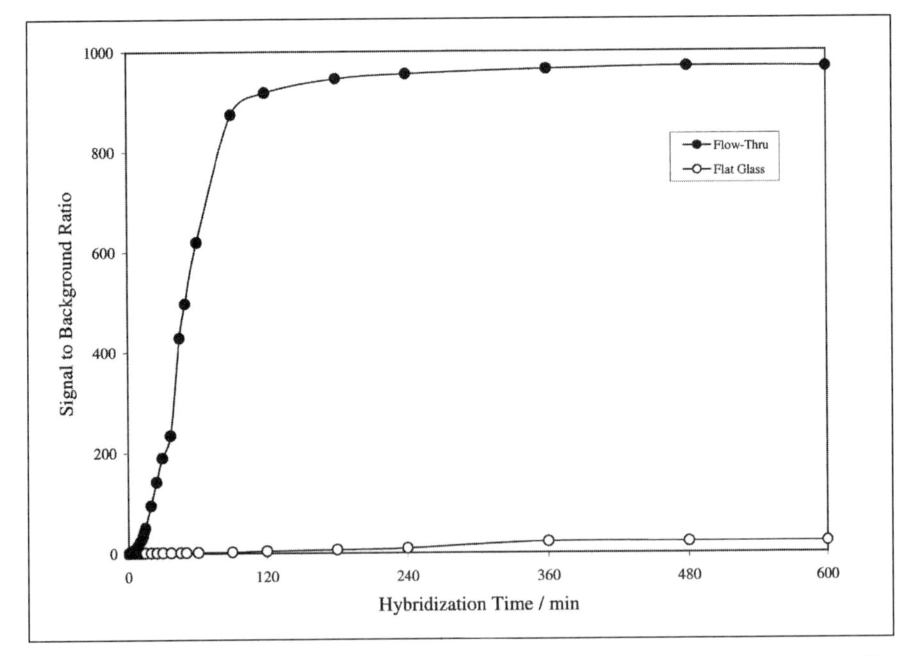

Figure 7. Comparison of hybridization signal on the Flow-Thru Chip and a traditional flat substrate. The signal-to-background ratio is plotted for the fluorescence intensity measured in situ. The filled circles are for the Flow-Thru Chip, and the empty circles are for the flat glass substrate. The ratio of the signal intensities at $t = 600$ minutes is 44:1 in favor of the Flow-Thru Chip. The hybridization rate is 5.6 times higher on the Flow-Thru Chip.

limiting step in the FT geometry if the bulk flow rate is too high. Therefore, we consider it reasonable that the relative rate of hybridization is approximately 6× faster on the Flow-Thru Chip, given the reduction of dimensionality in the three-dimensional chip geometry.

Probe distribution. An additional benefit of the Flow-Thru Chip substrate is afforded by the wetting properties of microporous materials. When a liquid comes in contact with a reasonably hydrophilic microporous material, the liquid is drawn into the material by capillary action. That same behavior is observed on the prepared capillary array glass. Such wetting has two desirable side effects: equivalent volumes have a smaller footprint on FT substrates and liquid evaporates from the channels at a much slower rate than on a flat surface. Images of an equivalent volume, 40 nL, of fluorescently labeled probe spotted on a Flow-Thru Chip and a flat glass substrate are shown in Figure 8. The uniformity of the spot on the FT substrate is quite good, with only a small amount of a diffusive halo at the edge of the spot due to either limited wetting of the edge capillaries or cross-talk of the fluorescence signal. It is interesting that the hexagonal packing of the capillaries determines the shape of the spot to a great extent. Also, the interior of the microchannels is brighter than the top surface, indicating that much of the signal detected was generated from within the microchannels. The spot on the flat substrate is quite different in appearance. The intensity is very nonuniformly distributed due to the rapid evaporation of the deposition mixture. Spot uniformity on flat glass substrates can be improved by incubation in a humidified chamber immediately following spotting. Rapid evaporation of the deposition mixture results in uneven probe deposition across the spot area, which is reflected in nonuniform intensity distribution across the spot after hybridization. Note

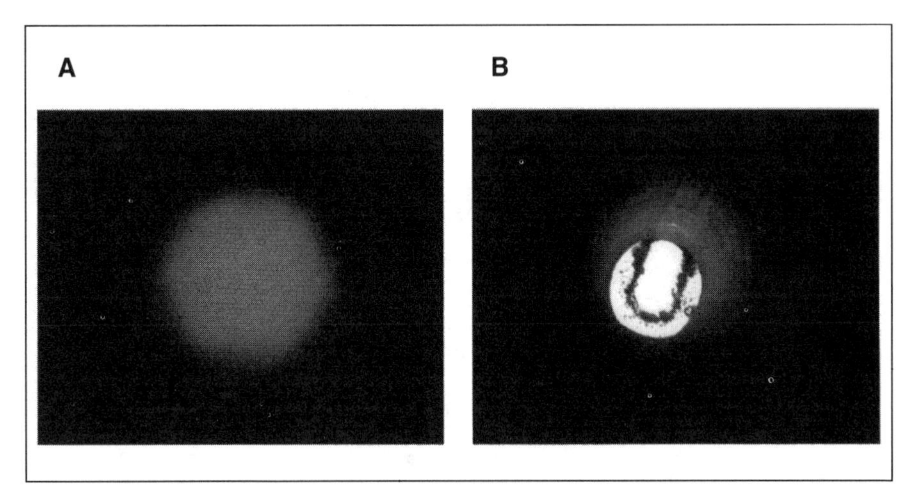

Figure 8. Fluorescence images of a 25-nL spot of fluorescently labeled probe. The spot was deposited on a Flow-Thru Chip (A) and on a flat glass substrate (B) prior to washing. The intensity is much more uniform on the three-dimensional substrate, resulting in superior hybridization results and more reliable image analysis.

that the contrast in the images is not the same. The flat glass image has been optimized to show the much dimmer outer ring; nevertheless, the integrated intensity from the two spots is identical. The diameter of the spots on the two substrates is also different. A plot of spot diameter versus deposition volume is given in Figure 9. Spots have a consistently larger diameter on the flat substrate for the same volume of deposited liquid. The spot diameter is about 60% larger on the flat substrate than on the Flow-Thru Chip. Smaller spot diameters mean that larger array densities can be realized on the three-dimensional substrate.

Flow-Thru Chip System Performance

System performance of the current configuration of the Flow-Thru Chip has been rigorously determined through a series of hybridization experiments. We will describe a series of experiments that demonstrate the reproducibility and potential for further development of the platform. One of the most instructive experiments for monitoring microarray system performance is to determine the uniformity of hybridization across the chip. Uniformity measurements give a reading for the entire system including spotting, fluid distribution, and detection in a single experiment. The relative contribution to nonuniformity of each element in the system can be determined by independent variation of a single parameter.

Uniformity of hybridization to a Flow-Thru Chip in the Gene Logic chip

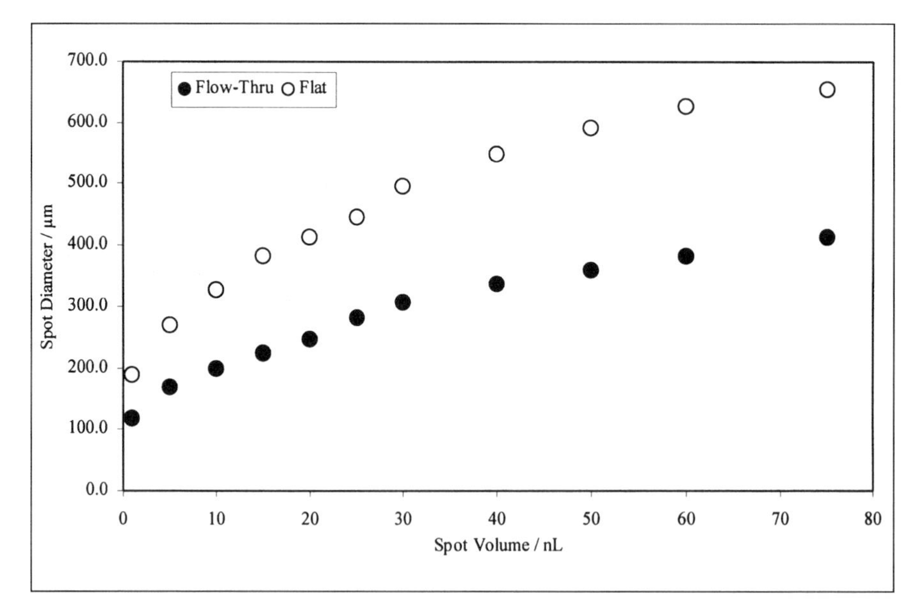

Figure 9. Spot diameter versus deposited volume on flow-thru and flat substrates. The spot diameter with increasing volumes of liquid deposited using a piezoelectric spotter on a Flow-Thru Chip (solid circles) or on a flat glass substrate (open circles) that were modified in an identical manner. The spot diameter is uniformly 60% larger on flat glass substrates than on the microporous substrate due to differences in wetting properties.

cartridge and fluid delivery system was determined by hybridizing a synthetic target mixture to an array covering nearly 75% of the active chip area. The array contained 256 spots divided into 16 identical 4 × 4 subarrays. Each subarray contained spots for four different probes (GL-P1, GL-P2, GL-P3, and GL-P4). The mixed target solution contained complementary targets to three of the probes, but not the fourth probe GL-P4, which served as a negative control. The concentration of each target was GL-T1 (2 nM), GL-T2 (10 nM), and GL-T3 (20 nM). The targets T1, T2, and T3 are complementary to probes GL-P1, GL-P2, and GL-P3, respectively. At 64 spots per probe type, the target concentration was 31, 160, and 310 pM, respectively, on a per spot basis. The target mixture was recirculated through the fluid handling system for a total of 4 hours prior to image acquisition. An image of the resulting hybridization is given in Figure 10. The columns in the figure from left to right correspond to GL-P1, GL-P2, GL-P3, and GL-P4. That pattern is repeated four times across the chip.

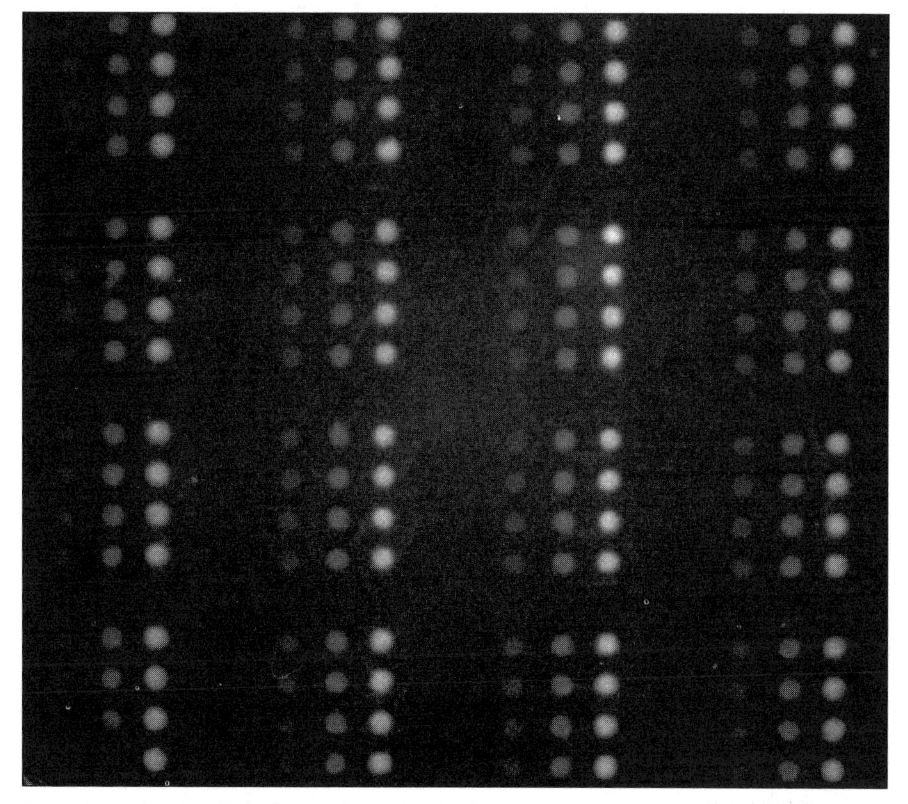

Figure 10. Uniformity of hybridization to microarray. The microarray contains 256 elements, divided into 16 identical 4 × 4 arrays. In each 4 × 4 array, the columns contain the same probe sequence. A target mixture containing 2 nM to probe 1, 10 nM to probe 2, 20 nM to probe 3, and no target to probe 4 (from left to right) was hybridized to the array for 4 hours. The response is nearly linear with concentration, and the average standard deviation for each probe type response is 10%.

The results are quite uniform for each probe. The standard deviation for the average of each probe is roughly 10% of the measured signal intensity. The relative spot intensities are in good agreement with the relative target concentrations. The ratio of the spot intensities for P3 and P2 to P1 are 9.3 and 5.6, respectively, whereas the relative target concentrations in the mixture were 10- and 5-fold higher. Hence, the spot intensity is proportional to the target concentration in solution after 4 hours of hybridization. The signal-to-background ratio at the lowest target concentration, 2 nM GL-T1 total or 31 pM per spot, is 12. This value suggests that the Flow-Thru Chip detection limit should be on the order of picomolar concentrations in the current configuration. A minimum detectable concentration of 3 pM has been realized in additional experiments that are not detailed here.

A detection limit of 1 pM target concentration translates to a sensitivity to 5×10^7 labels. In the case described here, there is a one-to-one correspondence between the number of targets and the number of labels. Strategies for multiple-label incorporation on the target are available and have the potential to enhance the sensitivity of the assay to a given target dramatically.

CONCLUSIONS

Applications

A large number of applications for DNA chips have been identified including, but not limited to, gene profiling/differential display, sequencing, diagnostics, pharmacogenomics, single nucleotide polymorphism detection, antisense activity determination, forensics, and genotyping. The role of the DNA chip in any application is to provide a quick and inexpensive means of determining the abundance of specific nucleic acid sequences in a sample in a reproducible manner.

Drug screening. The core of Gene Logic's Accelerated Drug Discovery system is its proprietary READS® (Restriction Enzyme Analysis of Differentially Expressed Sequences) technology for analyzing how genes are expressed in human cells or tissues. A schematic of how technologies converge in the Gene Logic drug discovery program is given in Figure 11. Gene Logic uses READS to generate a gene expression profile that provides a semiquantitative snapshot of the levels of expression of essentially all the genes expressed in a tissue sample. For instance, Gene Logic compares normal and diseased tissues through a series of such snapshots and creates a "molecular movie" that can be used to identify the changes in gene expression that occur as a disease develops and progresses and also to determine which genes are associated with the disease. By employing its READS technology in conjunction with its proprietary bioinformatics systems, proteins encoded by these disease-associated genes are prioritized as potential drug targets.

In its drug discovery process, Gene Logic uses its READS technology to identify the genes that are associated with a given disease. Once these genes are known, a customized Flow-Thru Chip incorporating probes specific to these genes is designed and used to test the effect of compounds in cellular assays. Test samples from treated cells are hybridized to the Flow-Thru Chip. Analysis of the microarray image permits correlation of the expression level in the treated sample with expression in the original sample. Compounds that have the desired effect on expression of the relevant genes, such as restoring the expression pattern to normal or mimicking the effect of a known therapeutic, are evaluated as drug leads. Each chip is configured to contain from just a few to as many as 100 genes identified using the READS technology; the number of genes included on each chip is expected to be increased to approximately 1000 in the future.

Other applications. Many other nucleic acid-based applications are amenable to Flow-Thru Chip analysis as well. We have concentrated primarily on nucleic acid analysis using this three-dimensional substrate but would like to stress that the advantageous concepts translate to other analyses as well.

Final Thoughts

The Flow-Thru Chip is a powerful analysis platform, with distinct advantages over traditional substrates for biological sensing applications. We have demonstrated advantages gained by using a Flow-Thru Chip over traditional flat glass

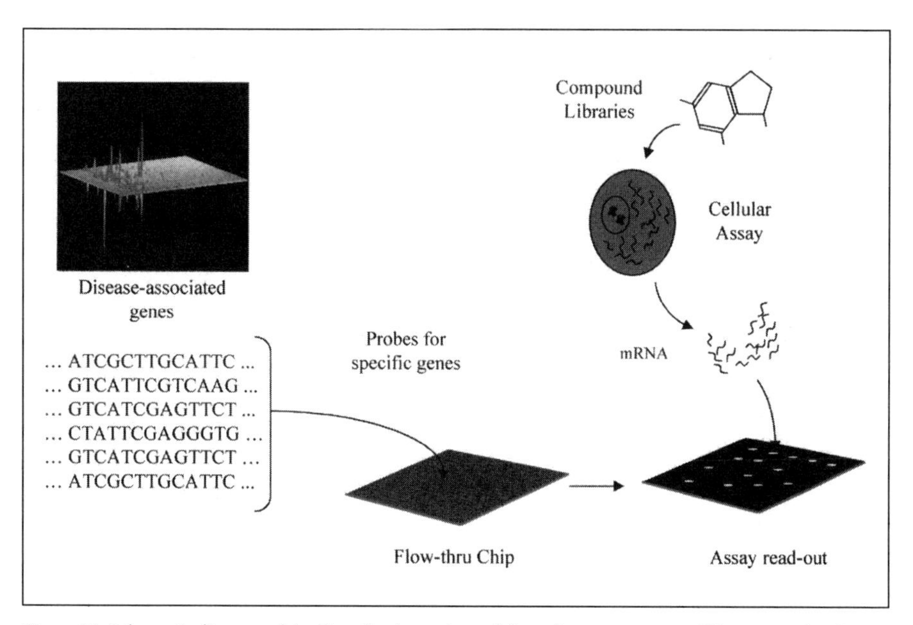

Figure 11. Schematic diagram of the Gene Logic accelerated drug discovery concept. Disease-associated genes are identified using the proprietary READS technology. Those gene are placed on a Flow-Thru Chip, which becomes the read-out mechanism for compound screening in cellular assays.

chips for DNA microarray applications. The advantages of using the FT substrate over a flat glass substrate included a 44-fold increase in target capture due to the larger surface area per spot, hybridization rates roughly sixfold higher due to reduction of distance between immobilized probe and the target, more uniform spots and probe deposition due to slower fluid evaporation from the microchannels, and higher array densities due to the 60% smaller spot diameter per unit volume on the Flow-Thru Chip.

We have demonstrated the potential power of using an FT substrate for microarrays even at this early stage of development. Although many challenges and obstacles are yet to be encountered on the path to realizing the full potential of microarrays for analysis, we believe that the Flow-Thru Chip will stand out as a leading platform in the next generation of analysis tools.

ACKNOWLEDGMENTS

We would like to thank our collaborators Prof. Anthony Cass and Dr. Catherine Halliwell of Imperial College for clarifying issues regarding silanization of glass and silicon, Prof. John Cooper and Mr. Vincent Benoit of Glasgow University for optical studies of FT substrates, and Dr. Kenneth Beattie and Dr. Mitch Doktycz of Oak Ridge National Labs for discussion of the Flow-Thru Chip.

REFERENCES

1. **Bamdad, C.** 1998. The use of variable density self-assembled monolayers to probe the structure of target molecules. Biophys. J. *75*:1989-1996.
2. **Bard, A.J. and L.R. Faulkner.** 1980. Electrochemical Methods. John Wiley & Sons, New York.
3. **Beattie, K.L.** 1998. Microfabricated, flowthrough porous apparatus for discrete detection of binding reactions. U.S. Patent 5,843,767.
4. **Beier, M. and J. Hoheisel.** 1999. Versatile derivatisation of solid support media for covalent bonding on DNA-microchips. Nucleic Acids Res. *27*:1970-1977.
5. **Bowtell, D.D.L.** 1999. Options available—from start to finish—for obtaining expression data by microarray. Nat. Genet. *21*:25-29.
6. **Chan, V., D.J. Graves, P. Fortina, and S.E. McKenzie.** 1997. Adsorption and surface diffusion of DNA oligonucleotides at liquid/solid interfaces. Langmuir *13*:320-329.
7. **Chan, V., D.J. Graves, and S.E. McKenzie.** 1995. The biophysics of DNA hybridization with immobilized oligonucleotide probes. Biophys. J. *69*:2243-2255.
8. **Cheung, V., M. Morley, F. Aguilar, A. Massimi, R. Kucherlapti, and G. Childs.** 1999. Making and reading microarrays. Nat. Genet. *21*:15-19.
9. **Doktycz, M.J. and K.L. Beattie.** 1997. Genosensors and model hybridization studies, p. 205-225. *In* T.J. Beugelsdijk (Ed.), Automation Technologies for Genome Characterization. John Wiley & Sons, New York.
10. **Hermanson, G.T.** 1990. Bioconjugate Techniques. Academic Press, San Diego.
11. **Janssen, K.** 1995. Current Protocols in Molecular Biology. John Wiley & Sons, New York.
12. **Lehmann, V.** 1990. Formation mechanism and properties of electrochemically etched trenches in n-type silicon. J. Electrochem. Soc. *137*:653-659.
13. **Lehmann, V. and U. Gruning.** 1997. The limits of macropore array fabrication. Thin Solid Films *297*:13-17.

14. **Lipshutz, R.J., S.P.A. Fodor, T.R. Gingeras, and D.J. Lockhart.** 1999. High density synthetic oligonucleotide arrays. Nat. Genet. *21*:20-24.
15. **Liu, J. and V. Hlady.** 1996. Chemical pattern on silica surface prepared by UV irradiation of 3-mercaptopropyltriethoxy silane layer: surface characterization and fibrinogen adsorption. Coll. Sur. B. *8*:25-37.
16. **Livshits, M.A. and A. Mirzabekov.** 1996. Theoretical analysis of the kinetics of DNA hybridization with gel-immobilized oligonucleotides. Biophys. J. *71*:2795-2801.
17. **O'Donnell, M.J., K. Tang, H. Koster, C.L. Smith, and C.R. Cantor.** 1997. High-density, covalent attachment of DNA to silicon wafers for analysis by MALDI-TOF mass spectrometry. Anal. Chem. *69*:2438-2443.
18. **O'Donnell-Maloney, M.J. and D.P. Little.** 1996. Microfabrication and array technology for DNA sequencing and diagnostics. Gen. Anal. *13*:151-157.
19. **Pease, A.C., D. Solas, E.J. Sullivan, M.T. Cronin, C.P. Holmes, and S.P.A. Fodor.** 1994. Light-generated oligonucleotide arrays for rapid DNA sequence analysis. Proc. Natl. Acad. Sci. USA *91*:5022-5026.
20. **Proudnikov, D., E. Timofeev, and A. Mirzabekov.** 1998. Immobilization of DNA in polyacrylamide gel for the manufacture of DNA and DNA-olignucleotide microchips. Anal. Biochem. *259*:34-41.
21. **Shriver-Lake, L.C.** 1998. Silane-modified surfaces for biomaterial immobilization, p. 1-14. *In* T. Cass and F.S. Ligler (Eds.), Immobilized Biomolecules in Analysis. Oxford Press, Oxford.
22. **Southern, E.M., S.C. Case-Green, J.K. Elder, M. Johnson, K.U. Mir, L. Wang, and J.C. Williams.** 1994. Arrays of complementary oligonucleotides for analysing the hybridisation behaviour of nucleic acids. Nucleic Acids Res. *22*:1368-1373.
23. **Southern, E., K. Mir, and M. Shchepinov.** 1999. Molecular interactions on microarrays. Nat. Genet. *21*:5-9.
24. **Special Report.** 1993. Genosensors: the next step in biosensor technology. Genesis Rep. *2*:6.
25. **Steel, A.B., T.M. Herne, and M.J. Tarlov.** 1998. Electrochemical quantitation of DNA immobilized on gold. Anal. Chem. *70*:4670-4677.
26. **Steel, A.B., R. Levicky, T.M. Herne, and M.J. Tarlov.** 1999. Electrochemical characterization and quantitation of DNA on gold. Proc. Electrochem. Soc. *99-5*:132-143.
27. **Tinland, B., A. Pluen, J. Sturm, and G. Weill.** 1997. Persistence length of single-stranded DNA. Macromolecules *30*:5763-5765.
28. **Williams, J.C., S.C. Case-Green, K.U. Mir, and E.M. Southern.** 1994. Studies of oligonucleotide interactions by hybridisation to arrays: the influence of dangling ends on duplex yield. Nucleic Acids Res. *22*:1365-1367.
29. **Yang, P., T. Deng, D. Zhao, P. Feng, D. Pine, B. Chmelka, G.M. Whitesides, and G.D. Stucky.** 1998. Hierachically ordered oxides. Science *282*:2244-2246.

6 Large-Scale Genomic Analysis Using Affymetrix GeneChip® Probe Arrays

Janet A. Warrington, Suzanne Dee, and Mark Trulson
Affymetrix, Inc., Santa Clara, CA, USA

INTRODUCTION

In this chapter we hope to enlighten, educate, and enthuse the inexperienced GeneChip® user regarding the Affymetrix technology. If you are an experienced GeneChip user, we hope this will be an informative review and stimulate you to think about new applications in your laboratory. The following aspects are covered: *(i)* a brief overview of the characteristics of the technology that distinguish it from other microarray technologies; *(ii)* an introduction to array design, probe selection, and array synthesis; *(iii)* a description of current applications including results from recent experiments; and *(iv)* a brief discussion on future applications.

Why the GeneChip System is Different

A number of critical aspects of the Affymetrix technology distinguish the GeneChip system from other microarray methods. First, the probes are photochemically synthesized on the chip. No cloning, spotting, or polymerase chain reaction (PCR) steps are required. With a minimum number of synthesis steps, our method produces arrays of hundreds of thousands of different probes with extremely high feature density. In designing and producing the arrays, probe selection is performed based on sequence information alone; consequently, the sequence of each probe on the chip is known. In expression applications, this enables one to distinguish and quantitatively monitor closely related genes by avoiding a sequence that is identical or highly similar among gene family members. Similarly, one can select probes from every exon in a gene and use the hybridization pattern information to identify splice variants in mRNAs from

Microarray Biochip Technology
Edited by Mark Schena

different tissues. In variation detection applications (i.e., polymorphism/mutation detection), probe selection based on known sequences allows one to use multiple strategies in a single experiment to identify different types of variation. For instance, one can simultaneously query single base pair mutations as well as specific insertions and deletions. In expression applications, multiple probes are selected to represent each gene or expressed sequence tag (EST). This redundancy allows one to tolerate polymorphisms, splice variants, hybridization inconsistencies, sequence similarities, and errors in sequence databases.

Because we synthesize multiple copies of each of the probes, we are able to collect quantitative information across a range of transcript abundance levels in expression applications. In variation detection applications, the redundancy in the number of probes interrogating a given base position aids in distinguishing homozygotes from heterozygotes, once again for reasons of signal intensity. GeneChip analysis uses intensity information as well as hybridization *pattern* information. The use of pattern information contributes to the robustness of the method and aids in measuring nonspecific hybridization. Robust assays and analysis algorithms have been developed to optimize the quality and amount of information obtained from the data. Also, we have created a standard for open analysis architecture [Gene Analysis Technology Consortium (GATC®)] in an effort to support expression and mapping customers who wish to create their own querying tools or wish to use tools developed by other companies such as SpotFire™, GeneLogic™, Pangea™, MAG™, and others.

The GeneChip system is an entire system, not just a chip. The photolithographic process is capable of generating exceedingly fine feature resolution over a large surface area. Thus the scanning system must be capable of meeting the mutually conflicting requirements of high resolution, large field of view, high sensitivity, and short scan times. The first generation of the microarray scanners involved adaptations of existing confocal fluorescent microscopes, which provided very high resolution over a wide field of view but became excessively slow when feature sizes shrank below 50 µm. The current generation of scanners, including the commercially available Hewlett-Packard (HP) system and custom research scanners developed at Affymetrix, utilize galvanometer scanning and achieve scan rates in excess of 20 lines/s. These systems generate confocal fluorescent images of the probe array consisting of 25 million pixels in less than 10 minutes.

In addition to the assay, software, arrays, and scanner, the GeneChip system includes an automated, software-controlled hybridization and wash station, the fluidics station. Controlled washing and staining is critical for achieving reproducible data.

ARRAY DESIGN, PROBE SELECTION, AND ARRAY SYNTHESIS

Because the oligonucleotides synthesized on the surface of the chip are used to interrogate or "probe" the sample, the oligonucleotides are called the *probes.*

The entire chip is called a *probe array*, and the material hybridized to the array is called the *target* or *sample.*

Generally, the arrays are all designed using the same basic strategy, the central feature of which is the match/mismatch probe strategy. Probes that match the target sequence exactly are referred to as *reference probes*, whereas probes that contain a single base mismatch in the center of each reference probe are known as *partner probes.* For each reference probe synthesized, a partner probe or probes containing a nucleotide mismatch at the central base position is synthesized. These partner probes are synthesized adjacent to one another on the array and allow us to measure the cross-hybridization that occurs due to fragments with similar sequence hybridizing indiscriminately under the experimental conditions (heat, salt, and pH). In other words, we have a reference built in for every probe on the chip. Because of the high density of the probes we are able to achieve by our synthesis method, ample space is available on the chip for partner probes. Additionally, we place the probe partners adjacent to one another to control for any differences in hybridization that may occur on different parts of the chip due to poor mixing during hybridization, washing, and staining as well as microscopic inconsistencies in the glass. The combined presence of partner probes and multiple probes per gene target avoids detection issues that arise from not having sufficient signal and not being able to distinguish specific signal from background.

There are two major categories of GeneChip probe array designs, expression and variation detection. Current expression probe arrays (Figure 1) contain enough probes to query more than 7000 genes or ESTs simultaneously, with a linear detection dynamic range of more than 500-fold (9,29,49). Generally, the expression arrays contain sets of 20 different 25-mer probes representing each EST or gene. Probes are selected using many of the same criteria that are used to design good amplimers, such as avoiding palindromes that may form hairpin loops. Additionally, we use melting temperature (T_m) clustering and empirically determined rules that have been found to correlate with desirable hybridization behavior. In expression designs, we have the luxury of sorting through sequences to select the probes that are predicted to behave optimally. To enhance our ability to collect information from samples prepared using poly (A) extraction methods, we often bias the probe selection to the 3′ ends of the genes. As mentioned earlier, we also preferentially select probes from regions that are unique among gene family members, allowing us to distinguish expression levels among members of gene families.

In expression designs, for each of the reference probes, referred to as perfect match (PM) probes, we synthesize a partner probe containing a single nucleotide (transversion) mismatch (MM) in the central position. RNAs are considered present in the target mixture if the signals of the PM probes are significant above background *after* the signal intensities from the MM probes have been subtracted (Figure 2). For transcripts not present in the target mixture, the average signal intensity for the MM probes is close to the average signal intensity for the PM probes. If the sum of the PM intensities minus the sum of the MM intensities is close to zero, we classify the corresponding RNAs as absent or undetectable.

Array design for genotyping applications differs from expression design in two ways. First, probes are not selected from various portions of the sequence, rather a continuous region of sequence is represented using sets of probes to interrogate each nucleotide position, as shown in Figure 3. Second, instead of

Figure 1. The HuGeneFL array. It contains probes representing more than 5000 full-length human genes. It is shown hybridized with human brain RNA. (**See color plate A11.**)

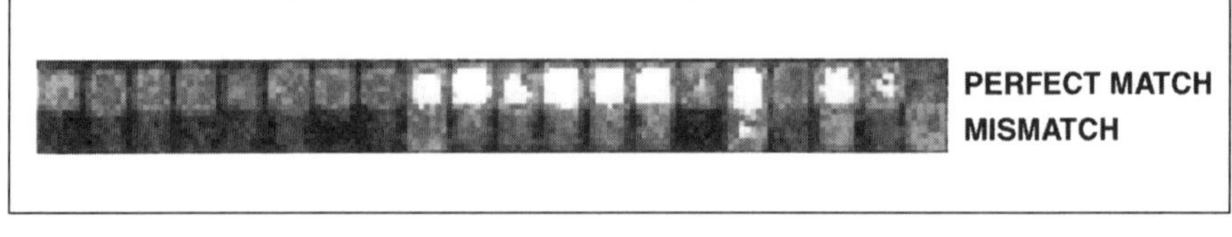

Figure 2. Twenty probe pairs representing a gene of interest. The top row contains probes that are perfectly complimentary to the target, and the bottom row contains the mismatch partner probes.

selecting a single MM partner for each nucleotide, all three possible MM partners are represented on the array. For instance if the reference sequence (PM) central position contains an adenine (A), the MM probes are represented by the precise same flanking sequence but with a thymine (T), guanine (G), or cytosine (C), respectively, at the central positions of the three MM probes. At the most basic level, for any given nucleotide position there are typically two sets of probes complementary to the sense and antisense strands of the reference sequence. This is usually referred to as the *standard tiling* design.

Including redundant probes on the chip for interrogation of a given nucleotide enhances the base-calling accuracy. For applications with a defined set of characterized polymorphisms or mutations, it is not necessary to query each position in the sequence, rather, a subset of nucleotide positions can be evaluated to determine whether they are different from a reference sequence. In this case, the array is designed such that there are sets, or "blocks" of probes that interrogate the position of interest and two flanking nucleotides. For each of these design types, variation is detected using intensity *and* pattern information. Specifi-

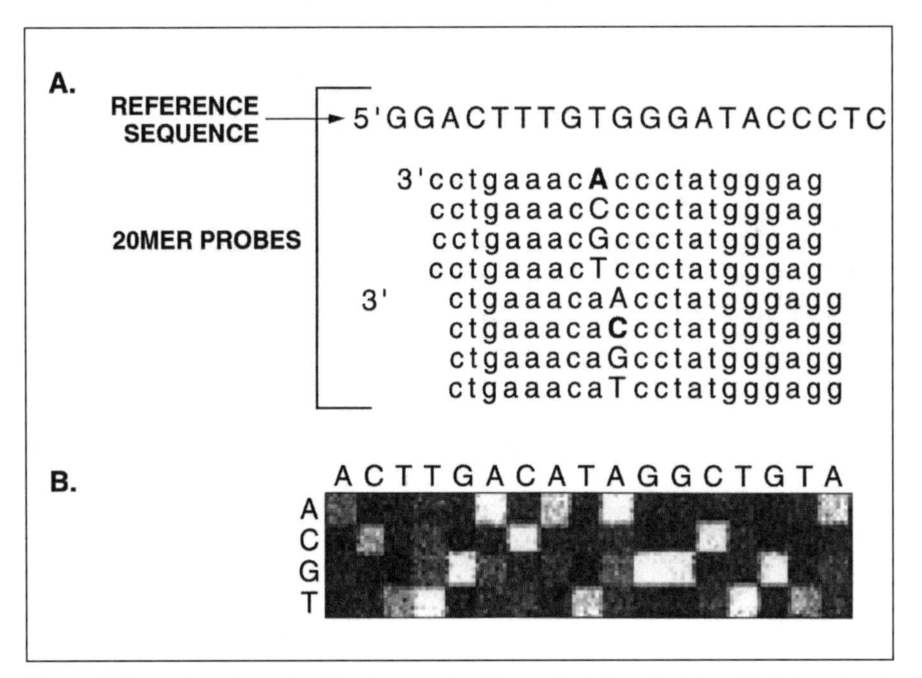

Figure 3. Variation detection probe tiling. An example of the standard tile probe design. (A) Each nucleotide in a target of known sequence (upper case sequence, top) is interrogated by a set of four probes (lower case sequences, below) that are identical to each other with the exception of a centrally located substitution position (upper case nucleotide) and complementary to the reference sequence. At the substitution, or interrogation site, all four possible nucleotides, A, C, G, or T are represented. The perfect match probe has the base at the substitution position shown in bold. Two probe sets are shown, querying adjacent bases in the target sequence. (B) An actual portion of probe sets on an array. For each nucleotide position, the probe perfectly complementary to the target sequence has the most intense signal.

cally, a fragment containing a single nucleotide variant will most strongly hybridize to a probe that is a reference sequence mismatch, and the sequence variation will disrupt hybridization in adjacent probes causing a characteristic hybridization bubble or footprint even though the adjacent probes are complementary to the reference sequence (Figure 4).

The GeneChip p53 array employs a modified standard tiling strategy for each nucleotide in the coding sequence, as well as the splice donor and acceptor sites flanking each exon. Standard-tile probe sets contain probes with each of the 4 possible bases in the substitution position, as well as a fifth probe designed to detect single base deletions, as shown in Figure 5. In addition to the standard tiling, the p53 array includes 12 additional probe sets (six probe sets each, complementary to the sense and antisense strands of the reference sequence, respectively) for more than 300 published substitution mutations. The redundant probes for a given site enhance the base-calling accuracy.

The GeneChip HIV PRT arrays are designed to detect known mutations in the protease and reverse transcriptase genes of human immunodeficiency virus-1 (HIV-1). The analysis region of the initial product, HIV PRT 440, included the entire protease gene (codons 1–99) and 243 amino acids of the reverse transcriptase gene. Two arrays were required to obtain information for both the sense and antisense strands of HIV-1. The product has been updated to include probes complementary to the sense and antisense strands on a single array and includes an additional sequence from the reverse transcriptase gene. The HIV PRT *Plus* array contains over 53 000 probes to interrogate coding sequences for amino acids 1 to 99 of the protease gene and amino acids 1 to 400 of the reverse tran-

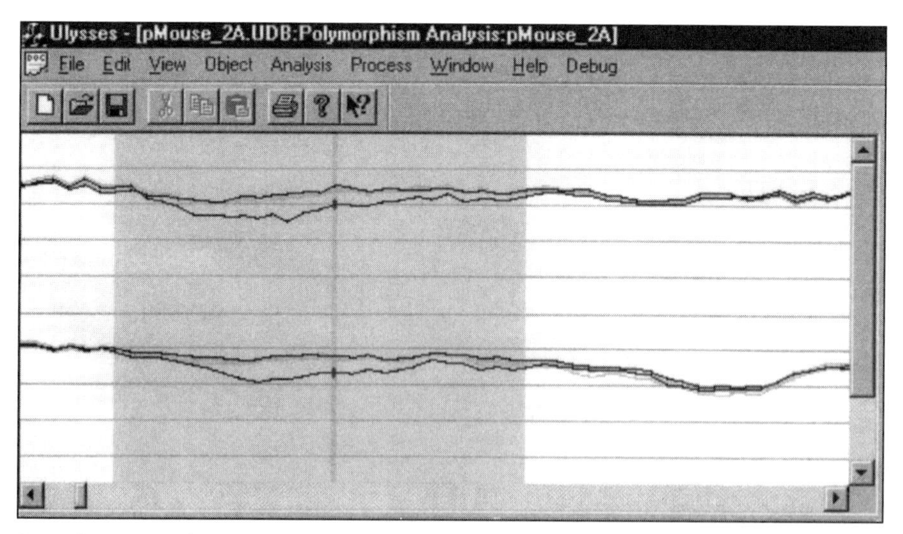

Figure 4. Footprint of a region containing a single nucleotide polymorphism. Detection of base differences is seen in a 45-bp region by comparing hybridization intensity patterns between the sample sequence and the reference sequence.

scriptase gene. The HIV PRT *Plus* array utilizes a standard tile design in which there are four probe sets contributing to the base call for each position in the sequence. There are additional probe sets for 155 short regions of the protease and reverse transcriptase genes. These are concentrated in the vicinity of known drug resistance mutations. The alternative tiles that incorporate polymorphisms into the probe design consist of additional probe sets that interrogate the mutant site and 4 flanking bases on either side. Probes interrogating bases that flank the mutant site differ from the standard tiles in that they are complementary to the mutant base instead of the wild-type base at the mutant site (Figure 6).

The GeneChip CYP450 probe array utilizes a block tiling strategy to detect polymorphisms in the 2D6 and 2C19 genes. For each characterized polymorphism site, there are five probe sets that are perfectly complementary (with the exception of the substitution position) to the target containing the wild-type base at the mutation site (W). Five additional probe sets are complementary to the target containing the mutant base at the mutant site (M). The wild-type and mutant probe sets are interdigitated such that each pair of probe sets interrogates the same position but is complementary to wild-type or mutant target (Figure 7).

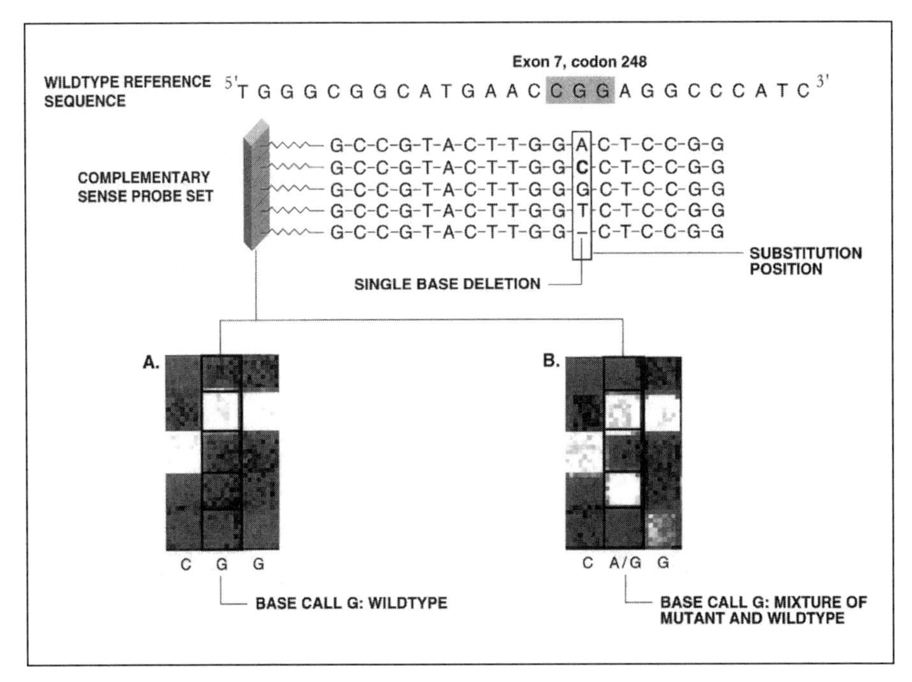

Figure 5. p53 probe array design. The p53 array employs a modified standard tiling design. Each probe set on the p53 array contains five probes complementary to a reference sequence, with the exception of the substitution position. In addition to the four probes with A, C, G, or T, there is a fifth probe complementary to a single-base deletion at each site (dash). (A,B) Two probe arrays were hybridized to target derived from wild-type p53 or p53 with a mutation in codon 248. The region of the arrays representing codon 248 is shown. The array hybridized with mutant p53 target shows the presence of a mixture (A/G) in the second base of codon 248.

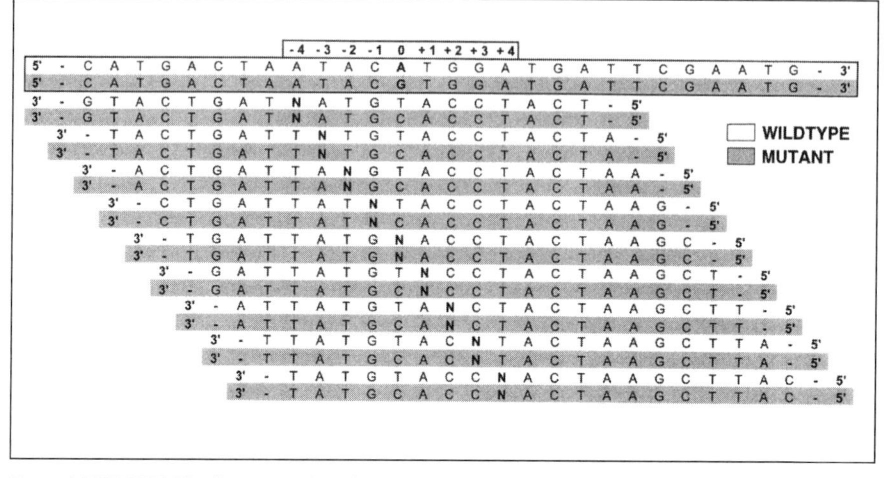

Figure 6. HIV PRT *Plus* alternative tiling design. The HIV PRT *Plus* alternative tile probe sets query the mutant site (position 0) and the four flanking bases on either side (-4 to +4 as shown). In this example, the wild-type base is A, and the mutant base is G. For each flanking base interrogated, each probe set is complementary to either a wild-type (unshaded) or a mutant base (shaded) at position 0. The substitution position is indicated by an N, which represents the four possible bases (A, C, G, or T).

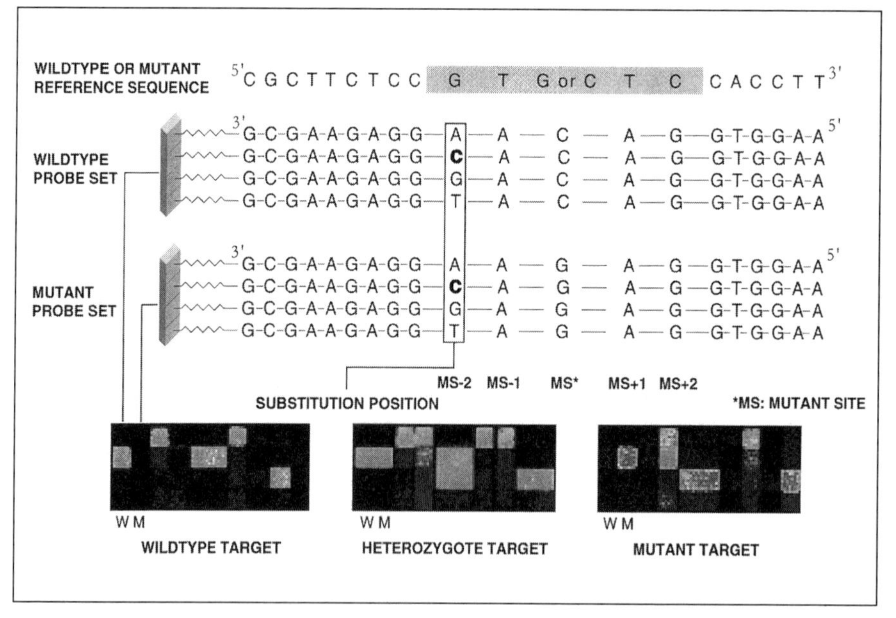

Figure 7. CYP450 block tiling. The CYP450 array has blocks of probes for characterized polymorphism sites. Each block contains probe sets that interrogate the target nucleotide sequence from 2 bases upstream of the mutant site (MS-2) to two bases downstream of the mutant site (MS+2). The columns are arranged in pairs that each interrogate the same position but are complementary to either a wildtype (W) or mutant (M) base at the mutant site (*MS). The array images show examples of hybridization patterns for homozygous wild-type, heterozygote, and homozygous mutant at the polymorphism site.

Conditions are optimized to ensure hybridization of the labeled target to the probe that best matches its sequence.

AUTOMATIC ANALYSIS

Automated image analysis software developed at Affymetrix converts signal intensity data to numerical output and data summaries. A brief summary of GeneChip analysis software follows.

Expression Analysis

GeneChip expression analysis is complex but is based on fairly elementary concepts that include computing background, probe usage, and the number of positive and negative scoring probes.

Background correction. In every hybridization reaction there are components and contaminants that contribute to some variability in overall fluorescence intensity/pixel intensity. This variability is called background. GeneChip software obtains fluorescence information from the array to measure the background and then automatically subtracts it to obtain intensity values.

Probe usage. To determine whether an RNA transcript is present or absent in the target/sample, the signal intensity is averaged over all probes. To be scored as "present," most of the probes representing a given gene must display a positive signal.

Positive probes. A probe pair is considered "positive" when the PM-MM intensity is greater than a calculated difference threshold and the PM/MM is greater than a set ratio threshold. The difference and ratio thresholds are set at default values in the software and can be changed by the user.

Negative probes. A probe pair is considered "negative" when the PM-MM intensity is less than the difference threshold and the PM/MM is less than the ratio threshold.

Determining the number of positive and negative probes. The probe pairs used in the expression analysis algorithms consist of the probes that are scored as positive *and* negative. As described above, the number of positive probes is computed after background subtraction. Probes that do not score as positive or negative are not used in the analysis.

Noise. Subtracting background from the specific hybridization serves to correct for the presence of nonspecific binding but does not remove random pixel intensity variation, or noise. The magnitude of the noise dictates the threshold above which a PM - MM difference is meaningful. The uncertainty in the mean signal for a given probe site is quantified by a quantity we call Q, which is given by the standard deviation of the pixel intensities over a synthesis site divided by the square root of the number of pixels averaged. For example, a 20-μm probe site is represented by the intensities of a 6 × 6 block of 3-μm image pixels. If the measured mean and standard deviation for this site are 500 counts and 150 counts, respectively, then the uncertainty in the mean is $150/(6 \times 6)^{1/2}$, or 25 counts. In this case, a mean intensity difference of >25 counts may be considered meaningful.

Variation Detection Analysis

A number of different variation detection analysis programs have been custom designed for specific arrays. Generally, the analysis methods share a number of common features including the use of intensity and pattern information to identify or "call" each base.

The GeneChip p53 algorithm compares all standard and redundant tiling probe sets for a given nucleotide position with the same probe sets from an array hybridized to a reference, or wild-type, p53 sample. Detection of mutations is based on differences in the pattern of hybridization intensities between the experimental sample and the control sample. Inclusion of redundant probe sets at known mutation sites heightens the ability to detect intensity pattern differences by increasing the number of times the base at the interrogation position is scored. Because probe sequence flanking the interrogation position is varied systematically, the influence of refractory, local sequence or structural problems is reduced. Data quality thresholds are built into the algorithm so that probe sets of poor quality will be eliminated from the analysis. If all probe sets for a given site do not exceed quality thresholds, an ambiguous (*N* call) will be given for that site. This can occur if the quality of the starting material is poor or if insufficient target is hybridized to the array. The other possible calls are "-" for a single base deletion, or a mixture call, where both the non-wild-type and wild-type bases are shown for a given position. For example, if the wild-type base is A, and the algorithm detects the presence of an additional base at that position, a possible call would be "G/A".

HIV PRT *Plus* base calls are computed at each position in the target sequence based on the hybridization intensities of the standard and alternative tile probe sets. The highest intensity probe determines the independent base call from any given set. Base calls from all probe sets for a given site are combined, and a consensus call is made. If the base call is non-wild-type, there are thresholds that must be exceeded to make the call.

The cytochrome p450 GeneChip probe array variation detection is performed utilizing a pattern discrimination method. Intensity information is extracted from each of the five paired probe sets on the hybridized array. A genotype call of "wild-type", "mutant", or "heterozygote" is determined based on the relative mutant/wild-type probe intensities within these blocks.

SCALEABLE ARRAY SYNTHESIS

We have adapted equipment used by the semiconductor industry to manufacture high-density oligonucleotide probe arrays (Figure 8) as described by Fodor et al. (12,13). The arrays are synthesized using photolithographic and solid-phase synthesis methods. Light-directed synthesis is carried out in a series of chemical steps that begins with the attachment of synthetic linkers modified with photochemically removable protecting groups. The first of a series of custom-designed masks is aligned with a 5 × 5-in piece of glass or wafer (Figure 9), and light is directed through the mask. Light passes through open regions of the mask to specific regions of the wafer, where it activates or de-protects the exposed linkers. A preselected single species of hydroxyl-protected deoxynucleoside (nucleoside chemically modified to carry its own removable protection group) is then flushed into the synthesis chamber and incubated with the surface. Chemical coupling of the deoxynucleoside with the linker occurs at the deprotected sites. The chemical coupling is followed by a capping step that acylates uncoupled active sites. Because of the efficiency of the coupling and the capping steps, uncoupled positions are not a problem. Following the capping step, the next in the series of masks is aligned, and the process is repeated (de-protection, chemical coupling, capping) until the probes are synthesized to full length at their discrete, designated positions on each microarray.

Each probe is synthesized in many identical copies in designated regions of the array called *features*. Having multiple copies of *each* probe in a given feature and many *different* probes per gene or transcript contributes to the robustness of this method and distinguishes it from other methods. Probes that are capped before

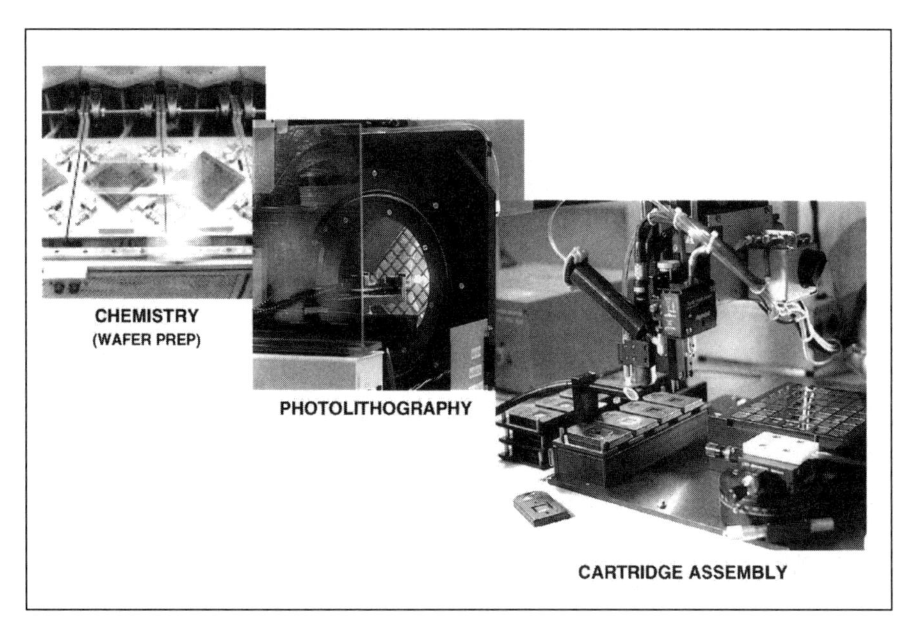

Figure 8. Equipment used for the manufacture of probe arrays.

reaching their complete, 25-mer length are sequence-specific truncations that do not introduce much nonspecific hybridization signal. The efficiency of the capping step ensures that the number of probes with sequence deletion errors is very small. Many identical arrays are produced in parallel on each glass wafer. Array synthesis is scaleable because the number of synthesis steps is dependent on the *length* of the probes being synthesized, not the *number* of different probes being synthesized. A complete set of 4^N oligonucleotides of length N can be synthesized in $4 \times N$ chemical steps whether producing an array to assay 70 genes or 7000 genes. The total number of probes synthesized on the array is limited by two practical considerations. The first is the density of the probe molecules per feature. If the probes are too close together on the chip surface, the hybridization efficiency is adversely affected, probably due to steric effects and probe-probe interactions. The second practical consideration is the physical size of the chip. We currently produce arrays that are 1.28 × 1.28 cm, 0.8 × 0.8 cm, and 0.5 × 0.5 cm. Small chips use small amounts of biological sample and are convenient to handle. The number of ESTs or genes queried on each array is restricted by the feature size, which is limited by the resolving power of the instrument used for signal detection, the scanner (see below). For expression applications, we routinely synthesize approximately 260 000 individual probes per 1.28 × 1.28-cm array, enough to query more than 7000 genes or ESTs. After synthesis each wafer is diced into individual arrays, and each is mounted into a plastic cartridge that provides a sample chamber and a convenient means to store and handle the array, as well as an accurate, repeatable positioning device of the array in the scanner. The sample chamber is 1 mm deep and holds a 200 μL volume of hybridization sample.

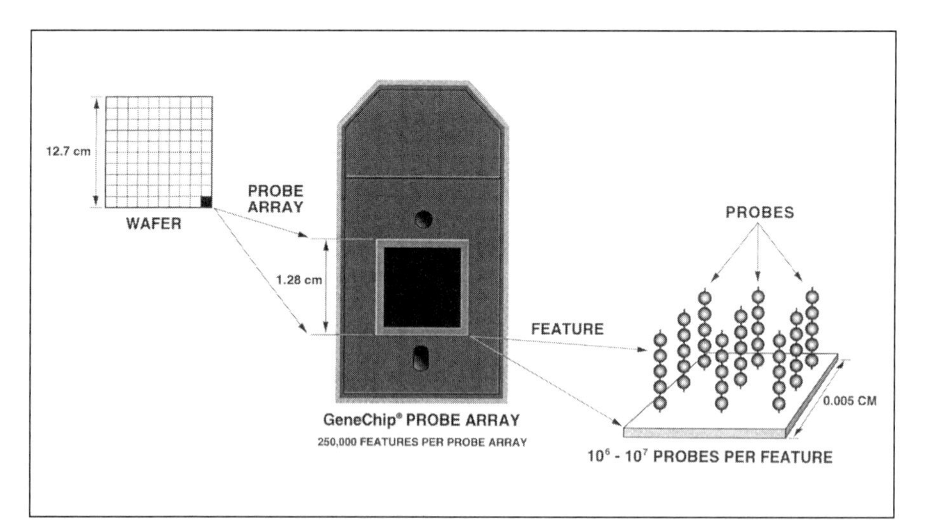

Figure 9. Wafer, chip, and feature. Many identical arrays are synthesized in parallel on a wafer, and the wafer is then diced and each array is placed in hybridization cartridge. Each HuGeneFL array contains approximately 250 000 different probes. Each probe is present in millions of copies at a distinct site on the array called a feature.

SAMPLE PREPARATION, HYBRIDIZATION, AND SIGNAL DETECTION

There are a variety of sample preparation methods for the different array types, but all share a few fundamental characteristics (3,9,15,29,30,40,46, 48,49). All methods require extraction of total RNA, poly(A), or genomic DNA (variation detection does not necessarily make cDNA, just amplifies genomic DNA in the case of p53), which is converted to cDNA or cRNA using enzymatic methods that modestly amplify the sample while tagging or incorporating biotinylated or fluorescinated nucleotides. In expression applications, the amplification must maintain the relative abundance levels of the different transcripts present, whereas for variation detection applications, relative abundance information is not usually as important. Depending on the particular application, the samples are hybridized to the array for 2 to 12 hours. After hybridization, the sample is removed from the hybridization chamber and in many applications can be saved for subsequent experiments because only a small fraction of the sample hybridizes to the chip in a given experiment.

If biotinylated nucleotides are used, the arrays are stained with a streptavidin-phycoerythrin conjugate that binds to biotin tags and emits fluorescent light when excited with a laser. Arrays are scanned using the GeneChip laser confocal fluorescence scanner, which focuses at the interface of the chip surface and the target solution (12,13). A lens collects the fluorescence and passes it through a series of optical filters to a detector (Figure 10). The presence of a slit aperture in the collection optics allows the system to discriminate molecules bound to the surface of the chip from those that are floating in the sample solution. The sensitivity of the instrument is sufficient to detect as few as 400 phycoerythrin molecules in a 20 × 20-μm probe site. Automated image analysis software measures fluorescence by calculating signal intensity units at each discreet probe site or feature on the array. The image is 4600 × 4600 pixels, corresponding to a scan area of 14 × 14 mm. The process of scanning and generating data output takes only a few minutes. Depending on the experiment, preparation of the labeled sample from cell lines requires 2 to 48 hours, and most expression and variation detection applications require 8 to 12 hours of hybridization time, although some assays require less. Without the aid of automation, an individual can process 8 to 12 arrays a day.

ARRAY RESULTS: QUANTITATIVE AND REPRODUCIBLE

How Quantitative are the Arrays?

In 1996, Lockhart et al. (29) demonstrated that hybridization signal intensity correlates with the expression level of the transcript. Thirteen labeled cytokine

RNAs were spiked into four 10-μg aliquots of murine B-cell RNA at 0.5, 5.0, 50, and 500 pM. The results show a linear correlation of concentration with hybridization signal intensity for 0.5 to 50 pM and then a flattening of the curve at 500 pM due to a saturation of probe hybridization sites (Figure 11). The hybridization intensity of the spiked transcripts correlated well with the expression levels of known mouse genes. It is important to note that sorting or binning by mRNA concentration is possible because the probes synthesized on the arrays are all selected using the same sets of rules; consequently, most of the probes hybridize with approximately the same efficiency under a single set of conditions. This is true regardless of the source of the target mRNA. The experiments reported by Lockhart et al. (29) were carried out using large (50 × 50 μm) features with a redundancy corresponding to hundreds of probe pairs per transcript.

In subsequent experiments, Wodicka et al. (49) also demonstrated the quantitative aspect of the arrays by hybridizing total yeast genomic DNA to *Saccharomyces cerevisiae* expression arrays. The authors reasoned that since most genes are present in the same copy number in the yeast genome, a good test of probe performance would be to hybridize the entire genome to the yeast microarray. If

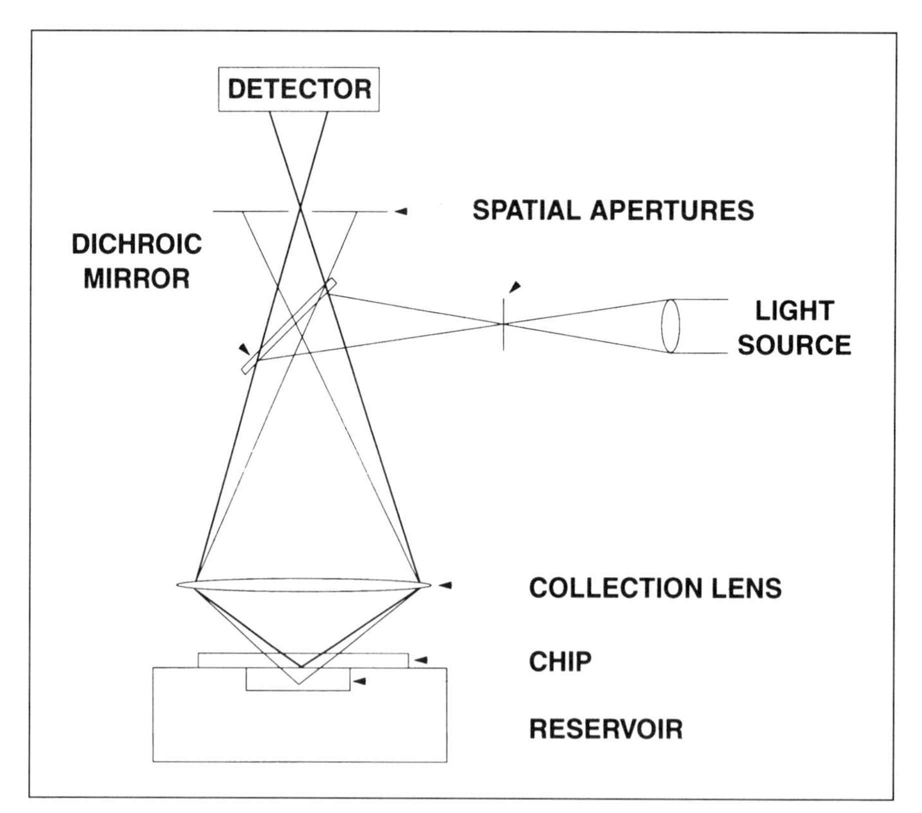

Figure 10. Laser confocal fluorescence detector.

the probes were unbiased in terms of hybridization efficiency, then most of the probes should produce a similar signal intensity on the yeast GeneChip probe arrays; indeed, 80% of the probes produced hybridization intensities that were within twofold of each other. Additionally, Wodicka et al. (49) correctly identified those genes present in more than one copy in the yeast genome. For these experiments, Wodicka et al. used arrays designed with large features (50×50 µm) and 20 probe pairs per transcript.

de Saizieu et al. (9) demonstrated that the arrays are quantitative and that the results were in good agreement with results obtained by conventional Northern blot analysis of selected genes. Using arrays designed with large features and 150 probe pairs per gene to measure expression levels in *Hemophilus influenzae* and *Streptococcus pneumoniae*, de Saizieu et al. hybridized labeled total RNA and reported sensitivity as high as 1 to 5 transcripts per cell. Furthermore, de Saizieu et al. performed retrospective analysis on subsets of probe pairs and found that the probe pair number could easily be reduced to 20 and still allow for quantitative accuracy for both high- and low-abundance RNAs. These early experiments demonstrated that the 50-µm feature arrays are quantitative. Recently, arrays

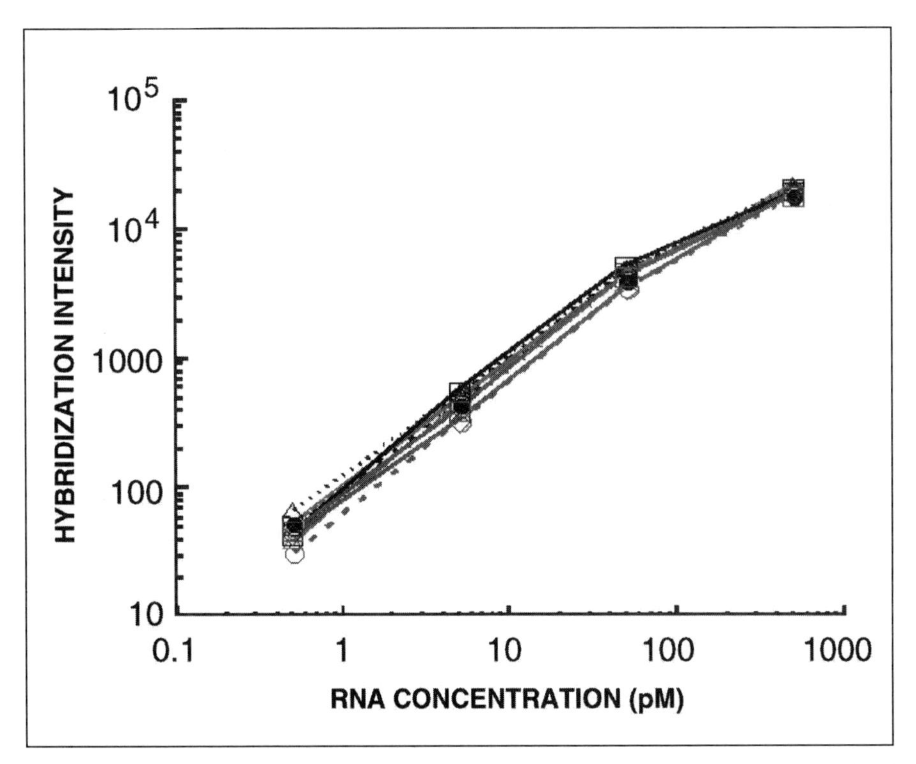

Figure 11. The relationship between hybridization signals and RNA concentration is linear. Log-log plot of the average hybridization intensities versus concentration for 11 different RNA targets spiked into labeled T10 RNA at 0.5, 5.0, 50, and 500 pM. (**See color plate A12.**)

have been produced with features reduced to 24 × 24 μm with 20 probe pairs per transcript. With modifications to the assay that include signal enhancement and a refined algorithm, we have found that the smaller feature arrays are as sensitive as those with 50-μm features (43).

Are the Data Reproducible?

The data obtained from the arrays are reproducible. In a set of experiments comparing the same sample hybridized to two different arrays, Wodicka et al. (49) demonstrated that method-associated experimental artifacts produce differences between two identical samples of less than twofold. In related experiments, Wodicka et al. demonstrated that the single largest source of variation is sample handling, specifically having different individuals prepare the samples. Over the past few years, dozens of laboratories have demonstrated the reproducibility of the data generated from GeneChip arrays (see Current Applications below).

CURRENT APPLICATIONS

The GeneChip system including the assays, arrays, fluidics station, and laser confocal fluorescence detection reader (scanner) provides a powerful set of tools for simultaneously investigating large sets of genes or entire genomes in a single experiment. Currently, most GeneChip probe array designs are for gene expression monitoring and variation detection applications. The arrays have fundamentally transformed the way experiments are designed and analyzed.

Expression Applications

Generally, gene expression experiments are designed to provide clues to gene product function, regulatory circuitry, and biochemical pathways. Experiments usually consist of comparing expression levels across a variety of tissues or comparing expression levels in a disease tissue versus an unaffected tissue, or investigating cellular response to some type of treatment including chemical or drug treatment, presence of an infectious agent, change in temperature, pH, or growth medium, or characterizing expression in transgenic animals.

Conventional methods of expression analysis for any one of these types of expression experiments have traditionally been limited to focusing on one or a few genes. In most cases, the data were qualitative and rapidly interpretable. By these methods, discovering the entire roster of genes involved in a biological process demanded a series of experiments that could require an entire scientific career and/or a cadre of graduate students and postdoctoral fellows to complete. With probe arrays, the questions addressed are broader because thousands of genes are queried simultaneously. Large-scale analysis of the genome makes it possible to address questions such as: Which genes coexpress with gene X? Which genes with adjacent loci are expressed simultaneously in particular tissues? Which of the genes expressed in the same tissue have mutant alleles that affect tissue function? Which genes expressed in this tissue have the same pattern of expression as gene X when expressed in this tissue and/or in other tissues? At a specific stage of a defined process, which genes are expressed after or before but not during the expression of gene X?

Many GeneChip probe array expression designs are currently available, and additional designs for a number of bacterial species and other model organisms are planned for 1999. The volume of data produced by a single experiment means that time spent in the lab per data point collected has become very small; similarly, because of the large volume of data produced in a single experiment, approaches to data analysis have changed remarkably. Many powerful querying tools have been developed and are under development to aid in data mining. Because of the open analysis architecture of GATC, GeneChip software customers are able to create their own query tools or use tools developed by other companies such as SpotFire, GeneLogic, Pangea, and MAG. The increased rate of expression information collection has made it feasible to establish whole genome expression databases (e.g., www.HUGEindex.org, and www.tigr.org/tdb/tbase.html).

Clues to Gene Product Function

In a set of experiments using human expression arrays performed on pools of normal human adult brain, fetal brain, adolescent liver, and fetal liver, we obtained transcript information for approximately 6500 genes and expressed sequence tags (47). Figure 12 shows an enlarged region of the same portion of two identical

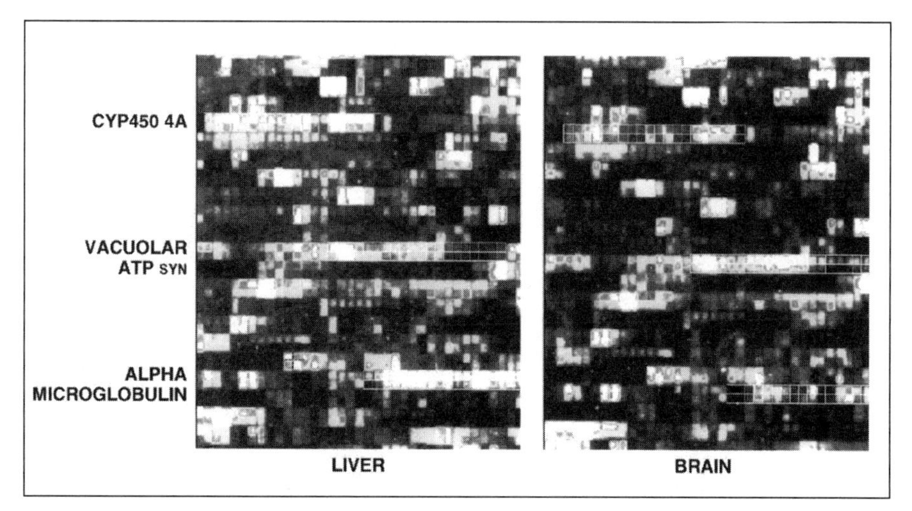

Figure 12. Enlarged view of one region of two identical arrays hybridized with human brain or liver mRNA. Cytochrome p450 4A and α-microglobulin are detectable in liver but not in brain. Vacuolar ATP synthetase is detectable in both tissues at different levels.

arrays hybridized with brain or liver. Many transcripts, such as the calcium channel receptor, synaptotagamin, calmodulin and acetylcholine receptor, are more abundant in brain; conversely, the very low-density lipoprotein receptor (VLDLR) is more abundant in liver (Figure 13). These data agree well with other methods. Not surprisingly, many transcripts of unknown function were also detected, such as the extracellular matrix protein represented on the far right of the bar graph in Figure 13. Measuring and comparing expression levels in many different tissues provides clues to gene product function and will undoubtedly produce additional information about genes for which some information is already available.

Recently, in collaboration with Steve Gullans at Brigham and Women's Hospital, we have begun building a public database of gene expression levels in normal human tissues (www.HUGEindex.org) funded by Affymetrix and the Merck Genome Research Institute. Our objectives are to create a database of normal gene expression of different tissues that can be accessed via the internet; perform a statistical analysis of expression within and among genes, tissues, and individuals; define an appropriate procedure for normalization of the data that will facilitate quantitative comparison of future investigations; and obtain feedback from the research community to improve future efforts.

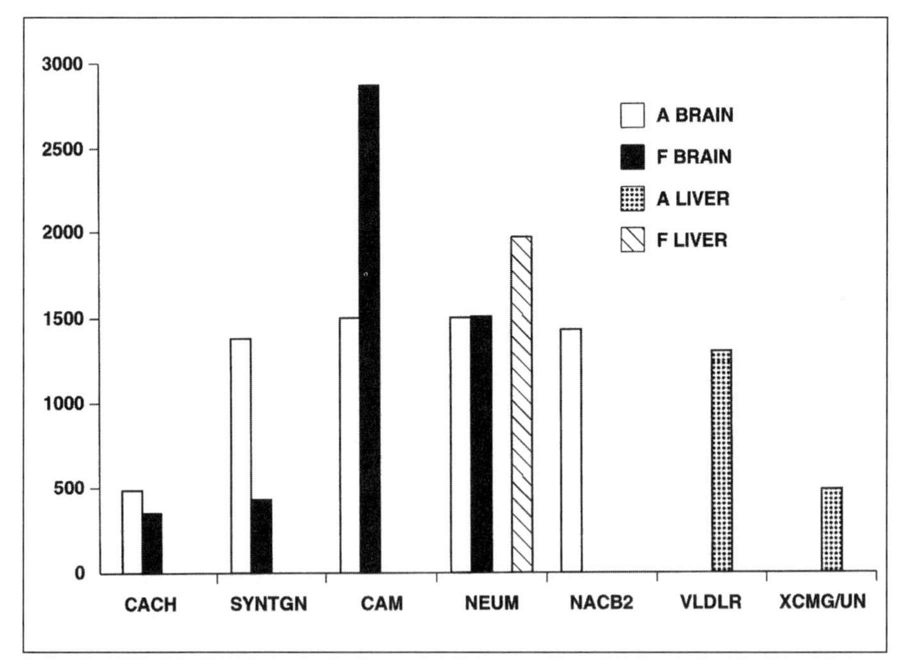

Figure 13. Human mRNA expression profiles. Calcium channel receptor (CACH), synaptotagamin (SYNTGN), calmodulin (CAM), neuromodulin (NEUM), acetylcholine receptor (NACB2), very low-density lipoprotein (VLDLR), and an extracellular matrix protein (XCMG/ON), in adult (A) brain, fetal (F) brain, adolescent liver, and fetal liver.

Clues to Regulatory Circuitry and Biochemical Pathways

In an effort to improve our understanding of the basic mechanism of human interferon (IFN) activity and to identify better markers for monitoring the effectiveness of IFN therapies, Der et al.(7) studied the mRNA profiles of human fibrosarcoma cell lines after IFN-α, -β, or -γ treatment. Forty genes known to be responsive to IFN treatment were identified as well as 82 genes not previously associated with the IFN-α, -β, or -γ response, including RAP46/Bag-1, phospholipid scramblase, and hypoxia-inducible factor-1α. In addition, many genes were identified that are preferentially upregulated by IFN-β but not IFN-α or IFN-γ. Stimulation by IFN-α or IFN-γ resulted in identification of about 100 genes upregulated by more than twofold, whereas twice as many genes, 268, were up-regulated by IFN-β stimulation, suggesting that IFN-γ may regulate a wider range of IFN-stimulated genes.

Genome-wide scans using GeneChip yeast arrays have been carried out by a number of groups. Holstege et al. (23) carried out a genome-wide study to identify key requirements of RNA Pol II transcription machinery in *S. cerevisiae* using GeneChip yeast arrays. Their goal was to elucidate the regulatory relationship between the transcription machinery and signal transduction pathways. Besides the expected regulation by gene-specific factors, they found a surprising level of regulation due to coordinate regulation of specific sets of genes induced by stress. Other *S. cerevisiae* genome-wide experiments have examined the effect of various growth conditions, including the differential expression response to rich and minimal media and expression level adaptation from fermentation to respiration (8,49). Jelinsky and Samson (25), studied the global response to methyl methanesulfonate, a known DNA-damaging alkylating agent. They reported finding 18 of 21 genes that were already known to be induced based on Northern blot analyses, as well as approximately 300 additional genes with increased transcript levels and approximately 76 genes with decreased levels. Most of the newly identified induced genes had transcription levels surpassing the levels of the 18 known genes. Additionally, many of the responsive genes fell into unexpected functional categories.

To determine whether chronic activation of a specific human G-coupled protein receptor (Ro1) expressed in mouse heart would cause dilated cardiomyopathy, Redfern et al. (38) measured mouse expression using GeneChip murine arrays. They showed that activation of Ro1 decreased heart rate by as much as 80% and caused severe congestive heart failure.

Innovative experiments designed to understand the mechanism of viral replication and pathogenesis better by examining host response to infection were carried out by Zhu et al. (50). Using primary human foreskin fibroblasts infected with HCMV AD169 or Toledo virions and hybridizing mRNA before and after infection at selected time points, the authors found 258 mRNAs changing expression level by more than fourfold when infected with human cytomegalovirus prior to the replication of the viral DNA.

In a tour de force, Winzeler et al. (48) used yeast expression arrays to address allelic variation within two isolates of *S. cerevisiae*. More than 3700 biallelic variants were identified. In a similarly creative experiment, Cho et al. (4) used yeast expression arrays to screen the genetic selection results of yeast two-hybrid experiments.

Screening Drugs

Probe arrays are used to identify drug-response targets by studying the expression of genes regulated by specific drugs or chemical compounds in carefully selected cell lines. In a study to examine the response of human promyeloid leukemia HL-60 cells to retinoic acid (RA), an agent used in the treatment of certain types of myeloid leukemia, Jing-Shan Hu (Affymetrix) identified more than 100 genes that were regulated by RA. Hu performed time point experiments to compare the expression levels of RA-treated and RA-untreated HL-60 cells (Figure 14). One of the genes repressed by RA treatment is *Lyl*-1, a gene previously found to contribute to the onset of T lymphoblastic leukemia (32). Another gene repressed is *bcl*-2, a proto-oncogene that prevents programmed cell death and is believed to play a role in leukemia. The genes induced by RA include *Mac*-1 (CD11b) and transglutaminase, markers of myeloid cell differentiation.

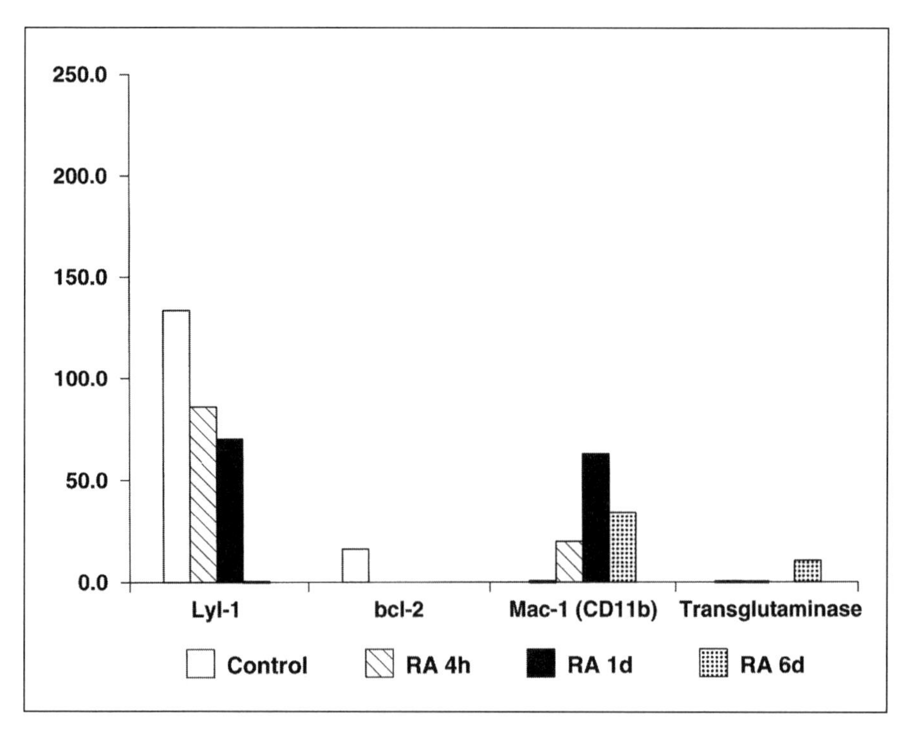

Figure 14. Gene expression levels. Expression levels for 5 of the 100 genes found to respond to retinoic acid (RA) treatment of HL-60 myeloid leukemia cells after 1 hour, 1 day, and 6 days of treatment.

In another approach to screening drugs, Giaever et al. (14) used diploid yeast cells and reduced the dosage of several yeast genes from two copies to one copy to profile drug sensitivities in haplo-insufficient yeast. These authors demonstrated that several haplo-insufficient strains which are sensitive to drugs grew more slowly than the controls. In a subsequent experiment, 12 uniquely tagged heterozygous yeast strains were pooled and grown competitively in the presence of tunicamycin. Every 2 hours, over a period of 12 hours, DNA was isolated from aliquots of cells removed from the pools, labeled, and hybridized to arrays containing probes representing each of the different strains tested. The resulting quantitative hybridization information was used to distinguish the strains that grew slowly in the presence of the drug, thus allowing identification of the genes targeted by the drug.

Variation Detection Applications

An improved understanding of the potential causative role of common genetic variants in human disease has the promise of improving diagnostics and therapeutics for a broad range of disorders. Of interest are mutations both sufficient to cause disease as well as polymorphisms associated with but not sufficient to cause disease. Affymetrix variation detection arrays are used to interrogate every base in a sequence of interest or, alternatively, to query a subset of characterized mutants or polymorphisms. Most applications are designed to identify changes in DNA sequence and establish relationships between genotype and phenotype. Affymetrix mutation detection arrays include GeneChip p53 and HIV PRT *Plus*, which interrogate and make base calls at each position in a specified region. Examples of arrays that interrogate only sites with characterized polymorphisms are GeneChip CYP450 and human single-nucleotide polymorphism arrays (HuSNP).

Commercial Arrays

p53. The p53 tumor suppressor gene is currently recognized as the most commonly mutated gene in human cancer (22,27). p53 is involved in cell-cycle regulation, and mutations in p53 are thought to have an impact on progression of disease and response to therapy. Gel-based mutation detection methods such as dideoxy sequencing and single-strand conformation polymorphism (SSCP) are labor-intensive and time-consuming. More rapid methods such as immunohistochemistry may fail to identify certain types of mutations such as those that cause protein truncations; also, these other methods do not provide information regarding the identity of the mutation. Our goal was to design an array that would permit rapid, accurate mutation detection analysis of the entire coding region of p53.

The starting material for the p53 assay is genomic DNA isolated from cells or tissue. A single-tube multiplex PCR is performed to amplify the 10 p53 coding exons (exons 2–11). The amplified target is fragmented with DNase I and end-labeled with terminal deoxynucleotidyl transferase and fluoresceinated dideoxy-

AMP. The fragmented, labeled target DNA is hybridized to the array, followed by washing and scanning. Assay duration starting from purified genomic DNA through data analysis is <5 hours. The p53 assay is designed to tolerate heterogeneous samples. The typical p53 sample derived from tumor tissue will contain a mixture of mutant and wild-type p53 sequences, where mutant p53 sequence is a minority population in the sample. Furthermore, p53 mutations frequently involve only a single base change. Thus, a very high level of base-calling accuracy is required. A mixture-detection algorithm utilizes a homogeneous sample with wild-type p53 sequence as a reference, which has been hybridized and scanned under the same conditions as samples with an unknown p53 sequence.

A concordance study was performed to compare mutation detection results from the GeneChip p53 assay with automated dideoxy sequencing (ABI) results. A blind test was carried out using the array assay with 60 genomic DNA samples derived from cell lines, white blood cells, or tissue. Exons 5 to 9 of p53 had been characterized previously by automated dideoxy sequencing. Mutations identified by chip assay were reamplified and rehybridized for confirmation. Dideoxy sequencing was repeated on samples with mutations identified by chip analysis but not by the first round of sequencing. Thirty-one of 34 mutations identified by dideoxy sequencing were correctly called as mutations using the p53 array, giving an analytical sensitivity for mutation detection of 91.2%. The base-calling concordance of the p53 assay compared with automated dideoxy sequencing was 99.94% (54 493 of 54 526 bases analyzed) for this sample set (Suzanne Dee, data not shown).

HIV. HIV-1 reverse transcriptase and protease genes are important targets of retroviral therapy. Combination therapy or use of drug cocktails inhibiting protease and reverse transcriptase activity of HIV-1 have been shown to slow disease progression and decrease mortality (20,21). HIV-1 mutations conferring resistance to inhibitors of protease and reverse transcriptase have been identified (5,11,41). The effectiveness of antiviral therapy is limited by variants resistant to reverse transcriptase and protease inhibitors; therefore, an improved understanding of the relationship between known mutations and different drug regimens may permit design of more effective therapeutic strategies.

The GeneChip HIV PRT *Plus* assay format includes a reverse transcriptase (RT)-PCR amplification of extracted viral RNA, with an optional nested PCR step for increased sensitivity for low-titer samples. The amplified DNA target is then fragmented with DNase I, end-labeled with terminal deoxynucleotidyl transferase and a biotinylated dideoxynucleoside, and hybridized to the array. Following hybridization and washing, there is an additional staining step with streptavidin-phycoerythrin before the scanning step.

Studies comparing dye-primer dideoxynucleotide sequencing and hybridization to HIV arrays were carried out for mutation detection in the protease and reverse transcriptase genes (17,26). Kozal et al. (26) analyzed the protease gene in 114 HIV-1 samples independently by dideoxy sequencing and array assay. A mutation detection concordance of 98.26% was observed for the 33 858 nucleotides analyzed by both methods. Gunthard et al. (17) examined both the protease and reverse transcriptase genes in 48 samples, including clinical samples and samples derived from viral stocks or clones. A concordance of 99.1% was found for a total of 30 865 bases compared.

Cytochrome p450. Pharmacogenomics is a term used to describe the study of the influence of DNA variation on drug efficacy. Variations are known to impair or augment drug effects or even result in drug toxicity (24,36). Several classes of enzymes are involved in drug metabolism, including members of the cytochrome p450 family of oxidases (10,33,34,44). Pharmacologists have evaluated the impact of variants in p450 genes on the metabolism of drug candidates. The GeneChip CYP450 array was designed to permit rapid genotyping of the 2D6 and 2C19 alleles of cytochrome p450 in order to determine metabolic efficiency for a number of important drug classes, including antiarrhythmic and neuroleptic drugs.

The CYP450 target preparation protocol employs a single-tube multiplex PCR to generate seven amplicons representing most of the 2D6 coding sequence and the sites of three characterized 2C19 alleles. The DNA target is fragmented and end-labeled with fluorescently tagged dideoxy-AMP, followed by hybridization, washing, and scanning of the array. A blind study carried out on 49 human DNA samples tested CYP450 array performance by comparing the results with genotypes determined by allele-specific PCR (28). A total of 486 allele-defining sites were compared, and the calls made using the array assay showed equivalent accuracy to allele-specific PCR (Figure 15). The analytical sensitivity observed for accurately

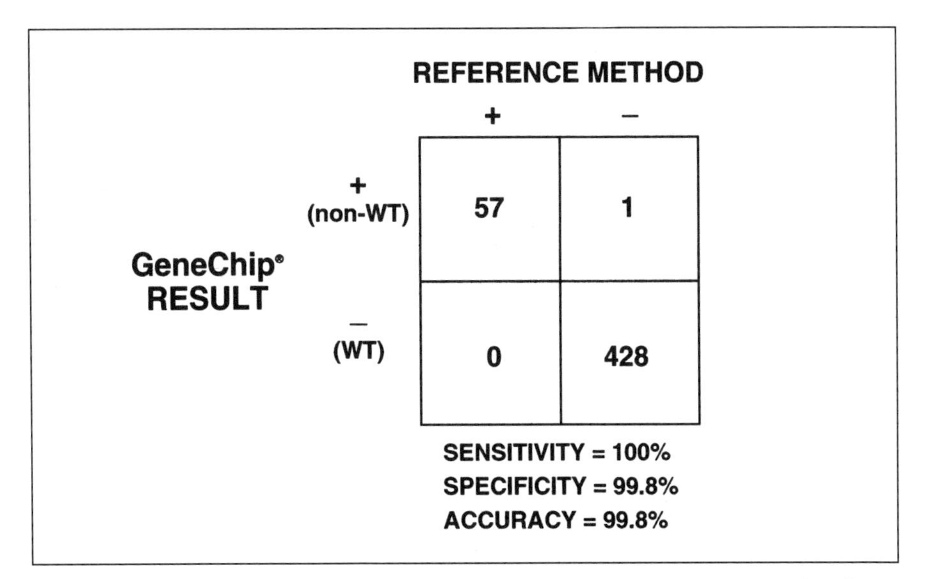

Figure 15. CYP450 results. The analytical performance of the assay based on comparison to the reference method is summarized. The true positive (+/+) cell reflects the number of allele-defining sites at which the GeneChip assay and the reference methods made the identical non-wild-type (WT) call. The true negative (-/-) cell shows the number of concordant wild-type calls.

identifying the presence of a polymorphism was 100%. The base-calling accuracy for sites evaluated was 99.8%. There was one discordant call, in which a non-wild-type call was made by the array assay but not by the reference method. One of the advantages of the array assay is the simplicity of the sample preparation procedure in which all 2D6 and 2C19 fragments are amplified in a single multiplex PCR.

SNP detection and mapping. The CYP450 array illustrates the utility of small-scale polymorphism analysis. There is also an ongoing program at Affymetrix to identify and map SNPs and to use the information to design probe arrays with thousands of polymorphisms, which can be used to carry out large-scale polymorphism analysis. An HuSNP array designed to enable the simultaneous genotyping of more than 1000 markers on a single array has recently been released. The HuSNP package contains an array representing more than 1000 SNPs, an assay composed of multiplex PCR amplification using primer pools (100+ primer pairs per reaction), and an automated analysis package. The HuSNP array design utilizes a variation on the block tiling strategy.

Four to five probe sets interrogate the polymorphism site and several neighboring bases. Unlike the standard tile design of four probes for each nucleotide interrogated with each of the four possible bases at the substitution position, the HuSNP design uses only two substitution bases complementary to perfect match and a mismatch base for each of the two expected alleles (Figure 16). As with CYP450, there are three possible calls, homozygous allele A, homozygous allele B, or an AB heterozygote. The hybridization intensities are expected to cluster into three distinguishable patterns representing the calls described above. As more SNPs are discovered, this type of array has potential usefulness for the study of disease susceptibility or population genetics research.

Research Applications

Microarray technology has the potential to streamline mutation and polymorphism detection significantly, by allowing the analysis of large amounts of sequence information simultaneously. As described above, several methods can be used for mutation detection. Cronin et al. (6) used two array designs for mutation detection in the cystic fibrosis transmembrane conductance regulator (CFTR) gene. Using a standard tiling design, each nucleotide in exon 11 of CFTR was interrogated. Alternatively, a second array was designed using a block tiling strategy to query samples for the presence of 37 characterized CFTR mutations. Using these arrays in a blind study with 10 characterized patient samples, the array-based assay gave results that were 100% concordant to a reference method.

Another method for variation detection involves a two-color labeling strategy and cohybridization of test samples and reference samples with known sequence to a single array. This permits the identification of sites where the hybridization pattern of the test sample differs from the reference, indicating the presence of a polymorphism or mutation. Chee et al. (3) used this approach to analyze 16.6 kb of the human mitochondrial genome in a single hybridization experiment. Analysis of the mitochondrial genome from 10 individuals resulted in the identification of 505 polymorphisms. Hacia et al. (18,19) demonstrated the utility of the array-based two-color assay for mutation detection/discovery in large genes that are challenging to analyze by conventional methods such as gel-based sequencing or allele-specific oligonucleotide hybridization. A screening assay is particularly useful for genes with multiple mutation types (substitution, insertion, and deletion) and for which there may not be a large number of characterized mutations. The coding region of the *BRCA1* gene, associated with hereditary breast and ovarian

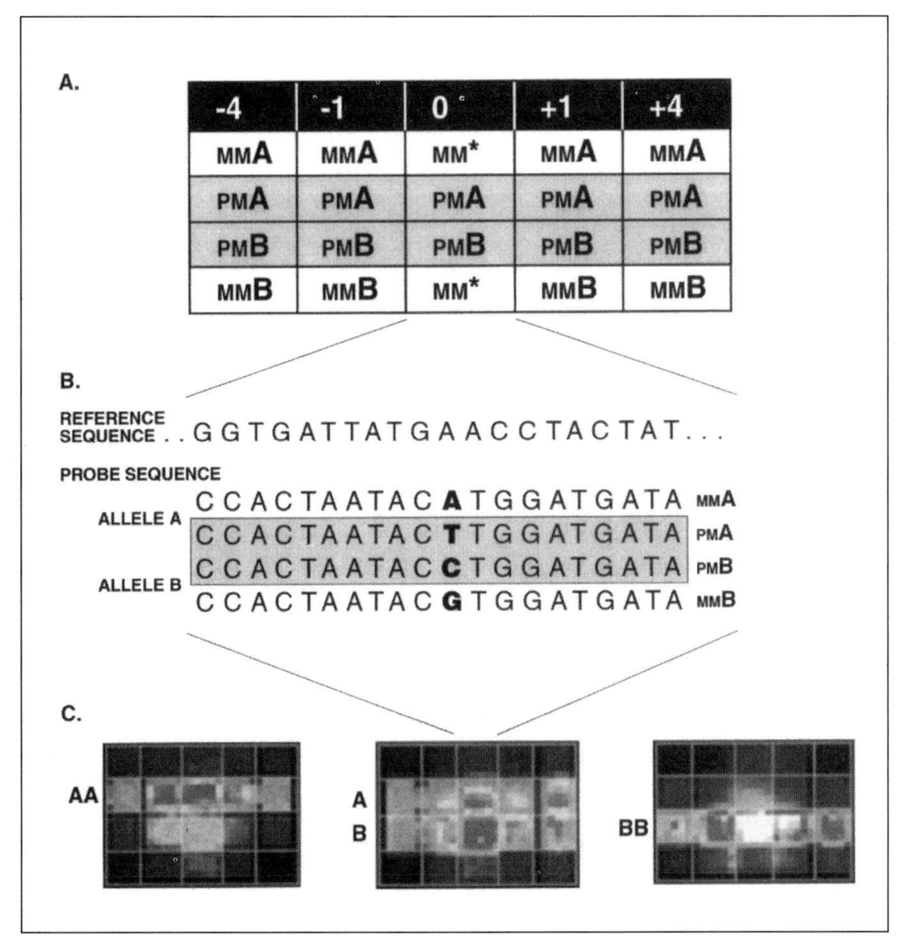

Figure 16. HuSNP array design. (A) The modified block tiling strategy used on the HuSNP array contains probe sets that query the site of a known polymorphism (0), and two flanking upstream (-4, -1) and downstream (+1, +4) sites. Rows 1 and 2 depict probe sets complementary to a mismatch or perfect match base to allele A, respectively. Rows 3 and 4 depict probe sets complementary to a perfect match or mismatch base to allele B, respectively. (B) The target reference sequence and probe sequences are shown. (C) Examples of hybridization patterns for homozygous allele A, heterozygous A/B, and homozygous allele B.

cancer, contains 5.5 kb in 22 exons (35). Thirty-five samples, including 15 patient samples, were evaluated on an array designed to interrogate the 3.45-kb *BRCA1* exon 11 sequence. Fourteen of the 15 characterized mutations were accurately detected, and 8 SNPs were identified in the chip assay. Furthermore, there were no false positives in the 20 control samples previously characterized by dideoxy sequencing (19). The feasibility of this approach was further demonstrated by screening for mutations in the ataxia telangiectasia (*ATM*) gene. Mutations in the *ATM* gene are associated with a predisposition to cancer, radiosensitivity, chromosomal instability, progressive neuromotor deterioration, and immunodeficiency (2,39). The coding sequence of *ATM* contains 9168 nucleotides in 62 exons (37). Twenty-five of 26 mutations were detected by analysis of the 9.1-kb *ATM* coding sequence from 22 genomic samples, with only 5 false positives (19). Another application with potential for clinical utility includes the use of arrays designed for genotyping and species identification of mycobacteria (15,45).

Generic Arrays

Universal or generic arrays containing all possible *N*-mers have been tested in a variety of applications. Instead of designing the arrays based on known sequences and the sequence order within the genome, the generic arrays are designed using a universal set of probes that consist of all or many of the possible sequence variants of a particular length in no particular order. For instance, Gunderson et al. (16) used an all 8-mer chip to demonstrate the feasibility of this approach for mutation detection. Analysis software sorts the probes into genome-relevant sets. This approach has been used in a number of applications including phenotypic analysis of yeast deletion mutants and mapping genomic clones (40,42).

Future Applications

Over the past 10 years, microarray technology has revolutionized the way we think about the genome and made rapid large-scale analysis of the human genome a reality. As the amount of sequence information made available by the Human Genome Project grows, so grows the power of the microarray technology. With current technology, we are able to measure tens of thousands of genes simultaneously (Figure 17). Genome programs dedicated to sequencing model organisms have made it possible to carry out experiments surveying the entire genome of *S. cerevisiae, E. coli, H. influenzae,* and *S. pneumoniae,* thereby providing a new view of genetic, biochemical, and biological processes. Increasingly, scientists are designing custom arrays with their favorite genes represented and tailoring probe number and feature size to the type of question they wish to address. For example, in 1998, in addition to standard designs for human, mouse, yeast, HIV, p53, and CYP450 arrays, Affymetrix produced 40 custom arrays for collaborators and partners and dozens of custom designs for internal research and development efforts. In 1999 we anticipate tripling these numbers. Custom variation detection arrays are being used in our high-throughput laboratory to scan the genome and detect SNPs from coding and noncoding regions. SNPs detected will be included in new designs, genotyping chips, and will make large-scale association and linkage studies a routine practice.

Advances in synthesis and scanner technology will enable us to distinguish smaller features, leading to an ability to read higher density arrays. Even now, adaptations of photoresist technology from the semiconductor industry have made the synthesis of higher density arrays possible (1,31). In addition, hardware and robotics modifications, such as an automated array loader (autoloader) for the GeneChip system will greatly simplify high-throughput handling and processing of arrays.

Figure 17. GeneChip probe array. The array contains probes for more than 40 000 expressed sequence tags (ESTs).

Along with a reduction in feature size, we are working to reduce the required amount of biological starting material required for the assays. Early sample preparation methods required millions of cells or milligrams of tissue. Recent developments have reduced that number to 50 000 cells or <1 mg of tissue (30). The development of methods requiring less material will make microarray studies of needle biopsies, neurons, or small clusters of specialized cells such as the α, β, or δ cells of the islets of Langerhans achievable. Simultaneously, we are working on new chemical and enzymatic labeling methods that may lead to alternative means of detecting signal.

REFERENCES

1. **Beecher, J.E., G.H. McGall, and M.J. Goldberg.** 1997. Chemically amplified photolithography for the fabrication of high density oligonucleotide arrays. Polymeric Mater. Sci. Engin. *76*:597-597.
2. **Canman, C.E. and D.S. Lim.** 1998. The role of ATM in DNA damage responses and cancer. Oncogene *17*:3301-3308.
3. **Chee, M., R. Yang, E. Hubbell, A. Berno, X.C. Huang, D. Stern, J. Winkler, D.J. Lockhart, M.S. Morris, and S.P.A. Fodor.** 1996. Accessing genetic information with high-density DNA arrays. Science *274*:610-614.
4. **Cho, R.J., M. Fromont-Racine, L. Wodicka, B. Feierbach, T. Stearns, P. Legrain, D.J. Lockhart, and R.W. Davis.** 1998. Parallel analysis of genetic selections using whole genome oligonucleotide arrays. Proc. Natl Acad. Sci. USA *95*:3752-3757.
5. **Condra, J.H.** 1998. Resistance to HIV protease inhibitors. Haemophilia *4*:610-615.
6. **Cronin, M.T., R.V. Fucini, S.M. Kim, R.S. Masino, R.M. Wespi, and C.G. Miyada. 1996.** Cystic fibrosis mutation detection by hybridization to light-generated DNA probe arrays. Hum. Mutat. *7*:244-255.
7. **Der, S.D., A. Zhou, B.R.G. Williams, and R.H. Silverman.** 1998. Identification of genes differentially regulated by interferon α, β, or γ using oligonucleotide arrays. Proc. Natl. Acad. Sci. USA *95*:15623-15628.
8. **DeRisi, J.L, V.R. Iyer, and P.O. Brown.** 1997. Exploring the metabolic and genetic control of gene expression on a genomic scale. Science *278*:680-686.
9. **de Saizieu, A., U. Certa, J. Warrington, C. Gray, W. Keck, and J. Mous.** 1998. Bacterial transcript imaging by hybridization of total RNA to oligonucleotide arrays. Nat. Biotechnol. *16*:45-48.
10. **Eichelbaum, M. and B. Evert.** 1996. Influence of pharmacogenetics on drug disposition and response. Clin. Exp. Pharmacol. Physiol. *23*:983-985.
11. **Erickson, J.W. and S.K. Burt.** 1996. Structural mechanisms of HIV drug resistance. Annu. Rev. Pharmacol. Toxicol. *36*:545-571.
12. **Fodor, S.P.A., J. L. Read, M.C. Pirrung, L. Stryer, A.T. Lu, and D. Solas.** 1991. Light-directed, spatially addressable parallel chemical synthesis. Science *251*:713-844.
13. **Fodor, S.P.A., R.P. Rava, X.C. Huang, A.C. Pease, C.P. Holmes, and C.L. Adams.** 1993. Multiplexed biochemical assays with biological chips. Science *364*:555-556.
14. **Giaever, G., D.D. Shoemaker, T.W. Jones, H. Liang, E.A. Winzeler, A. Astromoff, and R.W. Davis.** 1999. Nat. Genet. *21*:278-283.
15. **Gingeras, T.R., G. Ghandour, E. Wang, A.Berno, P.M Small, F. Drobniewski, D. Alland, E. Desmond, M. Holodniy, and J. Drenkow.** 1998. Simultaneous genotyping and species identification using hybridization pattern recognition analysis of generic *mycobacterium* DNA arrays. Genome Res. *8*:435-448.
16. **Gunderson, K.L., X.C. Huang, M.S. Morris, R.J. Lipshutz, D.J. Lockhart, and M.S. Chee.** 1998. Mutation detection by ligation to complete N-mer DNA arrays. Genome Res. *8*:1142-1153.
17. **Gunthard, H.F., J.K. Wong, C.C. Ignacio, D.V. Havlir, and D.D. Richman.** 1998. Comparative performance of high-density oligonucleotide sequencing and dideoxynucleotide sequencing of HIV type 1 pol from clinical samples. AIDS Res. Hum. Retroviruses *14*:869-876.

18. **Hacia, J.G., L.C. Brody, M.S.Chee, S.P.A. Fodor, and F.S. Collins.** 1996. Detection of heterozygous mutations in BRCA1 using high density oligonucleotide arrays and two-colour fluorescence analysis. Nat. Genet. *14*:441-447.
19. **Hacia, J.G., B. Sun, N. Hunt, K. Edgemon, , D. Mosbrook, C. Robbins, S.P.A. Fodor, D. Tagle, and F.S. Collins.** 1998. Strategies for mutational analysis of the large multiexon ATM gene using high-density oligonucleotide arrays. Genome Res. *8*:1245-1258.
20. **Hammer, S.M., K.E. Squires, M.D.Hughes, J.M. Grimes, L.M. Demeter, J.S. Currier, J. J. Eron, Jr., J.E. Feinberg, H.H. Balfour, Jr., L.R. Deyton, J.A. Chodakewitz, and M.A. Fischl.** 1997. A controlled trial of two nucleoside analogues plus indinavir in persons with human immunodeficiency virus infection and CD4 cell counts of 200 per cubic millimeter or less. AIDS Clinical Trials Group 320 Study Team. N. Engl. J. Med. *337*:725-733.
21. **Havlir, D.V. and J.M. Lange.** 1998. New antiretrovirals and new combinations. AIDS *12* (Suppl. A):S165-S174.
22. **Hollstein, M., D. Sidransky, B. Vogelstein, and C.C. Harris.** 1991. p53 mutations in human cancers. Science *253*:49-53.
23. **Holstege F.C., E.G. Jennings, J.J. Wyrick, T.I. Lee, Hengartner, C.J., M.R. Green, T.R. Golub, E.S. Lander, and R.A. Young.** 1998. Dissecting the regulatory circuitry of a eukaryotic genome. Cell *95*:717-728.
24. **Ingelman-Sundberg, M.** 1998. Functional consequences of polymorphism of xenobiotic metabolizing enzymes. Toxicol. Lett. *102-103*:155-160.
25. **Jelinsky, S.A. and L.D. Samson.** 1999. Global response of *Saccharomyces cerevisiae* to an alkylating agent. Genetics *96*:1486-1491.
26. **Kozal, M.J., N. Shah, N. Shen, R. Yang, R. Fucini, T.C. Merigan, D.D. Richman, D. Morris, E. Hubbell, M. Chee, and T.R. Gingeras.** 1996. Extensive polymorphisms observed in HIV-1 clade B protease gene using high-density oligonucleotide arrays. Nat. Med. *2*:753-759.
27. **Levine, A.J.** 1997. p53, the cellular gatekeeper for growth and division. Cell *88*:323-331.
28. **Liu, W.W., T. Webster, A. Aggarwal, M. Pho, M. Cronin, and T. Ryder.** 1997. Genetic mapping: finding and analyzing single-nucleotide polymorphisms with high-density DNA arrays. Am. J. Hum. Genet. *61*:1494-1494.
29. **Lockhart, D.J., H. Dong, M.C. Byrne, M. T. Follettie, M. V. Gallo, M.S. Chee, M. Mittmann, C. Wang, M. Kobayashi, H. Horton, and E.L. Brown.** 1996. Expression monitoring by hybridization to high-density oligonucleotide arrays. Nat. Biotechnol. *14*:1675-1680.
30. **Mahadevappa, M.** 1999. A high density probe array sample preparation method using 10–100 fewer cells. Nat. Biotechnol. *17*:1134-1136.
31. **McGall, G., J. Labadie, P. Brock, G. Wallraff, T. Nguyen, and W. Hinsberg.** 1996. Light-directed synthesis of high-density oligonucleotide arrays using semiconductor photoresists. Proc. Natl. Acad. Sci. USA *93*:13555-13560.
32. **Mellentin, J.D., S.D. Smith, and M.L. Cleary.** 1989. lyl-1, a novel gene altered by chromosomal translocation in T cell leukemia, codes for a protein with a helix-loop-helix DNA binding motif. Cell *58*:77-83.
33. **Meyer, U.A.** 1996. Overview of enzymes of drug metabolism. J. Pharmacokinet. Biopharm. *24*:449-459.
34. **Meyer, U.A. and U.M. Zanger.** 1997. Molecular mechanisms of genetic polymorphisms of drug metabolism. Annu. Rev. Pharmacol. Toxicol. *37*:269-296.
35. **Miki, Y., J. Swensen, D. Shattuck-Eidens, P.A. Futreal, K. Harshman, S. Tavtigian, Q. Liu, C. Cochran, L.M. Bennett, W. Ding, et al.** 1994. A strong candidate for the breast and ovarian cancer susceptibility gene BRCA1. Science *266*:66-71.
36. **Miller, M.S., D.G. McCarver, D.A. Bell, D.L. Eaton, and J.A. Goldstein.** 1997. Genetic polymorphisms in human drug metabolic enzymes. Fundam. Appl. Toxicol. *40*:1-14.
37. **Platzer, M., G. Rotman, D. Bauer, T. Uziel, K. Savitsky, A. Bar-Shira, S. Gilad, Y Shiloh, and A. Rosenthal.** 1997. Ataxia-telangiectasia locus:sequence analysis of 184-kb of human genomic DNA containing the entire ATM gene. Genome Res. *7*:592-605.
38. **Redfern, C.H., M.Y. Degtyarev, K. Desai, A.T. Kwa, E.K. Lee, P. Coward, N. Shah, J.A. Warrington, G. I. Fishman, D. Bernstein, et al.** Conditional expression of a G_i-coupled receptor causes a dilated cardiomyopathy. Cell (In press).
39. **Rotman, G. and Y. Shiloh.** 1997. The ATM gene and protein:possible roles in genome surveillance,

checkpoint controls and cellular defence against oxidative stress. Cancer Surv. *29*:285-304.
40. **Sapolsky, R.J. and R.J. Lipshutz.** 1996. Mapping genomic library clones using oligonucleotide arrays. Genomics 33:445-456.
41. **Shafer, R.W., D. Stevenson, and B. Chan.** 1999. Human immunodeficiency virus reverse transcriptase and protease sequence database. Nucleic Acids Res. 27:348-352.
42. **Shoemaker, D., D.A. Lashkari, D. Morris, M. Mittman, and R.W. Davis.** 1996. Quantitative phenotypic analysis of yeast deletion mutants using a highly parallel molecular bar-coding strategy. Nat. Genet. *14*:450-456.
43. **Sturniolo, T. E. Bono, J. Ding, L. Raddrizzani, O. Tuereci, U. Sahin, M. Braxenthaler, F. Gallazzi, M.P Protti, F. Sinigagalia, and J. Hammer.** 1999. In silico generation of tissue-specific and promiscuous HLA ligand databases using DNA chips and virtual HLA class II matrices. Nat. Biotechnol. (In press).
44. **Touw, D.J.** 1997. Clinical implications of genetic polymorphisms and drug interactions mediated by cytochrome p-450 enzymes. Drug Metabol. Drug Interact. *14*:55-82.
45. **Troesch, A., H. Nguyen, C.G. Miyada, S. Desvarenne, T.R. Gingeras, P.M. Kaplan, P. Cross, and C. Mabilat.** 1999. *Mycobacterium* species identification and rifampin resistance testing with high-density DNA probe arrays. J. Clin. Microbiol. 37:49-55.
46. **Wang, D.G., J.-B. Fan, C-J. Siao, A. Berno, P. Young, R. Sapolsky, G. Ghandour, N. Perkins, E. Winchester, J. Spencer, et al.** 1998. Large-scale identification, mapping, and genotyping of single-nucleotide polymorphisms in the human genome. Science *280*:1077-1082.
47. **Warrington, J.A., A.V. Nair, and D.J. Lockhart.** 1997. Expression profiles of 6500 human genes in poly(A) + and cDNA libraries. Am. J. Hum. Genet. Suppl. *61*:A36.
48. **Winzeler, E.A., D.R. Richards, A.R. Conway, A.L. Goldstein, S. Kalman, M.J. McCullough, J.H. McCusker, D.A. Stevens, L. Wodicka, D.J. Lockhart, and R.W. Davis.** 1998. Direct allelic variation scanning of the yeast genome. Science *281*:1194.
49. **Wodicka, L., H. Dong, M. Mittman, M-H. Ho, and D.J. Lockhart.** 1997. Genome-wide expression monitoring in Saccharomyces cerevisiae. Nat. Biotechnol. *15*:1359-1367.
50. **Zhu, H., J-P Cong, G. Mamtora, T. Gingeras, and T. Shenk.** 1998. Cellular gene expression altered by human cytomegalovirus: global monitoring with oligonucleotide arrays. Microbiology *95*:14470-14475.

7 Technology and Applications of Gene Expression Microarrays

Elisabeth Evertsz, Pascual Starink, Robert Gupta, and Drew Watson
Incyte Pharmaceuticals, Inc., Palo Alto, CA, USA

INTRODUCTION

The natural progression of sequencing entire genomes has been to develop technologies that allow researchers to experiment on large numbers of newly discovered genes. Only through highly parallel analysis can we hope to assign functions to the large numbers of expressed sequence tags (ESTs) that are being discovered in the course of advancing the Human Genome Project (2,18). Advances in Incyte's Gene Expression Microarray (GEM™) technology are enabling researchers to quantify the relative abundance of messenger RNA (mRNA) rapidly and accurately in diverse individual sequences obtained from complex DNA samples. These advances in GEM microarray technology, along with the increasing abundance of sequencing data obtained from high-throughput sequencing efforts allow researchers to analyze genome-wide patterns of mRNA expression across multiple samples and species.

Unlike oligonucleotide arrays, the sequence of genes on the cDNA microarray need not be known (17). This feature, plus the fact that the cDNA microarray is extremely easy to use and to analyze, makes the GEM microarray an ideal choice for studying expression levels of large numbers of genes and ESTs. The basic principles of current microarray technology are quite simple. DNA targets are immobilized on a glass surface, and labeled probe sample is hybridized to the DNA at high stringency. Gene expression is measured by detecting the amount of labeled probe hybridized to the immobilized DNA. Incyte GEM microarray technology employs a two-color system developed at Stanford (16,19) to hybridize both the control and experimental probe samples simultaneously. The advantage of this competitive hybridization is that it allows for a direct comparison

Microarray Biochip Technology
Edited by Mark Schena

between two different expression states in a particular sample and is independent of the DNA target density. This approach has been successfully used in the study of human, plant, and microbial gene expression studies (5,7,15).

The development of the gene expression microarray into a robust and reliable research tool has involved the optimization of a number of different processes. We describe here the GEM microarray fabrication process and other aspects of GEM microarray technology. We also discuss a study designed to characterize fully the reproducibility and accuracy of the GEM system, as well as an example of a typical GEM microarray technology application: a time-course experiment analyzing drug response.

GEM MICROARRAY TECHNOLOGY

Building a GEM Microarray

The GEM microarray fabrication process begins with the creation of a GEM microarray cDNA library—a collection of cDNA molecules containing genetic information from the specific biological system to be analyzed. A critical aspect of this method is that complete sequence verification is not needed for any aspect of GEM microarray technology. Generally, a normalized library is used to minimize redundancy in a population of known and unknown genes. To build the GEM microarray, individual cDNAs are isolated and amplified using vector-specific primers. Next, cDNA targets are purified in buffer conditions that maximize DNA deposition. A microsample of each cDNA molecule is deposited onto a prepared glass surface, with each gene (or amplified insert) occupying a unique location (Figure 1). The glass slides have been specifically prepared to

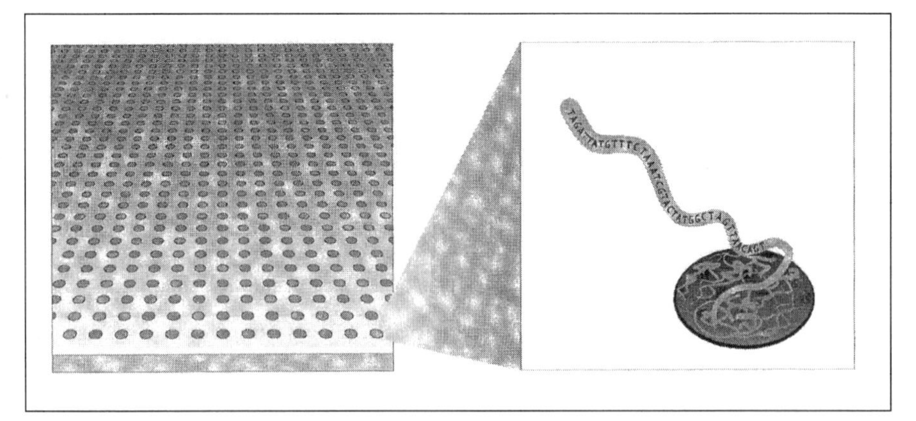

Figure 1. Schematic of a microarray showing individual cDNA elements deposited at unique locations in an array format. Incyte's microarray fabrication system tracks the location of each element, along with its clone identification.

facilitate uniform deposition of the DNA with a minimum of sample spreading. Current array technology allows for close spacing and robust deposition of distinct element samples, at an interelement distance of 175 µm. The arraying device used to create the GEM microarrays is a modified version of a format that can be found in many commercially available arraying devices today. The print head is mounted on an axis above the platen containing the slide. The motion of the head is perpendicular to the motion of the platen. The print head contains pins that load the clones from standard 96-well plates. After printing the slides on the platen, the tips are cleaned to eliminate cross-contamination before arraying the next set of clones. Once the GEM microarray has been completely fabricated, postprocessing steps ensure that the DNA is fixed on the glass.

Generating DNA Probes

To detect the differences in gene expression between two cell samples, the GEM microarray system uses a fluorescent labeling technique. The process begins by isolating mRNA populations from the tissues under consideration, using an oligo-dT-based purification technique. Kits are available from a variety of commercial sources that enable the user to isolate highly pure mRNA from tissue and cell samples (16). As little as 200 ng can be used in the labeling reaction, and no amplification protocol is used at any stage of probe generation. This ensures that differential expression ratios are not affected by any amplification artifacts. The resultant RNA populations are then converted to cyanine dye-labeled cDNA by a standard reverse transcription reaction (21). One cDNA population is labeled with Cy3 and the other with Cy5. These two fluorescent dyes are well matched at the cDNA generation step and maintain good fluorescent properties whether wet or dry. In addition, they have distinct excitation and emission spectra, facilitating subsequent scanning steps. Next, the two probe samples are mixed and simultaneously applied to a single microarray, where they competitively hybridize with the arrayed cDNA molecules (Figure 2). Competitive hybridization allows differential expression to be directly compared between the experimental and control samples. Ratios can also be calculated independent of target density. After an elevated temperature incubation, the microarray is rinsed

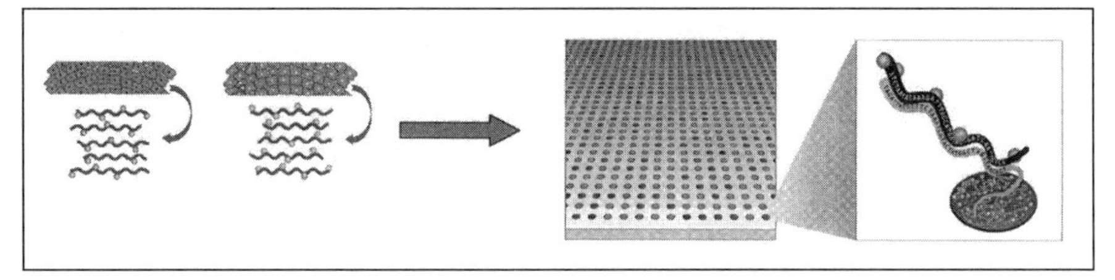

Figure 2. Fluorescent-labeling technique is used to visualize hybridized probe. An RNA sample from two experimental sources is purified and labeled with two fluorescent dyes in a standard reverse transcription reaction. Each probe pair is then mixed and hybridized to target elements on the GEM.

in two serial baths of decreasing ionic strength to eliminate unhybridized probe. The large length of cDNA targets (500–1000 nucleotides) and stringent wash conditions allow for highly specific hybridization.

Scanning and Analysis

After hybridization, the GEM microarray is read using a dual-laser point scanning and detection system. The slides are scanned twice to image both the Cy3 and Cy5 channel. The optical system is stationary, whereas the slide is mounted on an XY-stage and moves in front of the objective. In the optical path, HeNe (633 nm) and GreeNe (543 nm) laser beams are combined into a single beam and reflected by a beam splitter into the objective to excite the elements on the slide. The emitted light is collected by the same objective and passes through a beam splitter. This light then passes through a pair of band pass filters before a photon multiplier tube (PMT) translates the photons into an electrical signal. This signal is cleaned with an electronic filter before an analog-to-digital converter converts the data to a digital signal with 16 bits/pixel resolution, yielding a 64K-count dynamic range. This image is visualized in a false 16-color log scale. Currently, the scanning resolution is 10 μm/pixel.

The goal of the image analysis procedure is to separate and classify the true element signal pixels from background and unwanted pixels, mainly caused by noise, contamination, and nonspecific hybridization. With the resulting classification, an analysis is performed on an element-by-element basis to correct for the local background and estimate the expression level. The first step in the image analysis process is to generate a grid over the scanned image of the arrayed surface (Figure 3). Then an initial separation is performed to distinguish between signal and background through a sequence of segmentation algorithms. The resulting

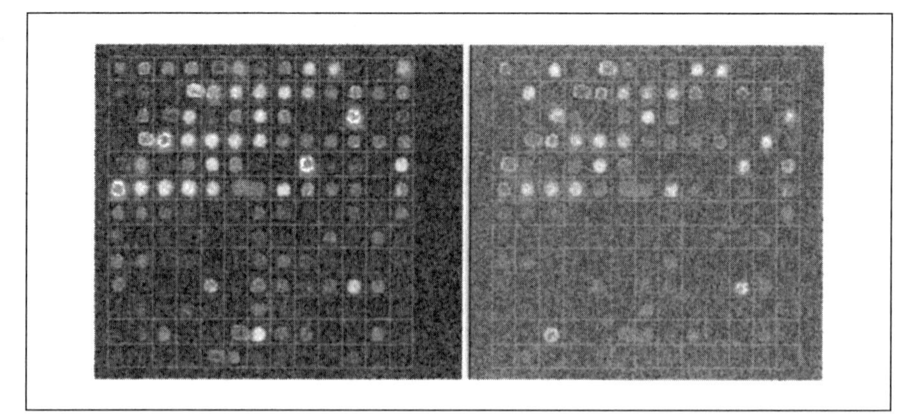

Figure 3. All arrays are gridded for analysis. Cy3 (right) and Cy5 (left) scans. Identical grids and element locations are calculated for the two images. Element identification occurs through a series of steps based on size and estimated location of each cDNA deposited within the grid. (**See color plate A12.**)

signal pixel regions are assigned to the array elements for classification. In the final stage of analysis, element intensity is calculated, background subtracted, and differential expression determined.

The segmentation procedure is used to delineate potential elements. First a composite image is generated from the Cy3 and Cy5 recordings, and then a noise peak elimination filter is applied to this image to reduce noise. This filter reduces single pixel spikes caused by electronic noise and very small contaminating particles. Next, a smoothing procedure reduces noise features that extend over more than one pixel. Groups of pixels with higher intensity than surrounding areas are designated as regions. The binary image is labeled, which assigns a unique number to pixels in the same region. The total number of regions resulting from the segmentation procedure will usually be four to eight times the actual number of elements. A further classification procedure is needed to assign the true elements correctly from a larger set of potential elements. This is accomplished by calculating the following statistical features for each region: the size in pixels, the extent (bounding box), and the "center of gravity." Based on the a priori knowledge that the elements are organized in a grid structure, the classification attempts to identify the true elements. The element regions are rejected based on three tests. The first test rejects regions that are simply too small or too big. The second test is based on the region locations, as determined by their "centers of gravity" and rejects the regions that are not in a regular grid structure. If an element has a neighbor directly left, right, up, or down, it is likely be a true element. The resulting element pairs are used to estimate the distance between the elements. The final test rejects elements that do not have neighbors at the estimated distance, within some tolerance. Each element region is assigned to a grid box, and grid boxes that remain empty are assigned an artificial circular element. The size of that element is the average size in pixels of all true element regions. Once the true elements have been successfully assigned to each grid location, analysis and calculation of differential expression can occur.

The analysis uses the delineated regions found by the segmentation and classification procedures for the final determination of differential expression. First, the background is calculated by averaging the signal from background pixels in a grid box and then fully subtracting that quantity from each grid pixel signal. Expression levels for each transcript are measured by determining the average signal in each fluorescent channel within the corresponding element area. The ratio of the two fluorescent intensities provides a quantitative measurement of the relative gene expression levels in the two samples. Only genes that show significant expression in at least one sample are considered for calculation of differential expression. For all elements whose signal does not measure at least 2.5 times greater than background signal in both channels, no differential expression ratio is calculated. This prevents ratios from being entered into a data-analysis expression database that do not reflect actual differences in transcription levels.

The overall expression level for a particular gene can be estimated by the signal intensity of each element. In general, the level of signal intensity is proportional to the number of fluorescent cDNA molecules bound to that gene. However, the absolute intensity of any particular element is also proportional to the arrayed DNA density. It is the ratio of the two fluorescent intensities in the competitive hybridization that provides a highly accurate quantitative measurement of the relative gene expression level in the two samples under analysis. Data are normalized to control for variations in Cy3 and Cy5 labeling by adjusting the Cy5 signal with a factor derived from the quotient of the total Cy3 and Cy5 signals. In addition, ratios are capped at 100:1 to ensure the biological relevance of the measurement. The statistics resulting from the image analysis are stored in the element table database, which can be converted to a tab delimited text file. Element statistics include Cy3 and Cy5 expression values, background signal, and differential expression ratios.

Quality Control

Before release, each Incyte GEM microarray is processed through a series of rigorous quality control filters. Using specially designed DNA control elements present on each microarray, and a known RNA mix that is added to each probe-labeling reaction, all GEM microarray hybridizations can be assessed for labeling, hybridization efficiency, minimum signal intensities, and expected Cy3/Cy5 ratios. The control elements are arrayed on each GEM microarray and are composed of DNA fragments derived from inter-open reading frame (ORF) regions in *Saccharomyces cerevisiæ*. These fragments are all approximately 1000 bp in length and are designed with minimal homology to any known gene. This ensures that no cross-hybridization to these elements occurs with any RNA sample.

A set of 14 control RNA molecules is generated from these fragments using polymerase chain reaction (PCR) and amplified with composite primers that include a T7 promoter and a 30-base oligo dT sequence. A simple transcription reaction using the PCR products as template generates artificial mRNA fragments. These RNA fragments are added to the two GEM microarray probe-labeling reactions in known amounts together with the expermimental samples. A set of four fragments (positions C1–D8 in Figure 4) is added in an concentration series in both the Cy3 and Cy5 reactions. These fragments control for signal sensitivity. Six other fragments (rows A and B) are added in known ratios to the two fluorescent-labeling reactions to control for expected ratios. Four RNA fragments (rows E and F) that have been previously converted to fluorescent DNA are either arrayed onto the control plate or are added to the experimental sample to control for overall scanning and hybridization signal, independent of the labeling reactions.

Additional elements composed of either housekeeping genes or mixed cDNA populations are also arrayed on the control plate (rows G and H). These elements will be hybridized by RNA in the experimental sample. The signal intensity of these elements is used as a measure of experimental RNA quality. All signal intensities in the control plate as well as total GEM microarray hybridization signals are collected. The values are then compared with a set of minimum intensity criteria for each particular control, which automatically grades each GEM microarray. Because the controls differentiate between processes in the labeling and hybridization steps, the quality of experimental RNA can be measured independently of labeling or hybridization efficiency.

VALIDATION OF GEM TECHNOLOGY

A set of experiments was performed to quantify and assess the performance of Incyte GEM microarray technology, in particular, to assess the performance with respect to assay reproducibility, accuracy, and precision. Experiments were also used to assess the effects of manufacturing variables on observed ratios for a range of probe concentrations and relative ratios. The dynamic range of the probe used in GEM microarray hybridization was also determined.

A useful index of reproducibility in hybridization signal is provided by the concordance correlation coefficient (8,9). Concordance correlation coefficients were calculated across all GEM microarrays under study using undifferentiated (1:1) probe samples. Results indicate that the degree of concordance between nondifferentiated Cy3 and Cy5 probe signals derived from the same RNA is approximately 98%, suggesting a high degree of agreement between probe signals (Figure 5). Additionally, concordance data can be used to calculate the limit of sensitivity of an assay. The limits of detectable differential expression (LDDE) may be calculated using statistical tolerance intervals for the fold change in differential expression. The LDDE may be defined as the smallest differential expression ratio that, with a high degree of confidence, ensures that the gene is differentially expressed. Using this approach, the LDDE was calculated to be approximately ±1.74. Consequently, any elements with observed absolute ratios

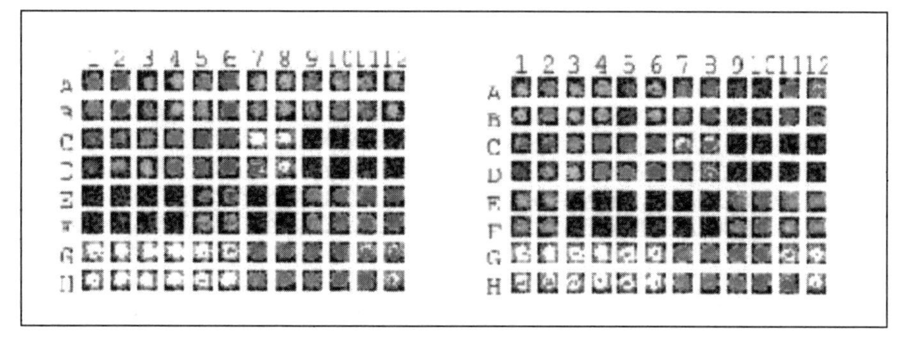

Figure 4. Every GEM contains a 96-well plate with controls for labeling and hybridization. In the Cy3 (right) and Cy5 (left) images shown here, the 96 wells are divided so that each control element is arrayed in quadruplicate. Intergenic yeast fragments are located in rows A–F. RNA sample quality is tested in rows G and H, where a heterogeneous population of DNA and individual housekeeping genes is arrayed. (**See color plate A13.**)

≥1.74 should be deemed to be differentially expressed. Based on results from this experiment, <0.5% of all nondifferentiated standards fell above this calculated 1.74-fold change, indicating that the GEM microarrays can be used to detect changes on the order of twofold in differential expression.

The level of accuracy of GEM microarray differential expression data is important in the design and analysis of all hybridization experiments. To measure the accuracy of GEM microarray technology, standards consisting of complementary RNA and DNA fragments were mixed at known probe concentrations and target densities. These provide the basis for obtaining estimates of the accuracy of ratios. For purposes of this experiment, standards were placed at differential expression ratios of 1:30, 1:11, 1:3, 1:1, 3:1, 11:1, and 30:1. Estimates of accuracy are obtained by comparing observed differential expression ratios with expected differential expression ratios from these standards. Analyses of these standards indicate that Incyte GEM microarray technology has only a slight bias in estimating weakly differentiated elements, that is, differential expression values near 1:1 are accurately reported (Figure 6). However, estimates of differential expression do appear biased at extremely high differential expression ratios. These results suggest that the bias increases as differential expression ratios increase; more precisely, differential expression ratios tend to be somewhat under-

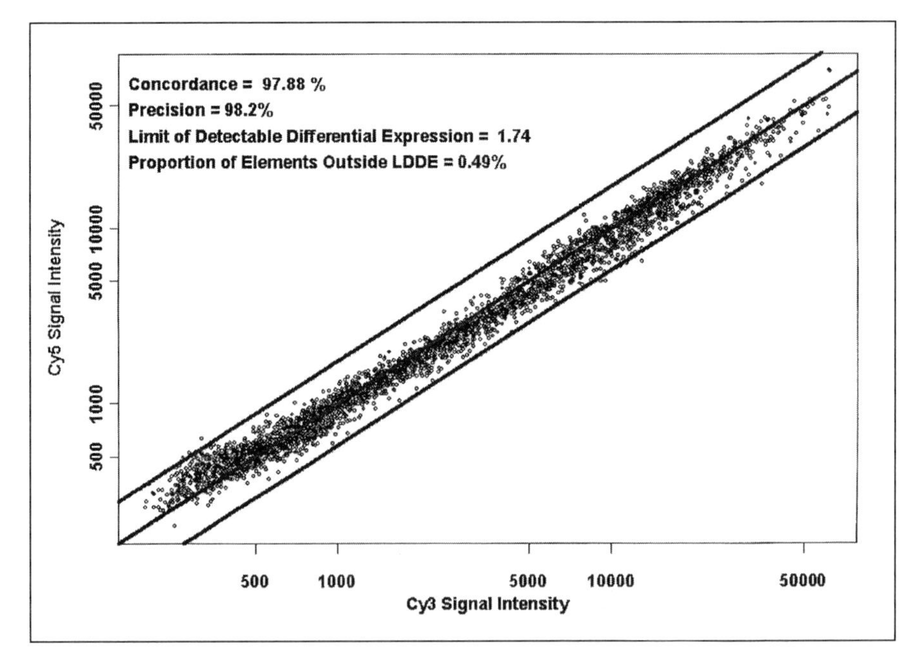

Figure 5. Concordance plot of undifferentiated probe samples. The limit of detectable differential expression is shown. Identical input RNA ranging from 2 to 250 pg was labeled with two fluorescent dyes. Each individual RNA sequence was input at a 1:1 ratio. Concordance and precision estimates indicate that 99.5% of all ratios are between −1.74 and +1.74.

stated (compressed) at extreme values.

The precision of differential expression results derived from GEM microarray hybridizations was also examined. The coefficient of variation (CV), or so-called relative standard deviation, provides a useful estimate of this parameter. Using a set of RNA fragments mixed at defined ratios, a precision profile of the CV for differential expression as a function of probe and target concentration was obtained. A profile of CVs is calculated over a two-dimensional surface of Cy3- and Cy5-labeled probe concentration using generalized linear models (14). The profiles were calculated for a range of DNA target concentrations representing typical PCR yields (0.05–0.2 μg/μL) and varied only slightly with this variable. Based on statistical analyses, the estimated CV for differential expression ratios is approximately 15% at low expected differential expression ratios. As differential expression values deviate from 1:1, CVs also increase. At very high differential expression levels (>30:1), CVs can be as high as 34%. The resulting contour plots are useful for assessing the precision of measurement of the GEM microarray. Figure 7 provides an example of a contour plot and response surface for the CV in differential expression as a function of probe concentration.

Analyses were also performed to determine the contribution of several GEM microarray manufacturing variables to the total variability in differential expression measurements, adjusting for the effects of different probe and target concentrations. These variables include glass lot, GEM microarray lot, arrayer, hybridization setup, postfabrication processing setup, and scan setup. Results indicate that the impact of these variables to the total variability in measurement is extremely small, <15%. The major components of variability in differential expression are scanning and microarray-to-microarray variability, and each

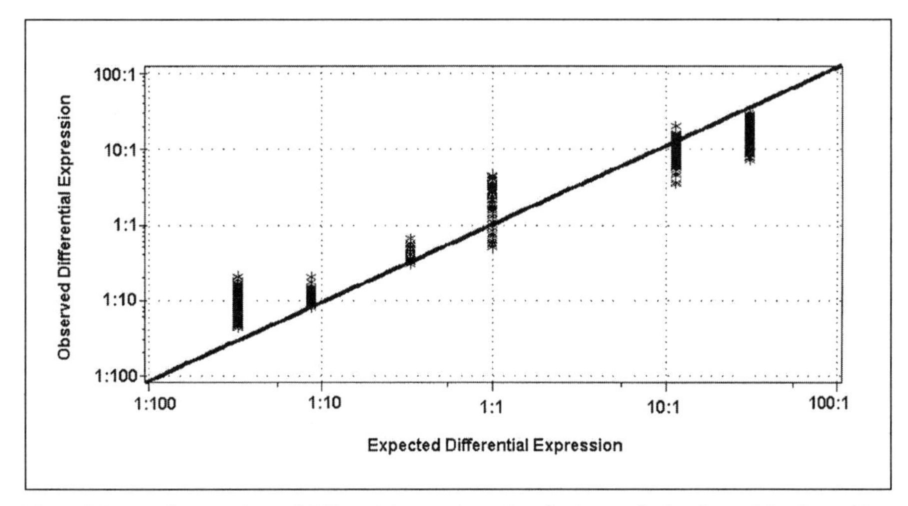

Figure 6. Expected versus observed differential expression values for the standard probe set. The diagonal line shows where perfect accuracy would be plotted. As input ratios increase, observed ratios tend to compress slightly.

accounted for <5% of the total variability in differential expression. Furthermore, there were no significant ($P = 0.64$) differences in differential expression due to different arraying robots. These results are the first of a series of statistical experiments intended to evaluate the performance of Incyte's GEM microarray technology.

In a separate study, we have determined the dynamic range of detectable level of RNA using Incyte's current GEM microarray hybridization and detection process. Incyte's probe-generation protocol requires a minimum of 200 ng of poly(A)+ mRNA per fluorescent-labeling reaction. Using these amounts, as little as 2 pg of an individual transcript can be detected. The dynamic range for probe detection using the described labeling and scanning system is between 2 and 2000 pg. To demonstrate the dynamic range, a series of yeast RNA fragments were added to a complex human RNA sample in a wide range of concentrations (Figure 8). The Cy3 probe generated from this sample was hybridized to a GEM microarray with yeast DNA target elements arrayed in a series of decreasing target densities. The DNA mass used to array target elements between 0.5 and 2.0 μg did not affect the observed dynamic range of the probe-generation reaction. The identical results were obtained for the Cy5-labeled probe.

Bioinformatics

Advanced data analysis packages are needed to extract relevant results from the huge datasets that microarray hybridizations generate. Incyte has developed several data management software tools that have been designed specifically for querying microarray data. Using these data analysis tools, a user has the capacity to ask questions and perform analyses across millions of data points derived from

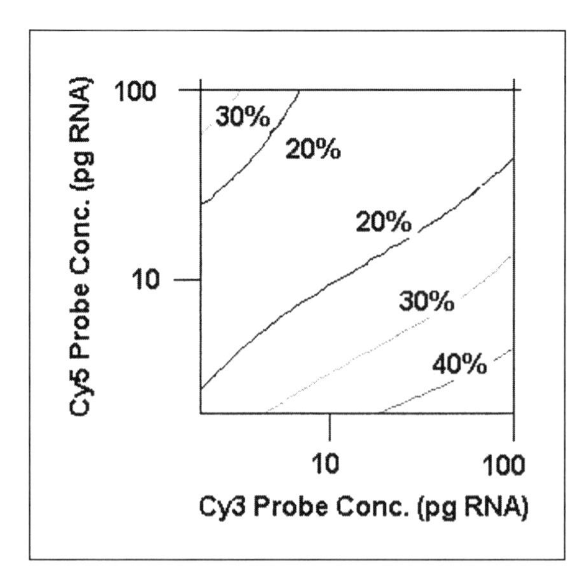

Figure 7. Contour plot for coefficients of variation in differential expression. The plot shows how the precision of ratios varies with input RNA concentration. Coefficient of variation is plotted as a function of RNA concentration (pg) in the Cy3 labeling reaction versus the Cy5 labeling reaction. Extreme ratios show the highest variation; lower ratios show relatively good precision.

GEM microarray hybridizations. Questions may be formulated using a query builder tool, and results can be displayed and further analyzed using a set of visualization tools. The query builder (Figure 9) enables the user to provide the specifications for a database search. The search can combine diverse constraints derived from input experimental parameters such as probe attributes and annotations, gene annotations, gene groups, and GEM microarray groups. In addition, GEM microarray results can be further narrowed based on numerical constraints associated with hybridization results such as differential expression and element signal. If sequence information exists for any clones contained in any GEM microarray, sequence analysis tools such as BLAST may also be used in conjunction with the query builder.

The results of such analyses can be saved and applied as additional constraints in related queries. Furthermore, the database search constraints can be combined using Boolean logic operators and grouping. The process works by translating queries into an optimized SQL query and submitting it to the database for execution.

GEM microarray query results can be viewed and further analyzed using customizable tables and graphs. The element display table permits sorting and display of data based on user-definable sorting criteria. The columns of the table can be enabled or disabled to produce custom displays and reports. Information for a result element such as details of hybridization probes, gene annotations, and element statistics may be accessed individually or in groups. Using a "drill down"

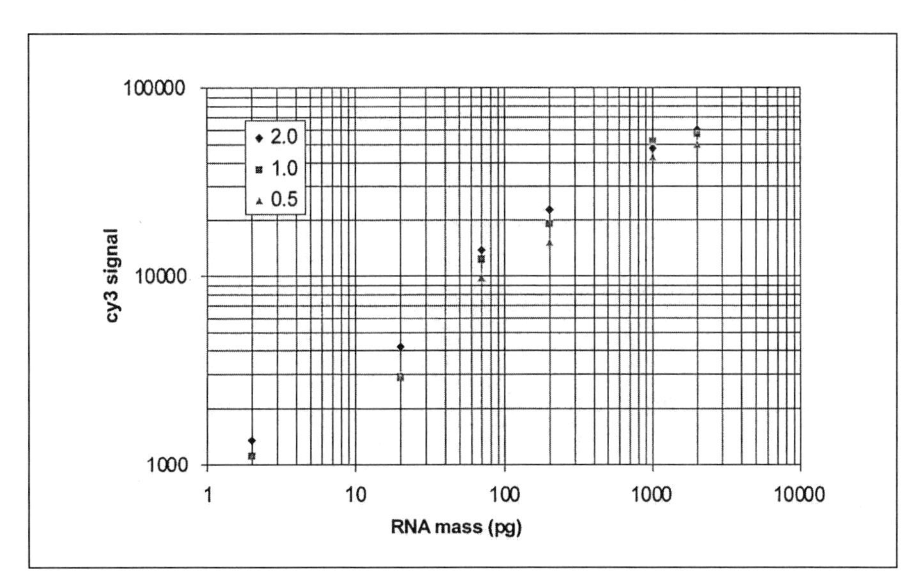

Figure 8. The dynamic range of RNA detection covers 3 orders of magnitude. As little as 2 pg of a single transcript is detectable under standard hybridization and scanning conditions. Data for three DNA element densities per well are plotted (♦, 2.0; ■, 1.0 μg; and Δ, 0.5 μg) as Cy3 signal versus input RNA mass. The dynamic range for mRNA detection is 2 to 2000 pg and is not affected by cDNA target density for target masses from 0.5 to 2 μg.

capability, a user can navigate via the World Wide Web to an external database such as GenBank. In addition, this feature enables researchers to access clone and gene information from a host of external (Internet) or internal (intranet) databases such as Incyte's LifeSeq™ database and related database products, internal proprietary databases, and external public databases.

Result sets can also be visualized by using a set of graph displays. Scatter graphs display the Cy3 intensities correlated with Cy5 intensities in standard or $\log_{10}/\log_{10}$ mode (Figure 10). As in the table viewer, detailed information for an element is accessible by drilling down on a data point. Data points may be colored/highlighted based on a range of differential expression, gene group membership, or protein function hierarchy classification. Many graph features such as the legend, axes text, colors, and orientation are customizable. Graphs may be printed, saved as a file, or copied to the Windows system clipboard and used in word processing or presentation documents.

Specialized groups of GEM microarray results from a single GEM microarray type may be combined into a series. Series groups are a set of ordered GEM microarray experiments that can involve a time course or drug concentration series. Specialized visualization tools are available to illustrate the differential expression pattern for each gene across a set of GEM microarrays over time or drug dosage.

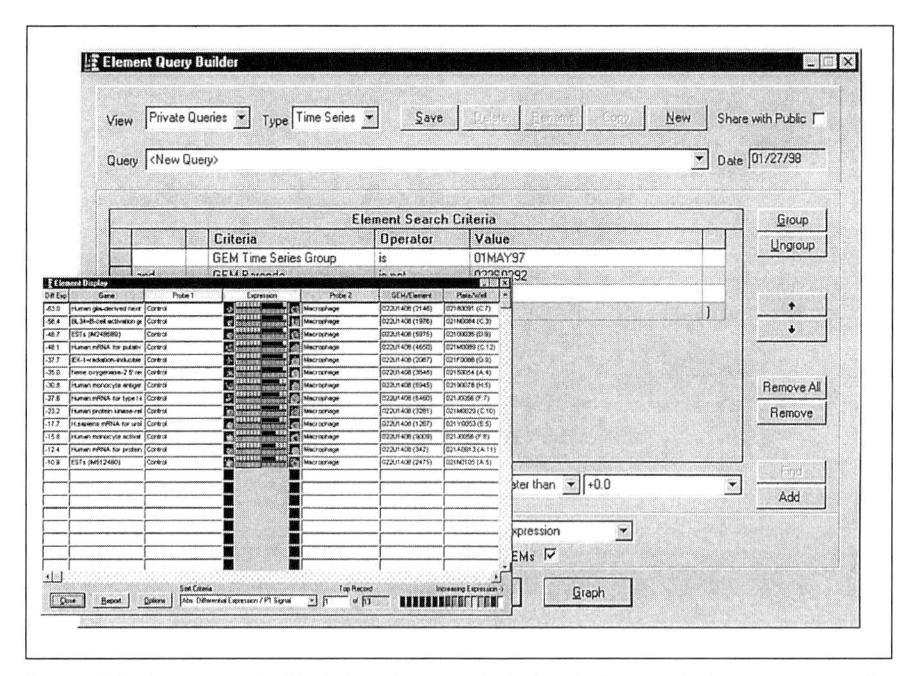

Figure 9. The element query builder. It is used to enter criteria for a database search. Query management features include query storage and retrieval and ability to share queries with colleagues. Query results can be displayed in an element report format in which differential expression, gene, probe and element image are linked in a single file.

These tools include scatter/line plot, 3D series charts, and bar charts with bars grouped by GEM microarray or grouped by gene. As described above, series graphs possess similar customization, printing, and copying features. To perform more rigorous GEM microarray data mining or analysis, result sets may be downloaded into mathematics, statistical, or custom bioinformatics software.

APPLICATION

Time Course Study with Microarrays

To examine the performance and analytic potential of the Incyte's GEM microarray technology, Incyte scientists designed an experiment aimed at maximizing the number of genes for which expression would be modulated. The experiment is based on the well-known responses of peripheral blood mononuclear cells (PMBCs) to treatment with phorbal myristate acetate (PMA) and ionomycin.

Samples were hybridized to UniGEM V, a microarray containing >7000

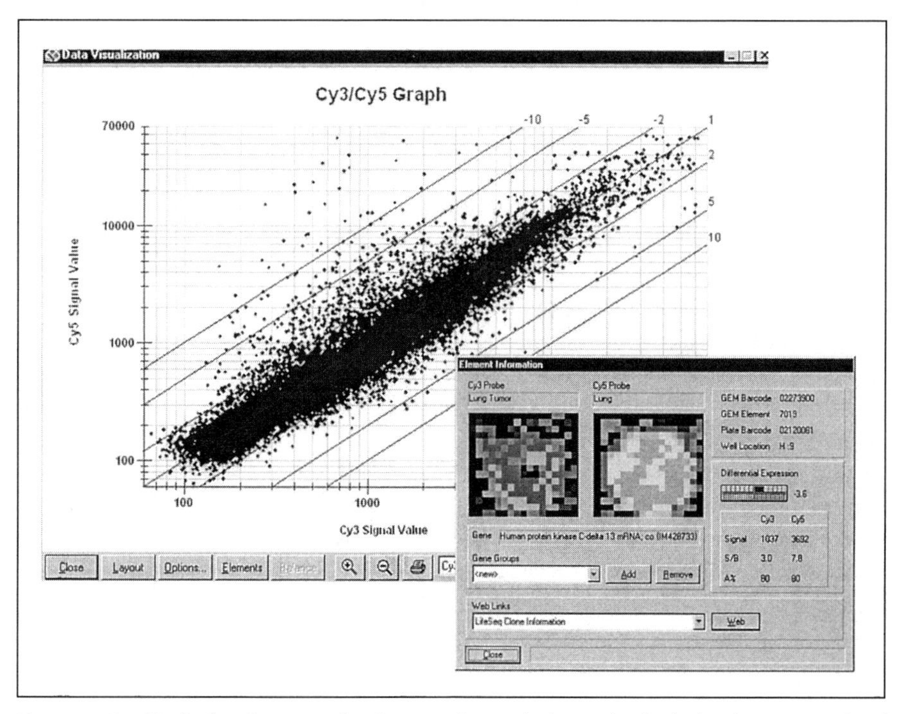

Figure 10. Graphic display of query results. Query results can also be graphically displayed in a scatter plot of the Cy3 and Cy5 signal for all genes in the query output. The graphs are presented in a $\log_{10}$ scale so that differential expression can be delineated with diagonal axis. Retrieval of detailed information on individual genes is available using "drill down" feature.

unique verified clones, composed of both known genes and ESTs. Hybridization results were generated from samples of PBMCs treated with PMA and ionomycin and collected over an 8-hour time course. The results proved consistent with the documented expression responses in this system. Furthermore, analysis of the gene expression on the microarray over five time points in series enabled Incyte scientists to identify expression patterns characteristic of particular gene groups. Indeed, using microarray to analyze group genes based on the temporal regulation of their expression may reveal critical associations that will help elucidate biological pathways for drug development.

The PBMC sample, which includes all major cell types of the immune system (T lymphocytes, B lymphocytes, monocytes, natural killer cells, and dendritic cells), was chosen to maximize the variety of expression responses after drug treatment. These cell types not only express common housekeeping genes but also tissue-specific genes such as cytokines, transcription factors, and membrane receptors that have been extensively studied and characterized. The combination of PMA and ionomycin triggers signaling events linked to the protein kinase C pathway and the intracellular calcium influx pathway. Because these two pathways represent the major route of signal transduction for most cells of the immune system, the combination of PMA and ionomycin reveals the full transcriptional potential of complex cell populations such as PBMC. Incyte scientists isolated PBMC samples from healthy donors, stimulated the cells in vitro with combined PMA (0.1 μM) and ionomycin (1 μg/mL), and then harvested the cells after 0.5, 1, 2, 4, and 8 hours of activation. For each of these five time points, samples of mRNA were purified, processed, and hybridized onto one microarray per time point. The control sample in all GEM microarray hybridizations was derived from untreated PBMC (0 hour time point). Using Incyte image processing algorithms, expression ratios were calculated only for those elements with sufficient signal in both the control and experimental channels. This includes all elements for which the signal in both channels was sufficiently above the background signal.

Of the 7075 elements on the UniGEM V microarray, differential expression results across the five time points were as follows: *(i)* 25% of the elements showed no differential expression results at any time point; *(ii)* 23% of the elements showed differential expression results at some time points; and *(iii)* 52% of the elements showed a differential expression result at all time points.

Focusing on the 3741 elements that showed expression data at all five time points hybridized, many changes in differential expression are observed. Among the expected responses is the change in expression for three interleukin genes that are known to have markedly different responses when stimulated with combined PMA and ionomycin. The levels of interleukin-2 mRNA detected by the UniGEM V microarray increased dramatically over time. In parallel, the levels of interleukin-13 mRNA increased initially but subsided at the end of the 8 hours of activation, whereas expression of interleukin-10 remained stable throughout the activation (Figure 11). All these changes correspond to published responses of these genes to this type of stimuli (1,6,22).

Marked differences in kinetics were observed for about 30% of genes with full expression profiles. These showed greater than twofold up- or down-regulation during the 8-hour course of the experiment. Using a simple clustering algorithm, the genes were grouped according to their behavior through time. Shown in Figure 12 are examples of early-, mid-, and late-activated genes. All the known genes identified as early-induced by hybridization on UniGEM V microarray correspond to those genes described in the literature as typical of early-activation events in PBMC (10–12). Highly parallel gene expression analysis combined with temporal clustering algorithms represents a powerful method to discover novel genes in critical biological pathways.

CONCLUSIONS

Gene expression microarrays have the potential to revolutionize how basic research and drug development are performed. The technology has evolved from a novel experimental tool into a robust and reproducible commercial method. Reliable and high-capacity probe generation methods have been developed that allow for the reproducible hybridization of RNA samples onto standard GEM microarray products. Incyte GEM microarrays derived from mouse, human, and rat are now commercially available, allowing for highly parallel expression studies without the generation of specialized cDNA libraries. Additionally, custom GEM microarray data can be generated from a cDNA library for any biological system. These systems include, but are not limited to, plant (15,16), yeast (7),

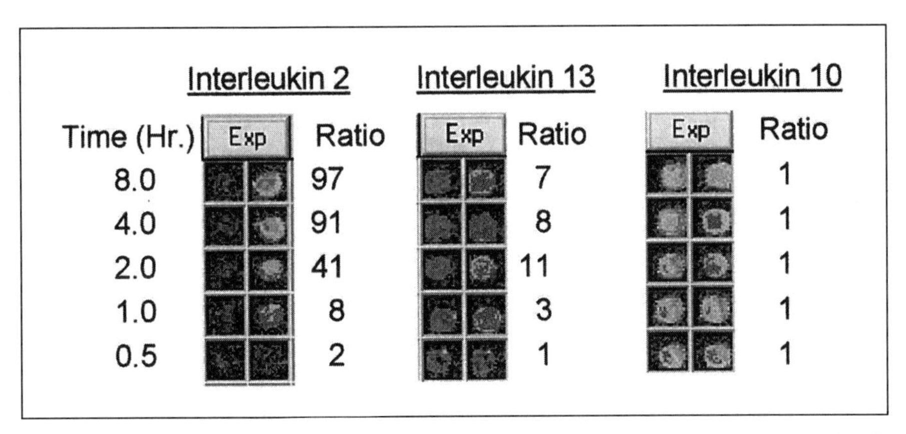

Figure 11. Differential responses in three interleukin genes. Responses are viewed in Incyte's GEMTools software over five time points. The control (left column) is the 0 treatment sample for all three genes. As time after PMA treatment increases, the three interleukins show markedly different responses (right columns). **(See color plate A13.)**

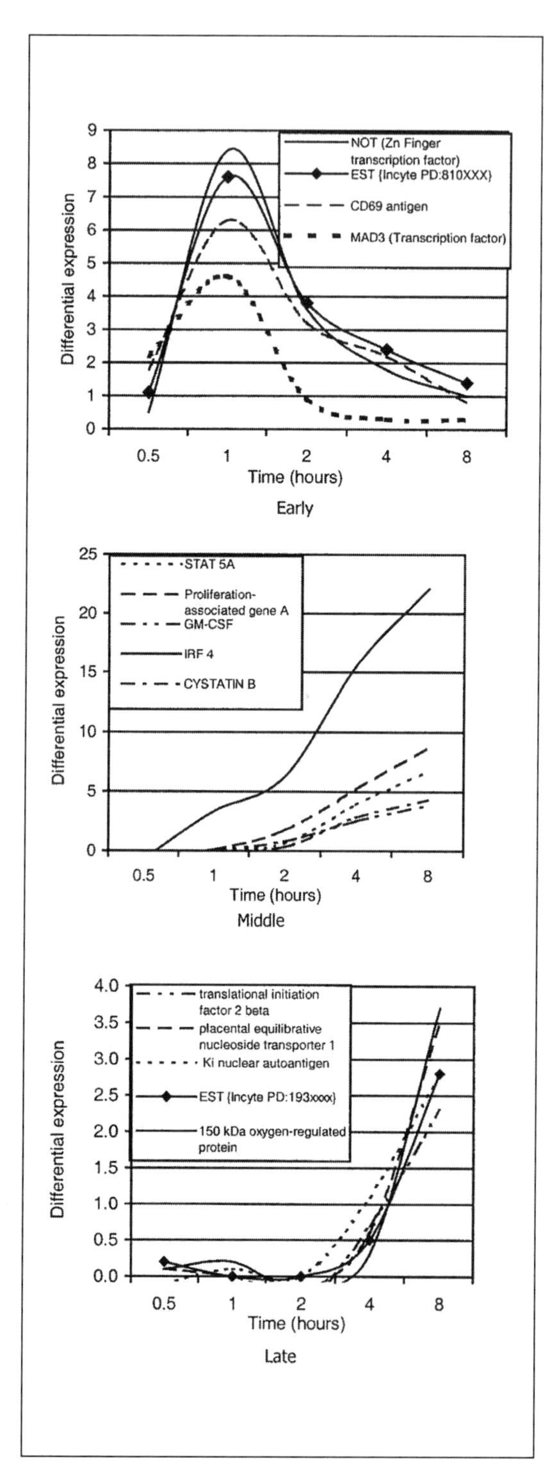

Figure 12. Clustering of temporal expression responses. Such clustering shows how genes can be grouped into early-, mid-, and late-activated groups. Genes in both the early- and late-activated groups show novel ESTs (♦), as do previously described genes.

and human studies (4,5,17). Advances in microarray manufacture and analysis have also allowed large-scale studies on the accuracy and reproducibility of this technology. These studies have demonstrated a robust probe generation protocol with a large dynamic range as well as a hybridization and analysis system that generates reliable and accurate data.

The development of hardware capable of creating microarrays that contain 10 000 elements precipitated the development of more sophisticated image analysis and database methods able to analyze and process the enormous amount of data generated by each GEM microarray experiment. Advanced image analysis techniques now permit the identification and localization of thousands of unique elements. These data are then filtered and refined to generate accurate expression data that can be transferred to a database. Software has been developed to manage these enormous sets of gene expression data and permit analysis with advanced query tools. With these advances, more and more researchers are using microarray technology as a research tool (3,5,7,13,20). In the future, highly parallel expression analysis will probably be crucial in many stages of disease treatment, including target discovery, toxicity analysis, and diagnosis. A wide range of applications for GEM s will continue to be developed as technology improvements proceed.

REFERENCES

1. **De Waal Malefyt, R., J.S. Abrams, S.M. Zurawski, J.C. Lecron, S. Mohan-Peterson, B. Sanjanwala, B. Bennet, J. Silver, J.E. de Vries, and H. Yssel.** 1995. Differential regulation of IL-13 and IL-4 production by humans CD8+ and CD4+ Th0, Th1, and Th2 T cell clones and EBV-transformed B cells. J. Immunol. *7*:1405-1416.
2. **Deloukas, P., G.D. Schuler, G. Gyapay, E.M. Beasley, C. Soderlund, P. Rodriguez-Tome, L. Hui, T.C. Matise, K.B. McKusick, J.S. Beckmann, et al.** 1998. A physical map of 30,000 human genes. Science *282*:744-746.
3. **DeRisi, J.L., V.R. Iyer, and P.O. Brown.** 1997. Exploring the metabolic and genetic control of gene expression on a genomic scale. Science *278*:680-686.
4. **Heller, R.A., M. Schena, A. Chai, D. Shalon, T. Bedilion, J. Gilmore, D.E. Woolley, and R.W. Davis.** 1997. Discovery and analysis of inflammatory disease-related genes using cDNA microarrays. Proc. Natl. Acad. Sci. USA *94*:2150-2155.
5. **Iyer, V.R., M.B. Eisen, D.T. Ross, G. Schuler, T. Moore, J.C.F. Lee, J.M. Trent, L.M. Staudt, J. Hudson, Jr., M.S. Boguski, D. Lashkari, D. Shalon, D. Botstein, and P.O. Brown.** 1999. The transcriptional program in the response of human fibrolasts to serum. Science *283*:83-87.
6. **Kumagai, N., S. Benedict, G. Mills, and E. Gelfand.** 1998. Comparison of phorbol ester/calcium ionophore and phytohemagglutinin-induced signaling in human T lymphocytes. Demonstration of interleukin 2-independent transferrin receptor gene expression. J. Immunol. *140*:37-43.
7. **Lashkari, D., J. DeRisi, J. McCusker, A. Namath, C. Gentile, S. Hwang, P.O. Brown, and R.W. Davis.** 1997. Yeast microarrays for genome wide parallel genetic and gene expression analysis. Proc. Natl. Acad. Sci. USA *94*:13057-13062.
8. **Lin, L.** 1989. A concordance correlation coefficient to evaluate reproducibility. Biometrics *45*:255-268.
9. **Lin, L.** 1992. Assay validation using the concordance correlation coefficient. Biometrics *48*:599-604.
10. **Lofquist, A.K., K. Mondal, J.S. Morris, and J.S. Haskill.** 1995. Transcription-independent turnover of I kappa B alpha during monocyte adherence: implications for a translational component regulating I kappa B alpha/MAD-3 mRNA levels. Mol. Cell Biol. *15*:1737-1746.

11.**Lopez-Cabrera, M., A.G. Santis, E. Fernandez-Ruiz, R. Blacher, F. Esch, P. Sanchez-Mateos, and F. Sanchez-Madrid.** 1993. Molecular cloning, expression, and chromosomal localization of the human earliest lymphocyte activation antigen AIM/CD69, a new member of the C-type animal lectin superfamily of signal-transmitting receptors. J. Exp. Med. *178*:537-547.
12.**Mages, H.W., O. Rilke, R. Bravo, G. Senger, and R.A. Kroczek.** 1994. NOT, a human immediate-early response gene closely related to the steroid/thyroid hormone receptor NAK1/TR3. Mol. Endocrinol. *8*:1583-1591.
13.**Marton, M.J., J.L. DeRisi, H.A. Bennett, V.R. Iyer, M.R. Meyer, C.J. Roberts, R. Stoughton, J. Burchard, D. Slade, H. Dai, et al.** 1998. Drug target validation and identification of secondary drug target effects using DNA microarrays. Nat. Med. *4*:1293-1301.
14.**McCullagh, P. and J. Nelder.** 1983. Generalized Linear Models. Chapman & Hall, London.
15.**Ruan, Y., J. Gilmore, and T. Conner.** 1998. Towards Arabidopsis genome analysis: monitoring expression profiles of 1400 genes using cDNA microarrays. Plant J. *15*:821-833.
16.**Schena, M., D. Shalon, R.W. Davis, and P.O. Brown.** 1995. Quantitative monitoring of gene expression patterns with a complementary DNA microarray [see comments]. Science *270*:467-470.
17.**Schena, M., D. Shalon, R. Heller, A. Chai, P.O. Brown, and R.W. Davis.** 1996. Parallel human genome analysis: microarray-based expression monitoring of 1000 genes. Proc. Natl. Acad. Sci. USA *93*:10614-10619.
18.**Schuler, G.D., M.S. Boguski, E.A. Stewart, L.D. Stein, G. Gyapay, K. Rice, R.E. White, P. Rodriguez-Tome, A. Aggarwal, E. Bajorek, et al.** 1996. A gene map of the human genome. Science *274*:540-546.
19.**Shalon, D., S.J. Smith, and P.O. Brown.** 1996. A DNA microarray system for analyzing complex DNA samples using two-color fluorescent probe hybridization. Genome Res. *6*:639-645.
20.**Spellman, P.T., G. Sherlock , M.Q. Zhang, V.R. Iyer, K. Anders, M.B. Eisen, P.O. Brown, D. Botstein, and B. Futcher.** 1998. Comprehensive identification of cell cycle-regulated genes of the yeast Saccharomyces cerevisiae by microarray hybridization. Mol. Biol. Cell *9*:3273-3297.
21.**Wang, A., M. Doyle, and D. Mark.** 1998. Quantitation of mRNA by the polymerase chain reaction. Proc. Natl. Acad. Sci. USA *86*:9717-9721.
22.**Yssel, H., R. de Waal Malefyt, M.G. Roncarolo, J.S. Abrams, R. Lahesmaa, H. Spits, and J.E. de Vries.** 1992. IL-10 is produced by subsets of human CD4+ T cell clones and peripheral blood T cells. J. Immunol. *149*:2378-2384.

8 Information Processing Issues and Solutions Associated with Microarray Technology

Yi-Xiong Zhou, Peter Kalocsai, Jing-Ying Chen, and Soheil Shams
BioDiscovery, Inc., Los Angeles, CA, USA

INTRODUCTION

Microarray technology provides an unprecedented means for carrying out high-throughput gene expression analysis experiments. With today's technology, a single hybridization experiment using one chip can yield expression profiles for tens of thousands of genes simultaneously. A microarray project, as with most genetic research projects, usually involves acquisition and validation of large datasets. These datasets contain a variety of different information, ranging from sequence data on the genes or clones placed on each slide to quantified expression values for each gene under different experimental conditions. Furthermore, each project typically requires iterations of series of processes, which compounds the amount of data needed to be handled. Figure 1 depicts a typical processing scheme. The process starts from experiment design and array fabrication, through array scanning, image analysis, and finally gene expression data analysis. The expression data analysis will likely lead to a new hypothesis, which will in turn require another iteration of experiment design, array fabrication, and so on. During each iteration, a process generates new data, which in turn can be used by other processes within the overall system. For example, data associated with the array fabrication process, containing information that links gene sequences to spots, can be exploited by the image processing process to automate many aspects of the operation. Due to the large volume of data generated in microarray projects, information processing issues become prominent.

In the maturation process of microarray technology, there are two kinds of challenges. One is to develop the hardware for fabricating, hybridizing, and reading

Microarray Biochip Technology
Edited by Mark Schena

arrays. The other is to manage the massive amount of information associated with this technology, so that the results can yield understanding of genomic functions in biological systems. With the steady progress in hardware technology development, currently available equipment can reliably produce and image >10 000 spots on single microscope slides, and 100 000 spots will be possible in the near future (e.g., http://arrayit.com). In other words, the fundamental challenge of microarray hardware has been mostly resolved. On the other hand, the informatics challenge has just begun to be addressed. There are three major issues involved. The first is to keep track of the information generated at the stages of the chip fabrication and hybridization processes. The second is to process microarray images to obtain the quantified gene expression values from the arrays. The third is to mine the information from the gene expression data. The goal of this chapter is to present these issues in detail and show some of the possible software solutions.

This chapter is organized into four sections. First, we present the informatics problems emerging from the array fabrication stage. The second section describes the problem of signal detection in image processing of the arrays. The third section illustrates the variety of ways to quantify the signal and to perform quality measurements for assessing the reliability of the data and revealing problems during the array fabrication and hybridization processes. Finally, the fourth section describes some of the issues and solutions for mining and analysis of gene expression data.

ARRAY FABRICATION INFORMATICS

Array Fabrication and Hybridization Process

The array fabrication process may be conceptualized in four steps and their subprocesses, each of which has its own set of data and/or parameters, as shown in Figure 2. Once a gene expression experiment has been planned, the first step is to design the microarray chips that will contain the clones of interest. The chips are

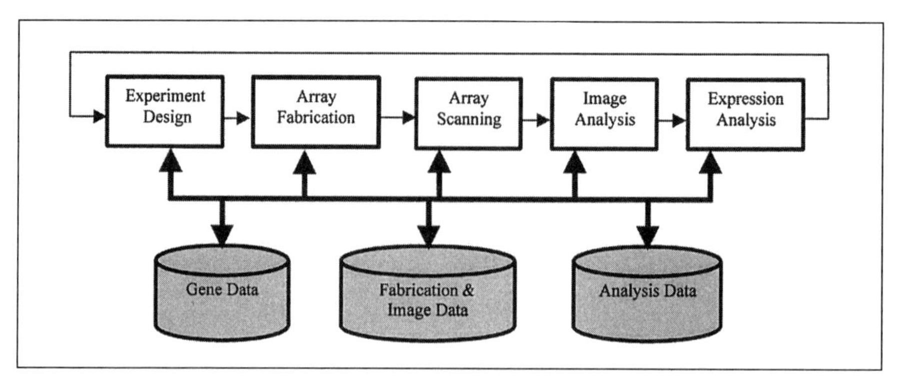

Figure 1. The overall microarray information processing system.

then manufactured using a robotic arrayer. Hybridizations are performed with either unamplified or amplified cDNA or RNA from sample tissues. Finally the florescence images of the hybridized chips are scanned and stored for image analysis.

In each of these four steps, a large amount of information is produced and needs to be handled appropriately. The use of this information is not limited to the array fabrication process. Some of it may be useful in the subsequent processes, such as the image processing and gene expression analysis. The informatics challenge in array fabrication is to maintain a vast amount of the information, organize it into a structured database for efficient access, and provide utilities for the design of microarray chips. These three informatics issues are the central theme of this section. After each of these issues are presented, solutions are proposed.

Keeping Information About the Process

The information generated in the stages of array fabrication and experimentation comes from multiple sources, as illustrated in Figure 2. It is essential to track the data as they pass through the various stages of the microarray fabrication process. Each step in this process has its own set of data and/or parameters. These are described below.

Plate data. Each plate may have a unique identification (ID), the user name, and the clone in each well.

Array data. Data related to an array may include the layout information such as the spot density, the spacing between adjacent spots, and, more importantly, information about the clones placed at each spot on the array.

Configuration of arrayer and array spotting. The arrayer configuration may

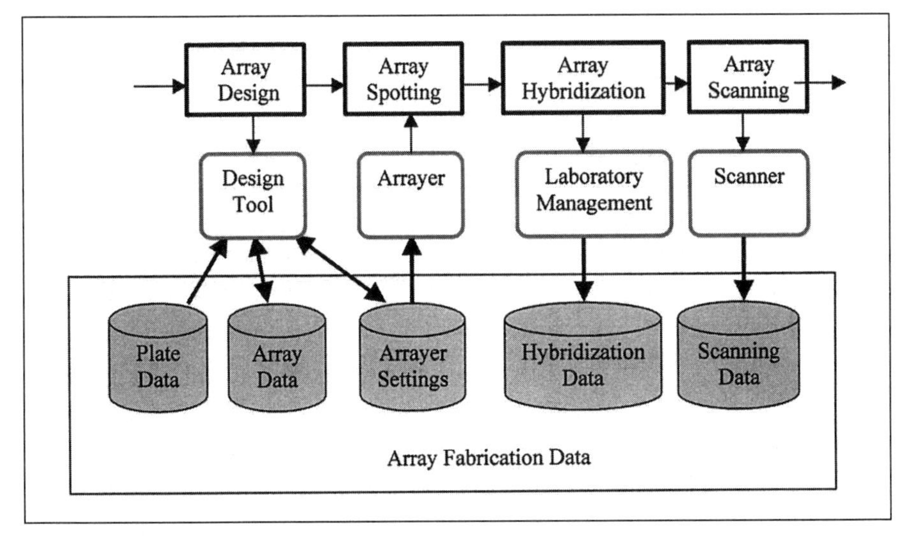

Figure 2. The microarray fabrication process.

include the number and type of pins, the number of pin cleaning cycles, and so forth. Some are parameters needed to configure the arrayer, and some are for tracking and logging purposes.

Array hybridization conditions. This may include the nature of the sample, the protocols, and other experimental conditions.

Array scanning conditions. This includes the type of scanner and the parameters used when scanning an array. From a conventional point of view, array scanning is not included in the array fabrication process, but from a laboratory information management point of view it makes sense to keep data regarding *physical* operations on arrays in the same place for tracking purposes.

The steady progress and increased density of microarray experiments is producing an explosion in the amount of data generated. The only solution is to employ an effective database management system.

Accessing the Array Information

The major purpose of building the database system for array informatics is to facilitate the access of information for experimental design, problem tracking and system tuning, replicating chips and experiments, and data sharing with other processes.

Experimental design. The ability to browse through a plate database or obtain spot densities of existing arrays can provide useful information for decision making, such as how many arrays to produce or the number of duplicates needed for each clone.

Problem tracking and system tuning. With an efficient database, the user can quickly identify a badly fabricated array or an erroneous data point, trace the source of the problem, and remedy it by fine tuning the array fabrication and experiment processes.

Replicating arrays and experiments. The user may want to replicate the same experiments on different arrays or vice versa. Having an efficient database system ensures that the information is available for replicating the processing steps.

Data sharing. The information in the database is useful not only for the array fabrication and experiments but also for subsequent processes, such as image processing and gene expression analysis. Image analysis software, such as ImaGene™ and AutoGene™, can take gene ID data as input and associate them with quantified values for each spot in the array.

Designing the Arrays

The informatics problem in the array design cannot be solved by a database alone. It involves the database infrastructure plus tools for navigating the database as well as software support for design of the array. The major components of this process are described below.

Laboratory information management. A management system is necessary that provides access interface to the underlying array fabrication database. This is used to monitor the overall fabrication process by letting the user query different aspects of the fabrication process, update data step by step according to the progress of the experiment, and track the various operations performed on each spot in each array.

Array design. An easy-to-use visual interface is needed that enables the user to design or adjust the array layout based on the type of mechanical arrayer used and the design objectives. The system could also export the various arrayer parameter settings, or even generate complete programs to control the arrayer robot directly.

Utilities for complex data manipulation. The information flow in the fabrication process involves several transformations that cannot be handled using simple database operations. One example is the transformation of clone data from a set of plates (plate data) into spot data on the array (array data). Generating the correct mapping is an essential component of array informatics since it relates the array data back to the plate data.

Solutions to Data Management

There are many solutions to the data management problem. Many companies and labs are developing their own specialized databases and user interfaces intended to work with internal projects. However, setting up an in-house microarray processing system is expensive and difficult to integrate with other modules such as image processing and gene expression data analysis. The ideal solution would be a standard, industry-wide scheme that is sufficiently flexible to handle diversities arising from the needs of different users. Such a solution would allow direct exchanges of microarray data between different labs and would foster collective efforts among multiple labs.

There are two kinds of diversities among users. One is due to the size of the labs. Users in small labs may have limited or no database tools in place, whereas more sophisticated users have already invested in database systems and are only interested in incorporating some useful functionality into their systems. The other kind of diversity is related to differences in the type of data being stored and tracked. Different users may use different kinds of fabrication methods or hybridization protocols, or they may store the plate/array data in different tables or formats (which may depend on the manufacturer). To deal effectively with these diversities, we have proposed and implemented a simple, yet complete, relational database schema that is itself immediately useable but can also be customized to suit specific needs of different users.

As shown in Figure 3, our scheme classifies the data into categories according to different stages in the experimental process. At the top level, there is a global structure recording management information. In each category, there is a *skeleton* table that links to the data in that category, and the top-level structure refers only to these skeleton tables. For each data category a set of tables is provided, but the user may augment or replace them as desired.

Software support is needed to provide simple navigation and management interfaces to the database. When browsing user-defined data, special interfaces are required to ask the user for the tables information.

Tools for Array Design

One of the major challenges in the design of high-density arrays is establishing a map between the wells in the plates to the spots on the chip. We have developed a software system, called CloneTracker™, to address this problem. The software provides several tools to help the user in different stages of the fabrication process. In addition, the software provides database maintenance and query tools for tracking and maintaining information about plates, arrays, and sets of arrays. All the tools are integrated into a central user interface, as shown in Figure 4. The key features of this system are described below.

Visual array design interface. The layout of an array can usually be characterized by a handful of parameters. That is, it can be determined given the chip size,

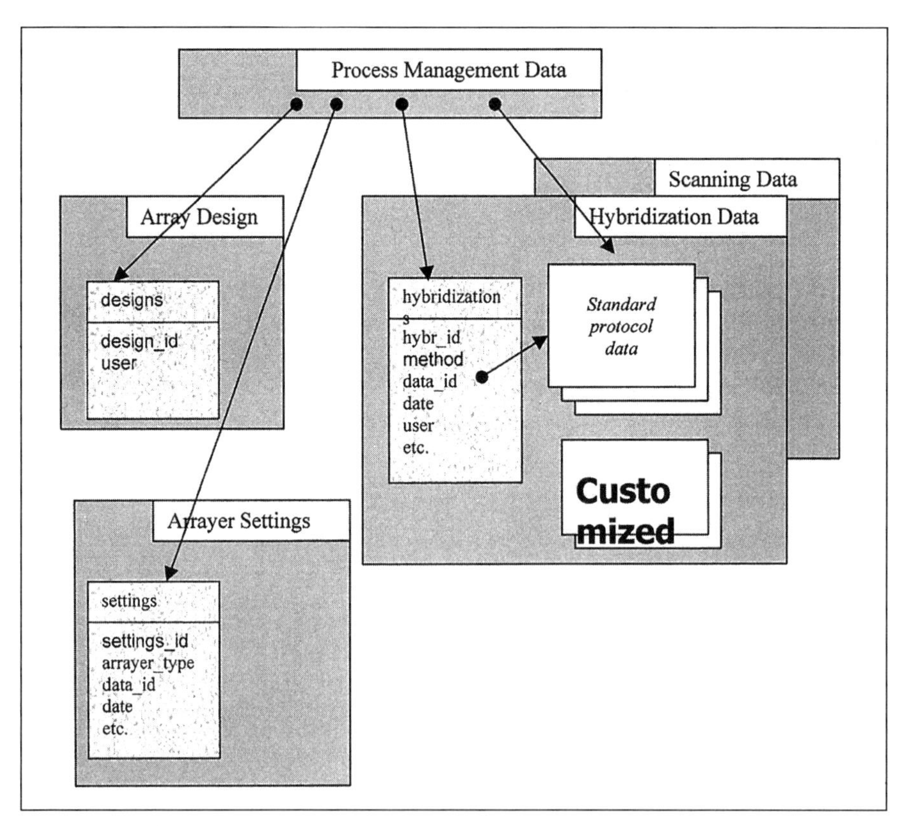

Figure 3. The informatics database.

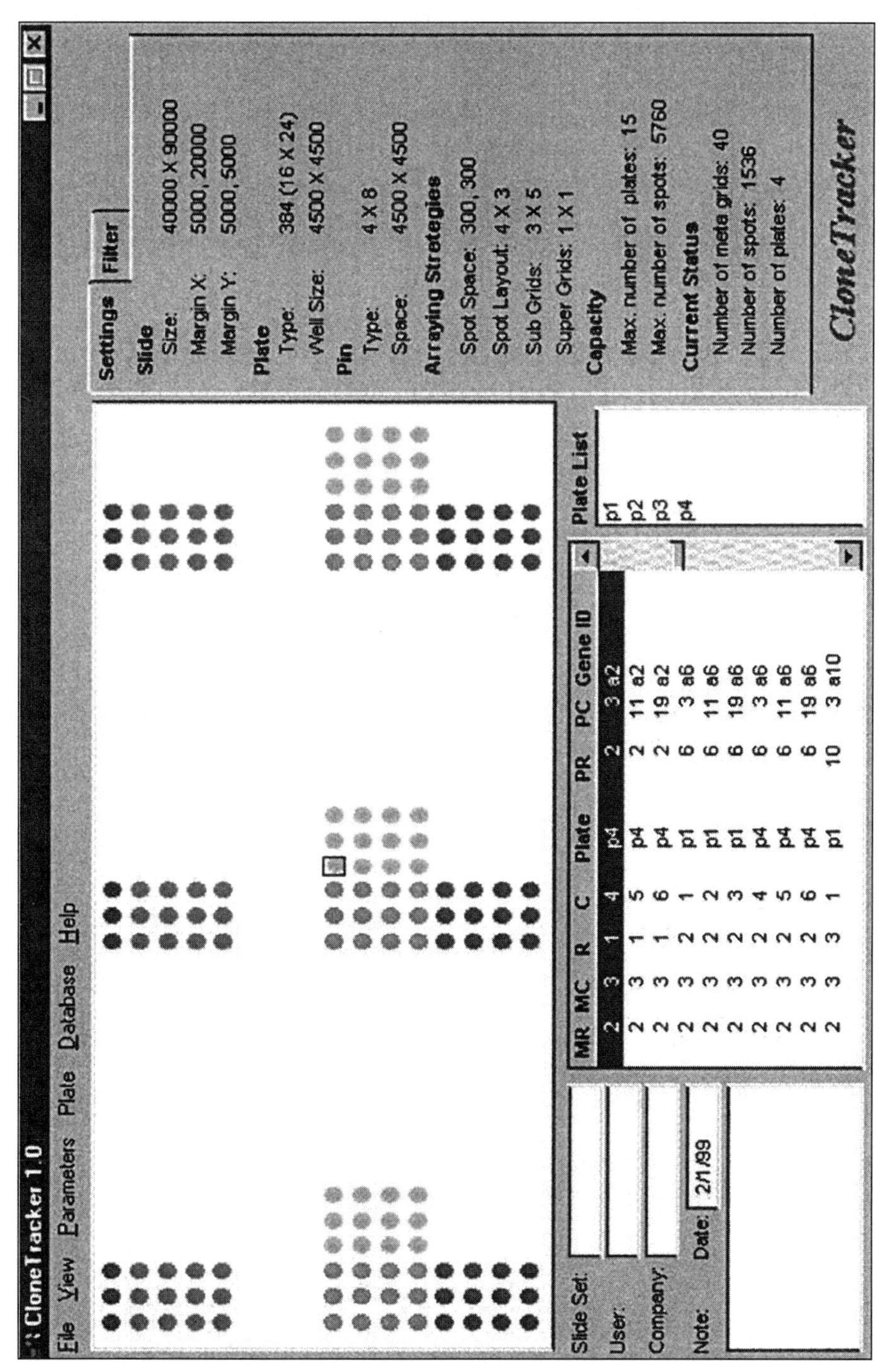

Figure 4. Array fabrication application user interface. (See color plate A14.)

spot spacing, pin number, arraying orientations, etc., regardless of the actual arrayer that manufactures it. By identifying a set of parameters, we can provide a simpler interface for array design without digging down too much into arrayer programming details. This has two different uses. First, multiple arrays can be "virtually" designed and visualized before a specific arrayer program is written. The user can adjust different parameters, see the effects, and obtain useful information such as the spot density or the numbers of rows and columns. The parameters used in the final "optimum" design can then be used to set up the arrayer. Second, if the user already has a prearrayed slide, the software can assist in establishing the necessary association of gene IDs to each spot.

Generation of plate/array mapping. Given an array design, a map between wells of microtiter plates to spots on an array can be generated. With this mapping, it is straightforward to derive the plate and well location from a spot on a given array and vice versa, hence making the tracking of plate/array data easy.

Database maintenance and tracking. Data about all available plates, including plate numbers, size, and IDs for each clone in each well, as well as information about different arrays produced, are all maintained in the database for tracking and archival purposes.

Generation of programs for specific arrayers. Given a generic arrayer settings plus some arrayer-specific parameters, it is possible to generate programs for different arrayers. This is highly arrayer specific and also requires support from vendors. (They need to provide public interfaces.)

ARRAY IMAGE ANALYSIS I: SPOT DETECTION

The fundamental goal of array image processing is to measure the intensity of the arrayed spots and quantify their expression levels based on these intensities. In a more sophisticated and complete approach, the array image processing will also assess the reliability of the quantified spot data and generate warnings for possible problems during the array production and/or hybridization phases. Before these operations can be performed, the spots in the images must be identified. This is the theme of this section.

We will start with enumerating the properties of DNA spots and issues in the microarray images. They are the central concerns in the design of image processing systems for microarray images. These concerns can be seen throughout this section and the next. In fact, any effective image processing system must be designed based on the specific properties of the images to be dealt with.

This section proceeds with a discussion of a range of solutions to the spot detection problem, describing their advantages and disadvantages under various conditions of the images. The section ends with a brief description of the existing approaches used in software systems currently available. The issues of signal intensity measurements and quality measurements are presented in the next section.

Properties and Problems of Microarray Images

Microarray images consist of arrays of spots arranged in grids. All the grids have the same number of rows and columns of spots. These grids, called subgrids, are arranged with relatively equal spacing to each other, forming a complete array. Figure 5 shows an example of a complete grid structure in a microarray image. There is a 2 × 2 array in this image, with each subgrid having 20 × 16, or 320, total spots. This microarray structure is formed with a robotic arraying system that employs pins to create the array. Each pin deposits DNA material in each of the subgrids. The microarray shown in Figure 5 has been produced with

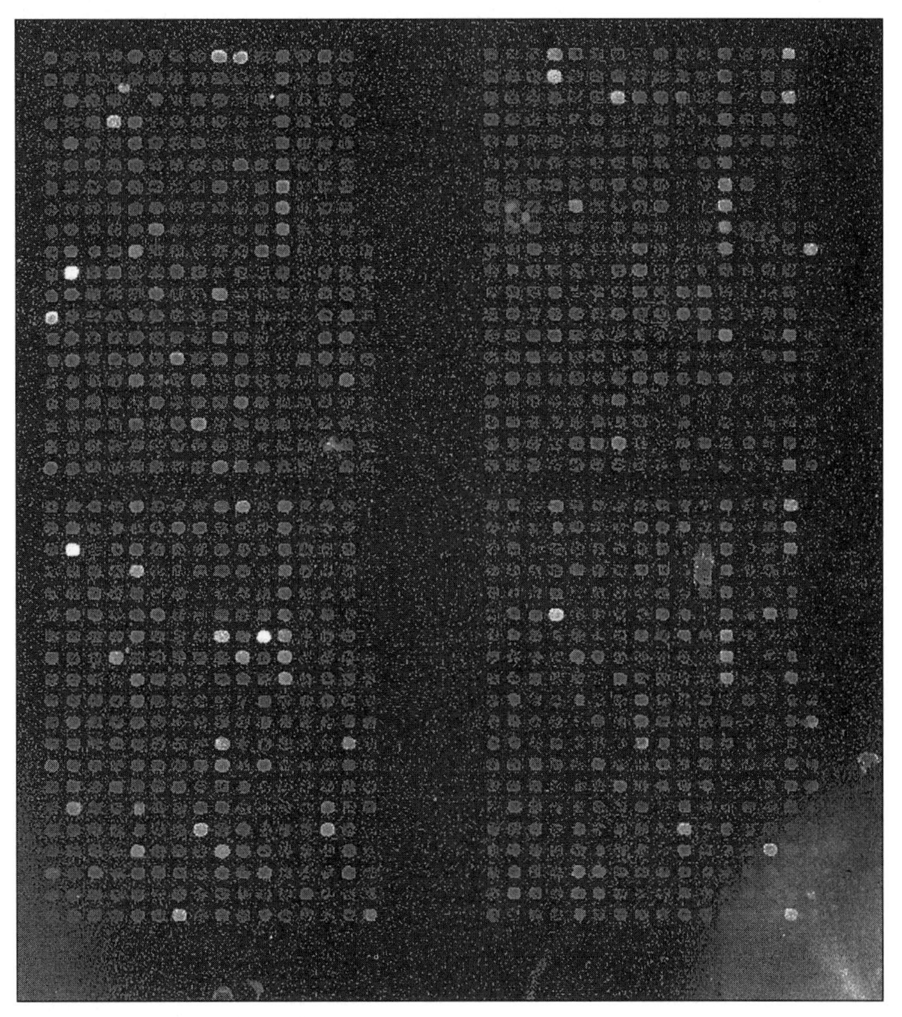

Figure 5. Microarray scanned image containing four subgrids.

a four-pin assembly organized into a 2 × 2 configuration.

Ideally, the microarray images should have the following properties: *(i)* all the subgrids are the same size; *(ii)* the spacing between the subgrids is regular; *(iii)* the distance between rows and columns should be the same; *(iv)* the location of the spots should be centered on the intersections of the lines of the rows and columns (the canonical position); *(v)* the size and shape of the spots should be perfectly circular and the same for all the spots; *(vi)* the location of the grids is fixed in images for a given type of slides; *(vii)* no dust or other contamination is on the slide; and *(viii)* minimal and uniform background intensity across the image.

If all these idealized conditions were satisfied, the image processing task could be easily accomplished by a simple computer program. An array of circles with the defined dimensions and spacing could be superimposed on the image. The pixels falling inside these circles would be considered signal and those outside would be background. However, some of these idealized conditions are often violated in a real-world setting. These violations may be conceptualized in four categories: spot position variation, spot shape and size irregularity, contamination, and global problems that affect multiple spots.

Spot variation is caused, in part, by the mechanical limitations in the arraying process. Researchers are motivated to put as many spots as possible onto a single array to increase the data content of a given hybridization. The fundamental concern in the fabrication of the array is to make sure that each spot is spatially distinct within the arrays. Whether the spots are spaced perfectly is a secondary concern. There are several sources contributing to the nonuniform positioning of the spots within the array, including inaccuracies in robotic systems, the printing apparatus, and the platter that holds the slides. The spatial mapping between slides and scanned images may not be perfectly linear, causing the spacing between the adjacent columns to be unequal across the arrays. Some of the spots may be "missing" from the grid due to lower than measurable expression level or empty wells in the plates used for array fabrication. Some of these issues are observed in Figure 5. Because the spot sizes are on the scale of 100 to 200 μm, fixing the location of the quantitation grids on each image means the location of spots would have to be accurate to ±10 μm (condition *vi* above). This requirement is unrealistic to expect from many research laboratories. Also, because the spots in a subgrid are typically printed with multiple pins, aligning the subgrids accurately would mean that the pin's spacing would have to be ±10 μm as well (conditions *i–iv* above), which is another difficult requirement for many research laboratories. The existence of these positional variations in microarray images demands an intelligent spot localization operation, which must be done either by human intervention or by computer vision algorithms.

In addition to the spot position variation, the shape and size of the spots may fluctuate significantly across the array. The sizes of the droplets of DNA solution may vary, which can cause the size of the spots to vary. Concentrations of the DNA and salt in solution may change over time, making the shape of spots deviate from a typical circle; furthermore, the density of DNA can vary within a given spot. The surface properties of low-quality slides can vary across the surface. All these factors perturb the shape and size of the spots from the ideal. Figure 6 illustrates some of the irregularities of the spots, such as the shape and intensity variation inside of the spots. To obtain accurate measurements of the hybridization signals, quantitation of pixels needs to be done intelligently to take account of these problems.

Contamination is also a source of difficulty in microarray image processing. From the procedure of spotting to the hybridization step, airborne dust can contaminate the array and produce fluorescent signal in the scanned images. Impurities on the glass surface can also cause speckles in the image. Uneven drying of the arrayed samples on the surface may cause unevenness in the images. Figure 7 shows an example of contamination in both the background areas and inside the spots, some of which produces intense fluorescence. The solution to these problems is to identify the contaminants and remove them from the image before applying measurements.

There are also other factors affecting the quality of data in a larger area of the slides, such as glass quality, temperature nonuniformity across the slide, scratches, or bubbles that occur during hybridization. Figure 5 demonstrates a few global problems that can occur in microarray images. The two lower corners of the image have elevated background, some of which obscures the array. Also, three spots in the fourth subgrid are smeared into each other.

All these global and local quality problems need to be handled either by human inspection or by software systems. In this section, we present a number of solutions to spot localization and the methods of signal pixel determination.

Design of the Signal Detection Process for Microarray Technology

The above four issues in microarray image analysis may be resolved in two steps. The first step is to localize the position and size of each spot, which addresses the spot position variation problem. The second step addresses the shape irregularity and contamination problems through a signal pixel identification process. The last step is to deal with the spatially global problems. The first two steps are discussed in this section. The process in the third step requires global quality information and is discussed in "Importance of Quality Measurements" in the next section.

Spot finding methods for resolving the spot localization errors. The goal of the spot finding operation is to locate the signal spots in images and estimate the size of each spot. There are three different levels of sophistication in the algorithms for spot finding, corresponding to the degree of human intervention in the process. These are described below in order of most to least amount of manual intervention.

Manual spot finding. This method is essentially a computer-aided image processing approach. The computer itself does not have any visual capabilities to "see" the spots, rather, it provides tools to allow users to tell the computer the location of each signal spot in the image. Typically, a grid frame is given that the user can place manually on the image and then manipulate to fit the spots in the image. Because the spots in the image may not be spaced evenly, the user may need to adjust the grid lines individually to align with the arrayed spots. The size of each circle may also need manual adjustment to fit the size of a particular spot. To conduct an accurate measurement, the manual spot finding method is prohibitively time-consuming and labor-intensive for microarray images with

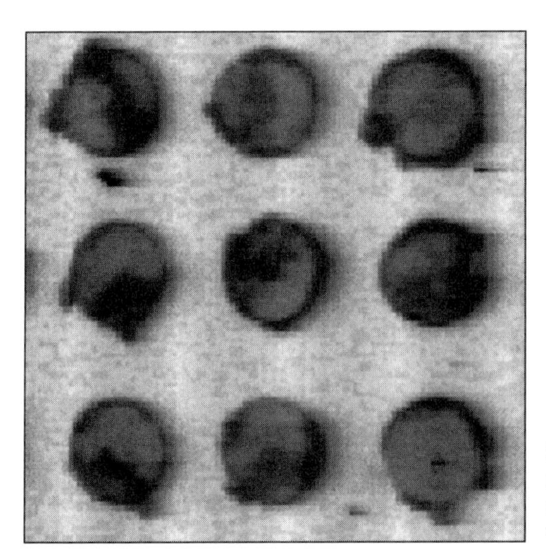

Figure 6. Irregularity of the spot shape. The shape of the spot may deviate from a perfect circle, and the intensity inside the spot may vary considerably.

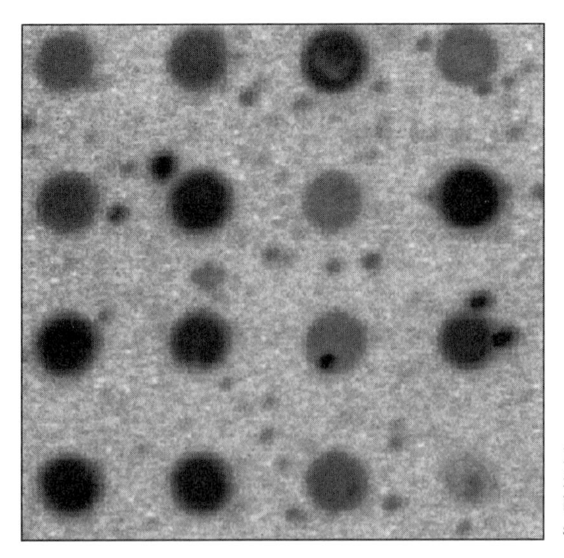

Figure 7. Contamination in microarray images. Punctate contamination is visible both in the background and inside of the spot signal regions.

thousands of spots. Thus, considerable imprecision may be introduced due to human errors, particularly with arrays having irregular spacing between the spots and large variation in spot sizes.

Semiautomatic spot finding. The semiautomatic method requires some level of user interaction. This approach typically uses algorithms for automatically adjusting the location of the grid lines, or individual grid points after the user has specified the approximate location of the grid. What the user needs to do is to tell the program where the outline of the grid is in the image. For example, the user may need to put down a grid and adjust the size of it to fit on the array of the spots, or to tell the program the location of the corners of the subgrids in the images. Then the spot finding algorithm adjusts the location of the grid lines, or grid points, to locate the arrayed spots in the image. User interface tools are usually provided by the software to allow for manual adjustment of the grid points if the automatic spot finding method has not correctly identified each spot.

This approach could potentially offer great time savings over the manual spot finding method since the user needs only to identify a few points in the image and make minor adjustments to a few spot locations if required. The key issue for spot finding algorithms here is to identify the spots correctly even at very low levels of intensity. In addition, the algorithm must deal effectively with two opposing criteria. First, due to variation in spot position, as described earlier, the algorithm must tolerate a certain degree of irregularity in the spot spacing. At the same time, the algorithm must not be "distracted" by contaminants that could be adjacent to a true arrayed spot. Such an algorithm has been developed and implemented in BioDiscovery s ImaGene™ software.

Automatic spot finding. The ultimate goal of array image processing is to build an automatic system that utilizes advanced computer vision algorithms, to find the spots without the need for any human intervention. Such a method would greatly reduce human intervention, minimize the potential for human error, and offer a great deal of consistency in the quality of data. One such system, called AutoGene™, has been developed by BioDiscovery. This tool only requires the user to specify the configuration of the array (i.e., number of rows and columns), and will automatically search the image for the correct grid position. In many tests, the software has been able to locate grids as accurately as those obtained by manual placement of the grid.

Methods of spatial segmentation of signal and background pixels. After the spot location is determined in the image, a small patch around that location (target region) can be used to quantitate the spot intensity level. The next step is to determine which pixels in the target region are signal and which are background. This operation is called signal or image segmentation in computer vision terminology. At this stage, size and shape irregularities of the spots and any contamination problem in the images are the major concerns to the algorithm design. A number of methods have been developed with different levels of sophistication. Their advantages and disadvantages are described below.

Pure space-based signal segmentation. Methods of this class use purely spatial information from the result of spot finding to segment out signal pixels. After the spot finding operation has been completed, the location and size of the spot are determined. A circular mask is placed in the image at the determined position and size to separate the signal from background. It is assumed that the pixels inside the circle are due to the true signal and that those outside are background. Measurements are then performed on these classified pixels. These types of methods are optimal when the spot finding operation is effective (i.e., when spots have been correctly located and sized, the spot shapes are close to perfect circles, and no contamination is present).

However, irregularities of the spot shape and sizes are rules rather than exceptions in many microarray images. Whenever these conditions occur, the accuracy of the measurements is largely compromised. In addition, spot contamination is still an issue in many microarray images. An example is shown in Figure 8 to illustrate the problem encountered with this method. In this example, nearly 50% difference is observed between mean measurements made before and after the removal of contamination from the image.

In addition to quantification inaccuracies generated due to the presence of contaminants in the image, problems arise due to variation in the shapes of the spots. There are two broad classes of spot shapes that will cause problems with pure space-based signal segmentation. First, doughnut-shaped spots, which are often seen with many arraying systems, contain many nonhybridized pixels within the circular spot area. These pixels will be mistakenly classified as signal pixels using this segmentation approach. Second, noncircular spots (e.g., more elliptical spots) cannot be fit perfectly with a round circle, thus causing some signal pixels to be considered as background and vice versa.

This method of quantification is used by a number of available microarray image analysis software tools and would be appropriate for "quick and dirty" look at the image data. However, it is not sufficiently sophisticated to ensure accuracy of the quantified data in most cases.

Pure intensity-based signal segmentation. Methods of this class use intensity information exclusively to segment out signal pixels from background. They assume that the brightness intensities of signal pixels are statistically higher than that of the background pixels. As an example, suppose that the target region around the spot taken from the image consists of 40×40 pixels and that the spot is about 20 pixels in diameter. From a total of 1600 pixels in the region, about 314 ($\pi \times 10^2$) pixels or approximately 20% are signal pixels that are expected to have intensity values higher than the background pixels. To identify the signal pixels, all the pixels from the target region are ordered in a one-dimensional array from the lowest intensity pixel to the highest one $\{p_1, p_2, p_3, \ldots p_{2500}\}$, where p_i is the intensity value of the pixel of the *i*-th lowest intensity among all the pixels. If there are no contaminants in the target region, the top 20% of the pixels in the intensity rank are classified as signal pixels.

The advantage of this method is its simplicity and speed. The method works well on computers with moderate computing speed and when the spots are high intensity relative to the background pixels, but it works less well on spots of low intensities or noisy images.

When the signal intensity is low, the intensity distribution of the signal overlaps largely with background. The signal and background pixels are not separable based on their intensity values alone. Applying this method will produce biased estimates of the signal and background intensity values. The shortcoming of the intensity-based segmentation method can be remedied by exploiting spatial information. One simple method of obtaining spatial information is to segregate the target region into potential signal and background regions before conduct-

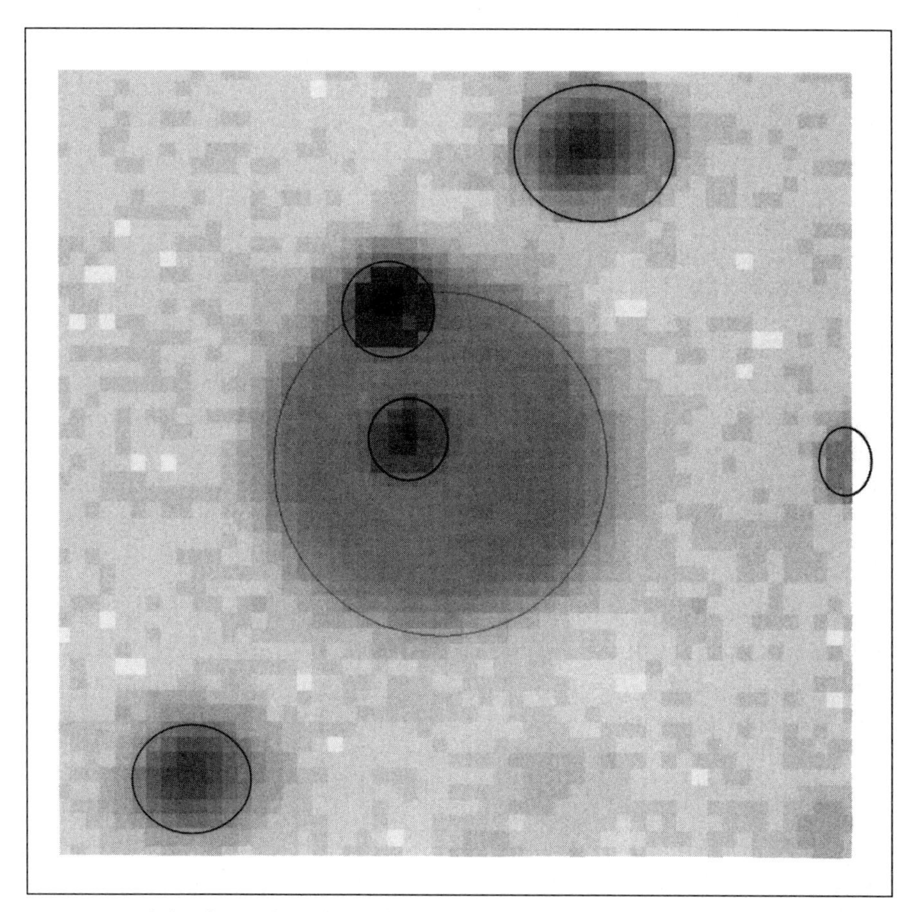

Figure 8. Signal identification from the spot finding operation. All the pixels inside the large circle are considered signal according to this method. When contaminants exist (small circles), measurement errors are introduced into both the signal and background values. The mean signal intensity using purely spatial segmentation is 1230. After the contaminant pixels are removed, the signal mean is 834 or a 48% difference. The mean background values are 113 and 28, respectively, or a 300% difference before and after the contamination removal.

ing intensity-based segmentation. A circle can be placed in the target region based on the result of the spot finding. Pixels inside the circle are then temporarily classified as signals, and those outside are background pixels. Statistical differences in the two regions are then assessed and compared for signal segmentation. In the following discussion, we describe two such methods: the Mann-Whitney test and the trimmed measurement method.

Mann-Whitney segmentation. Chen et al. (1) introduced a nonparametric statistical method, the Mann-Whitney test, to segment out signal pixels from target regions. After the spot finding operation, a circle is placed in the target region to demarcate the spatial region of the spot. Because the pixels outside the circle are assumed to be background, the statistical properties of these background pixels can be used to determine which pixels inside the circle are signal pixels. The Mann-Whitney test is the method used to obtain a threshold intensity level. Pixels inside the circle that exceed the threshold intensity are identified as signal. This method works very well when the spot location is found correctly and when there is no contamination in the image. However, when contaminated pixels exist inside of the circle, they are incorrectly scored as signal pixels. If there are contaminated pixels outside the circle or if the spot location is not found correctly such that some of the signal pixels are outside the circle, these high-intensity pixels will raise the intensity threshold level; consequently, signal pixels with intensities lower than the threshold will be incorrectly scored as background. Applying the Mann-Whitney test to the example in Figure 8, using α value of 0.0001, the mean intensities are 1230 for signal pixels and 33 for background pixels. These values deviate 50% and 20% from the true values, respectively.

This method also has its limitations when dealing with weak signals and noisy images. When the intensity distribution functions of the signal and background are largely overlapping, classification of pixels based on an intensity threshold is prone to errors, resulting in measurement biases. This scenario is similar to what has been discussed in the pure intensity-based segmentation method.

Trimmed measurements method. This approach is another method that combines both spatial and intensity information in segmenting signal pixels from background. The logic of this method proceeds as follows. After the spot is localized and a target circle is placed in the target region, *most* of the pixels inside the circle are signal pixels, and *most* of the background pixels are outside the circle. Due to shape irregularities, some signal pixels may lie outside the circle, and some background pixels may lie inside the circle. These pixels are considered outliers in the sampling of the signal and background pixels. Similarly, contaminant pixels can also be considered as outliers in the intensity domain. These outliers will severely change the measurement of the mean and total signal intensity. To remove the impact of the outliers on these measurements, one may simply "trim off" a certain percentage of signal and background pixels.

Typically, a certain percentage of pixels from the high end of the intensity distribution, say 10%, is trimmed off from the pixels inside the circle, due to the high possibility that such pixels may be contaminants. A certain percent of pixels at the low end of the intensity distribution, say 20%, is trimmed off from the pixels inside the circle, due to the high possibility that such pixels may be background. Whatever remains inside of the circle after trimming is used for quantification. The exact amount to be trimmed depends on the effectiveness of the spot finding process and the quality of the image, such as the extent of size and shape of irregularities. A good estimate for these thresholds can be determined empirically. As long as the image quality does not significantly change (e.g., major increase in contamination), this method is effective.

The major advantage of the trimmed measurement method is the robustness of the measurement against outliers, at the expense of accuracy. When there are no outliers, trimming will reduce the measurement accuracy by reducing the number of pixels used in the calculation of the mean signal value. However, if there is a significant number of pixels per spot and only a small percentage of the pixels are removed, this method can yield highly accurate values. Furthermore, by averaging the measurements from replicate spots, this method should achieve very good performance. Figure 9 illustrates the application of such a method used on the data in Figure 8. The measurements derived from this method deviate about 30% from the true value, which is less than the other methods discussed above.

The major limitation of this method is the loss of the geometric information of the spots due to trimming. However, this information is mainly important if it is used as part of the quality measurement score associated with a spot. Its considerable computational needs demand a fully automated system for practical implementation. In semiautomatic systems, computing speed is a major concern. The trimmed measurement method provides an optimal blend of speed and accuracy.

Integrating spatial and intensity information for signal segmentation. The two methods discussed above use a minimal amount of spatial information. They use target circle derived from the spot localization process to improve the detection of signal pixels. Their design priority is to conduct the spot intensity measurements with minimal computation. These methods are useful in semiautomatic image processing because the speed has high priority and the user can inspect the data visually.

In a fully automated image processing system, the accuracy of the signal pixel classification becomes a central concern. Not only does the correct segmentation of signal pixels offer accurate measurement of signal intensity, but it also permits multiple quality measurements based on the geometric properties of the spots. These quality measures can be used to ask for human intervention for spots of questionable quality that are flagged after the completion of an automated processing run. The correct classification of the signal pixels can be realized by algorithms that use information from both spatial and intensity domains. Such an approach has been implemented in the AutoGene software mentioned earlier.

ARRAY IMAGE ANALYSIS II: DATA QUANTIFICATION AND QUALITY MEASUREMENT

Methods of Quantitation

On a single microarray chip, the expression levels of many genes are in parallel. Under the proper conditions, the total fluorescence intensity of a spot is proportional to the expression level of a gene. These conditions are as follows:

1. The preparation of the probe cDNA (through reverse transcription of the extracted mRNA) solution is performed such that the probe cDNA

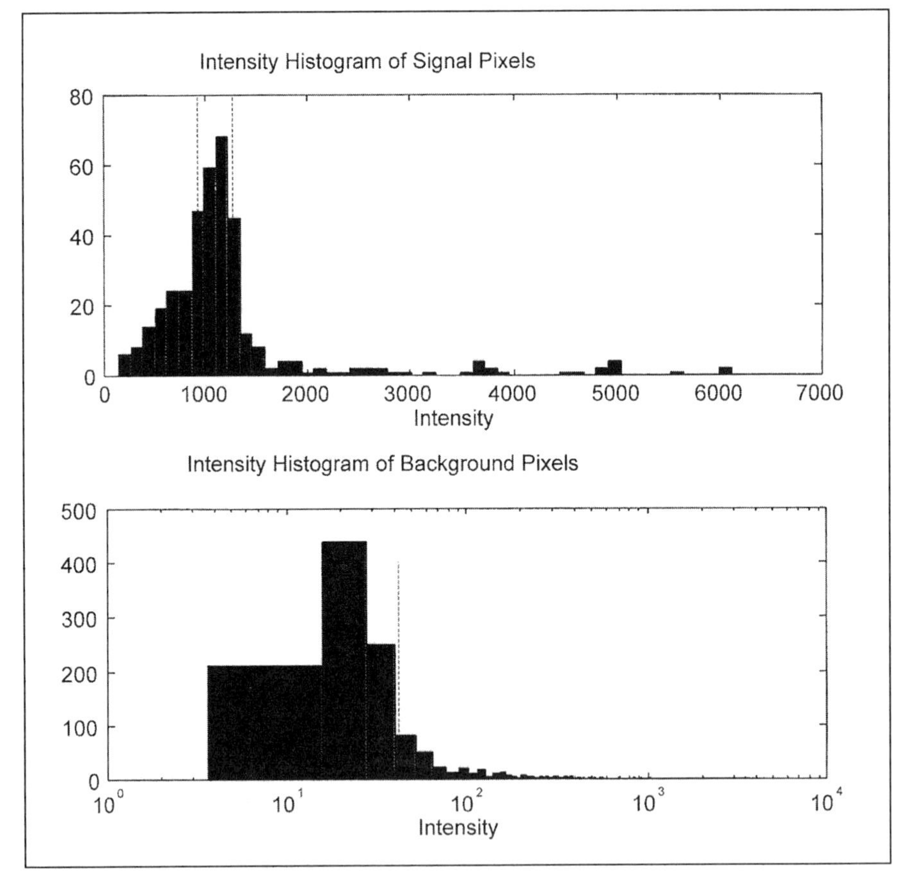

Figure 9. Trimmed measurement method for signal pixel identification. Top, plot of the intensity histogram of the pixels inside the target circle. Thirty percent of the pixels are trimmed off at the lower intensity side (left of the left vertical line). Twenty percent of the pixels are trimmed off the high-intensity side (right of the right vertical line). Bottom, plot of the histogram of the pixels outside the target circle. Thirty percent of pixels are trimmed off at the high-intensity side of the histogram (right of the vertical line). The abscissa is a log scale to show the detail on the lower intensity side. The mean intensities are 1100 for signal and 20 for background. They differ by 32% and -29% from the true value. (**See color plate A15.**)

concentration in the solution is proportional to the mRNA in the tissue.

2. The hybridization experiment is performed such that the amount of cDNA binding to each spot is proportional to the partial concentration of each cDNA species in the probe solution.
3. The amount of cDNA target deposited at each spot during the chip fabrication is constant and in approximately 10-fold excess relative to the most abundant species in the probe solution.
4. There is no contamination on the spots.
5. The signal pixels are correctly identified by the image processing.

In the following discussion, we assume that conditions 1 and 2 are satisfied. Whether these two conditions are truly satisfied is determined through the design of the experiments. For the quantitation measurements, the more closely conditions 1 to 5 are followed the better. Often, conditions 3, 4, and 5 are violated to varying degrees. The DNA concentrations in the spotting procedure may vary from time to time and spot to spot. Higher or lower concentrations may result in altered signals. When adjacent spots overlap, the signal intensity corresponding to the contaminated region is not measurable. The image processing may not correctly identify all the signal pixels; thus, the quantification methods should be designed to address these problems. The commonly used methods are total, mean, median, mode, volume, intensity ratio, and the correlation ratio across two channels. The underlying principle for judging which one is the best method is based on how well each of these measurements correlates to the amount of the DNA probe hybridized to each spot location.

Total. The total signal intensity is the sum of the intensity values of all the pixels in the signal region. As has been indicated above, total intensity is sensitive to variations in the amount of DNA deposited on the surface, the existence of contamination, and anomalies in the image processing operation. Because these problems occur frequently, the *total* may not be an accurate measurement.

Mean. The mean signal intensity is the average intensity of the signal pixels. This method has certain advantages over the *total*. Very often the spot size correlates with the samples and pins used in the arraying step. Measuring the mean will reduce the error caused by the variation of the amount of DNA deposited on the spot. With advanced image processing allowing for accurate segmentation of contaminated pixels, the mean is perhaps the best measurement method.

Median. The median of the signal intensity is the intensity value that splits the distribution of the signal pixels such that the number of pixels above the median intensity is the same as the number below the median intensity. The median is a landmark in the intensity distribution profile. The advantage of choosing this landmark as the measurement derives from the resistance of the median value to outliers. As discussed in the last section, contamination and problems in the image processing operation introduce outliers in the sample of signal pixels. The mean measurement is very vulnerable to these outliers. When the distribution profile is unimodal, the median intensity value is very stable and close to the mean. In fact, if the distribution is symmetric in both high and low intensity sides, the median is equal to the mean. Thus, if the image processing operation is not sophisticated enough to ensure the correct identification of signal, background, and contaminated pixels, the *median* is a better choice than the *mean*. An alternative to the median measurement is to use a trimmed mean, as was discussed in the of "Trimmed measurements methods" section above. The trimmed mean estimate is obtained by trimming a certain percentage of pixels from the high- and low-intensity sides of the distribution.

Mode. The mode of the signal intensity is the "most likely" intensity value and can be measured as the intensity level corresponding to the peak of the intensity histogram. It is also a landmark in the intensity distribution. Thus, it enjoys the same robustness against outliers offered by the median. The tradeoff is that the mode will be more unstable than the median when the distribution is multimodal. This is because the mode value will be equal to one of the modals in the distribution, depending on which one is the highest. When the distribution is unimodal and symmetric, mean, median, and mode measurements are equal. Often the difference between mode and median values can be used as an indicator of the degree to which a distribution is multimodal.

Volume. The volume of signal intensity is the sum of the signal intensity above the background intensity. It may be computed as:

$$(\text{signal mean} - \text{background mean}) \times \text{signal area}.$$

This method adopts the argument that the measured signal intensity has an additive component due to the nonspecific binding, and this additive component is the same as that of the background. This argument may not be valid because the nonspecific binding in the background is different from that in the spot. Perhaps the better way is to use blank spots for measuring the strength of nonspecific binding inside spots. It has been shown that the intensity on the spots may be lower than it is on the background, indicating that the nature of the nonspecific binding is different between what is on the background and what is inside of spots. Perhaps the background intensity should be used for quality control rather than signal measurement.

Intensity ratio. If the hybridization is two-color and the scanning measurements are taken in two channels, then the intensity ratio between the channels is often an important quantified value of interest. This value will be insensitive to variations in the exact amount of DNA spotted since the ratio between the two channels is being measured. This ratio can be obtained from the mean, median, or mode of the intensity measurement, obtained as discussed above, for each channel.

Correlation ratio. Another way of computing the intensity ratio is to perform a correlation analysis across the corresponding pixels in two channels on the same slide. It computes the ratio between the pixels in two channels by fitting a straight line through a scatter plot of intensities of individual pixels. This line must pass through the origin, and the slope is the intensity ratio between the two

Table 1. Measurements Using Different Quantification and Image Processing Methods

Method	Mean	Median	Mode	Total
Full AutoGene	834	935	998	384 450
Trimmed	1103	1109	1109	207 375
Pure spatial	1233	1076	1063	464 765

The pure spatial method is discussed in "Pure space based signal segmentation." It places a circle in the signal region and identifies the pixels inside of the circle as signal and outside as background. The trimmed method in the table is discussed in "Method of trimmed measurements." Thirty percent of the signal pixels were trimmed from the low-intensity side of the distribution and 20% from the high-intensity side. For background pixel identification, pixels in a 3-pixel-wide ring outside the signal circle were discarded due to their potential of containing signal pixels "bleeding" into this region. In addition, 30% of the remaining background pixels were trimmed from the high-intensity side. The measurements were performed after trimming. Full AutoGene identified the signal, background, and contamination pixels automatically, which are considered the true values based on visual inspection of the segmentation results. The measurements were performed after the regions were segregated.

channels. This method may be effective when the signal intensity is much higher than the background intensity. The motivation behind using this method is to avoid the signal pixel identification process. However, for spots of moderate to low intensities, the background pixels may severely bias the ratio estimation of the signal toward the ratio of the background intensity. The only remedy to this problem is to identify the signal pixels prior to performing correlation analysis. Then the advantage of applying this method becomes unclear, and the procedure suffers the same complications encountered in the signal pixel identification methods discussed above. Thus its advantage over the intensity ratio method may not be warranted.

Table 1 lists the results of mean, median, mode, and total quantification measurements using three signal pixel identification methods. The image data are displayed in Figure 8. Because of the existence of contamination in the image, the mean method with full AutoGene is optimal. The median and mode measurements with full AutoGene are about 10% to 15% deviated from the mean. With the trimmed method, the three measurements are very similar and represent about 30% deviation from the mean with full AutoGene. Without trimming, the mean measurement produces a 50% deviation; however, median and mode estimations are essentially unchanged.

We derive the following recipe for selecting the measurement methods using different signal detection processes. With full AutoGene and the trimmed method, the mean is the best measurement. Without trimming, the median is the best choice. Although mode provides the same value as median in this case, because it is not stable when the intensity distribution is multimodal, median is a better choice.

Importance of Quality Measurements

An important step in processing microarray images is to assess the reliability of the data obtained and report the problems in the images that may be arising from array fabrication process and experiments. These tasks may be conducted by human intervention or by image processing software. The importance of these tasks cannot be overstated. Without assessing the reliability of the data, the conclusions drawn from analyzing such data may be misleading. Very often such errors can lead to a false hypothesis or overlooking important findings.

The reliability of the data is affected by multiple factors, ranging from problems in array fabrication to problems in image processing. In a high-throughput gene expression analysis system, the reliability assessment needs to be done with a software system. Figure 10 shows an example in which quality measurements can be used to improve the reliability of the data. The data were obtained from an

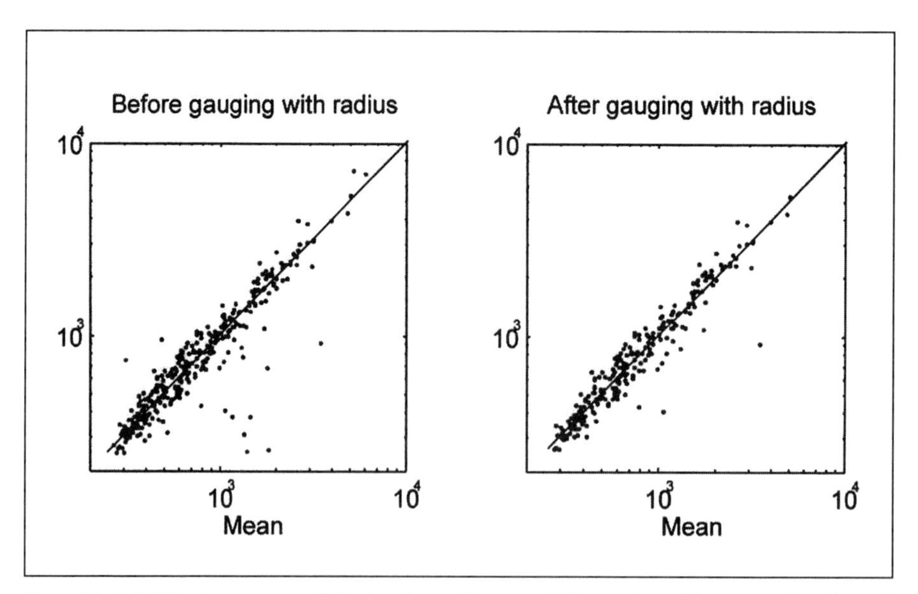

Figure 10. Reliability improvement following the quality control. Scatter plots of the mean values obtained from duplicated spots. The axes represent the mean pixel intensity values. The image was processed with AutoGene. Left, data from all the spots in the image. Right, plot of 75% of the spots in the left plot. The 25% discarded spots have an excessively large or small radius. The coefficient of variance is 27% for the left plot and 19% for the right, a reduction of 30%.

image file processed with AutoGene. The coefficient of variance [(standard deviation)/mean] was 27%. Using spot size as the quality measurement, we discarded about 25% of the spots that had excessively large or small diameters; the coefficient of variance was then reduced to 18.5%, suggesting an improvement in the reliability of the data after discarding low-quality spots.

DATA ANALYSIS AND VISUALIZATION TECHNIQUES

The image processing and analysis steps produce a large number of quantified gene expression values. The data typically represent thousands or tens of thousands of gene expression levels across multiple experiments. To make sense of this many data, it is essential to use multiple visualization and statistical analysis techniques. One of the most typical microarray data analysis goals is to find statistically significant up- and down-regulated genes. Another goal is to find functional groupings of genes by discovering a similarity or dissimilarity among gene expression profiles, or predicting the biochemical and physiological pathways of previously uncharacterized genes.

In the following sections, visualization approaches and algorithms implementing various multivariant techniques are discussed that could help realize the above-mentioned goals and provide solutions to the microarray research community.

Scatter Plot

Probably the simplest analysis tool for microarray data visualization is the scatter plot. In a scatter plot each data point represents the expression value of a gene in two experiments, one assigned to the x-axis and the other to the y-axis (Figure 11). In such a plot, genes with equal expression values in both experiments fall along a diagonal "identity" line. Genes that are differentially expressed fall off the diagonal. The larger the deviation from the identity line, the larger the difference in the expression of a given gene between two samples. Absolute expression levels can also be readily visualized in this plot in that the greater the expression of a given gene the further the data point lies from the origin.

Principal Component Analysis

It is easy to see how the scatter plot is an excellent tool for comparing the expression profile of genes in two samples. Three samples could be plotted and compared in a three-dimensional scatter plot. Three-dimensional plots can be rendered and manipulated on a computer screen. However, when more than three samples are analyzed and compared, the scatter plot cannot be used. In the case of 20 samples, for example, we cannot draw a 20-dimensional plot. Fortunately, there are mathematical techniques available for dimensionality reduction, such as principal component analysis (PCA), in which high-dimensional space can be reduced down to two or three dimensions, which can be plotted. The goal of all dimensionality reduction methods is to perform this operation while

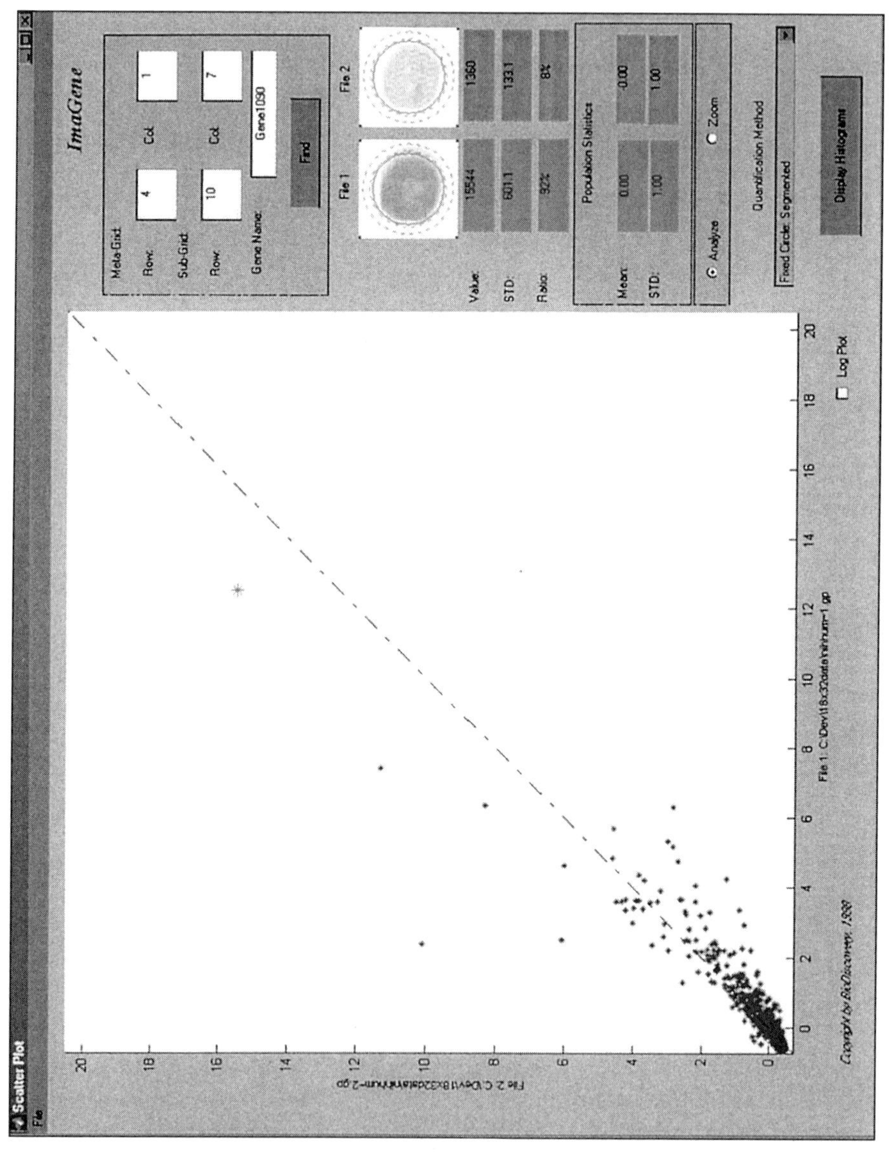

Figure 11. Scatter plot of two experiments. Each point in the plot represents a single gene. The position of the point on the graph reveals the absolute expression level of the gene, such that highly expressed genes lie far from the origin. Genes that are expressed at the same level in two samples fall along the identity (45° angle) line.(**See color plate A16.**)

preserving most or all the variances of the original dataset. In other words, if two points are relatively "close" to one another in the high-dimensional space, they should be relatively close in the lower dimensional space as well. In general, it is not possible to maintain this relationship perfectly. Imagine if a three-dimensional spherical orange peel were rendered flat on the two-dimensional surface of a table. Most of the points lying on the surface of the orange peel would keep their relationship to their neighbors. However, we must tear the orange peel in several places, thus breaking the neighboring relationship between some points on the orange peel to lay it flat.

PCA is a method that attempts to preserve the neighboring relationships as much as possible. Figure 12 is a three-dimensional plot of 600 genes in 21 experiments indicating the position of all 600 genes with respect to the first three principal components. Each "principal component" is made up of a linear combination (summation) of all the 600 genes each weighted by a different value. This multivariate technique is frequently used to provide a compact representation of large amounts of data by finding the axes (principal components) on which the data vary the most. In PCA the coefficients for the variables are chosen such that the first component explains the maximal amount of variance in the data. The second principal component is perpendicular to the first one and explains the maximum of the residual variance. The third component is perpendicular to the first two and explains the maximum of the still remaining variance. This process is continued until all the variance in the data is explained. The linear combination of gene expression levels on the first three principal components could easily be visualized in a three dimensional plot (Figure 12). This method, just like the scatter plot, provides an easy way of finding outliers in the data such as genes that behave differently than most of the genes across a set of experiments. It can also reveal clusters of genes that behave similarly across different experiments. PCA can also be performed on the experiments to find out possible groupings and/or outliers of experiments. In this case, every point plotted in the three-dimensional graph would represent a unique microarray experiment. Points that are placed close to one another represent experiments that have similar expression patterns. Recent findings show that this method should be able to detect moderate-sized alterations in gene expression (3). In general, PCA provides a rather practical approach to data reduction, visualization, and identification of outlier genes and/or experiments.

Parallel Coordinate Planes

Two- and three-dimensional scatter plots and PCA plots are ideal for detecting significantly up- or down-regulated genes across a set of experiments. These methods, however, do not provide an easy way of visualizing the progression of gene expression over several experiments. These types of questions usually come up in time series experiments in which, for instance, gene expression is measured at 2-hour intervals. The important question in this case is how gene expression values vary over the duration of the entire experiment. The parallel coordinate planes plotting technique is an excellent visualization tool to answer these types of questions. By this method, experiments are ordered on the *x*-axis and expression values plotted on the *y*-axis. All genes in a given experiment are plotted at the same location on the *x*-axis; only their *y* location varies. Another experiment is plotted at another *x* location in the plane. Typically the progression of time

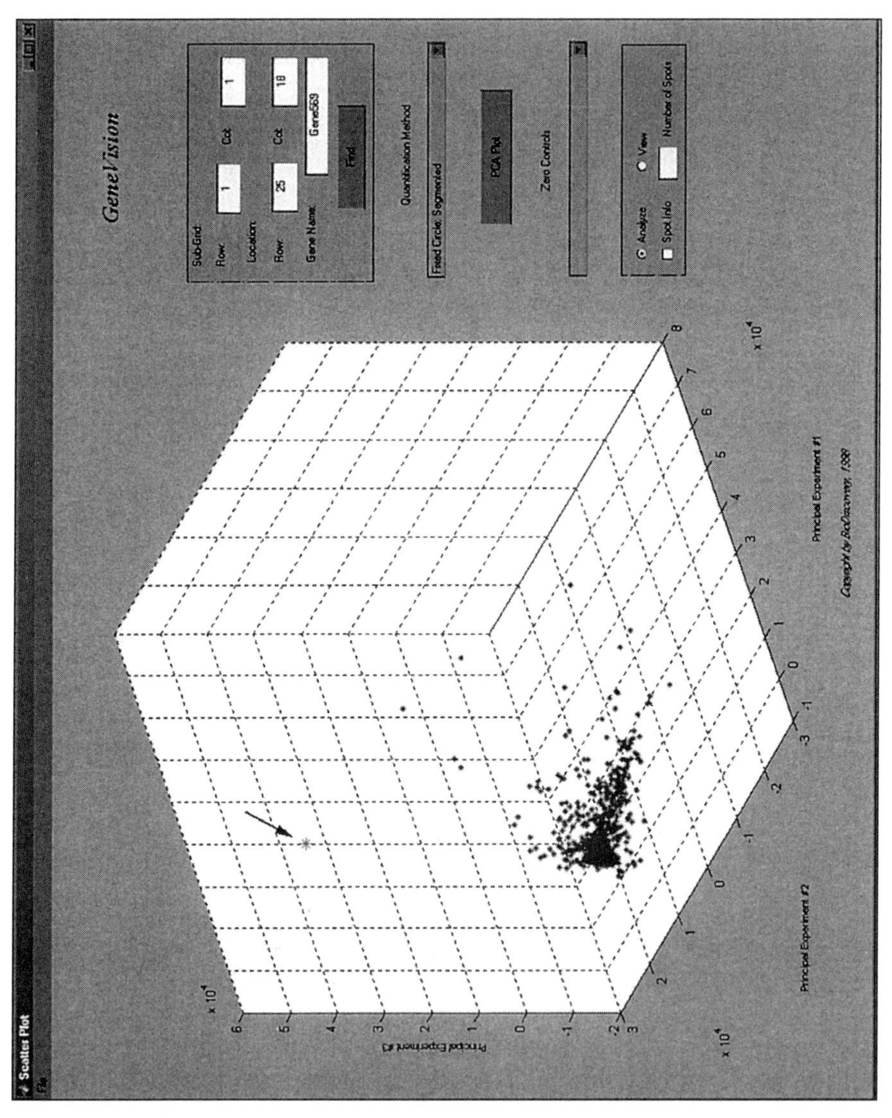

Figure 12. Principal component analysis on 600 genes across 21 experiments. Gene scores are plotted on the first three principal components. The arrow indicates a possible outlier. (**See color plate A17.**)

would be mapped into the x-axis by having higher x values for experiments done at a later time. By connecting the expression values for the same genes in the different experiments, one obtains an intuitive depiction of the progression of gene expression (Figure 13).

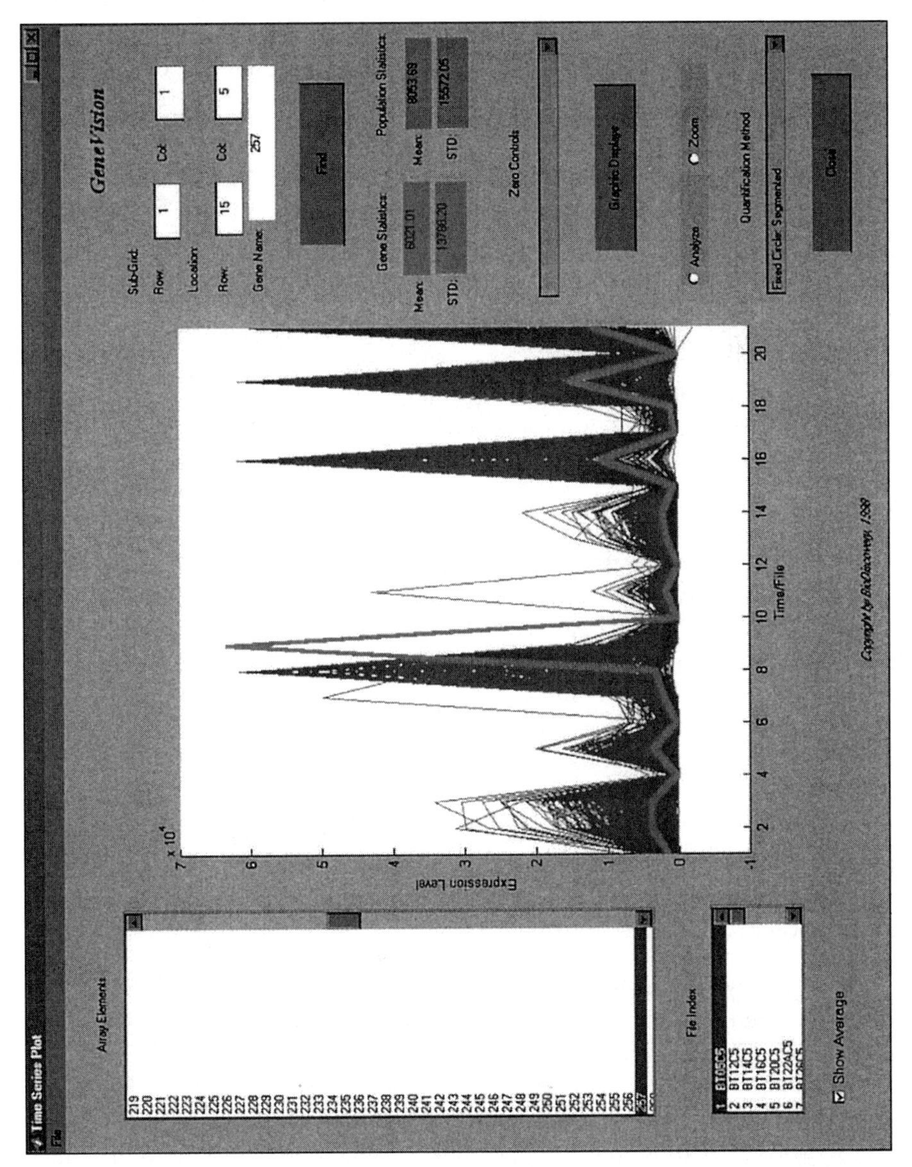

Figure 13. The parallel coordinate plot. The plot displays the expression levels of all genes across all experiments/files in the analysis. On the x-axis, the experiments or experimental files are ordered. The y-axis reveals the expression level of all genes across all experiments. **(See color plate A18.)**

Cluster Analysis

Another common microarray issue involves finding groups of genes with similar expression profiles across a number of experiments. The most commonly used multivariate technique to find these groups is cluster analysis. Cluster analysis orders the data by grouping genes with similar expression patterns close to each other. Because genes are often expressed or induced only when they are required for a given process, cluster analysis can help establish functionally related groups of genes or predict the biochemical and physiological roles of previously uncharacterized genes.

The clustering method that is most frequently used in the microarray literature is hierarchical clustering. This method attempts to group genes and/or experiments in small clusters and then group these clusters into higher level clusters and so on. As a result of this clustering or grouping process, a tree of connectivity of observations emerges that can easily be visualized as dendrograms. For gene expression data, not only the grouping of genes, but also the grouping of experiments is important. When both are considered, it becomes easy to search for patterns simultaneously in gene expression profiles across many different experimental conditions, as shown in Figure 14. For example, a group of genes behaving similarly (e.g., all up-regulated) can be seen in a particular group of experiments (Figure 14).

Every colored block in the middle panel of Figure 14 represents the expression value of a gene in an experiment. The 600 genes are plotted horizontally, and the 21 experiments are plotted vertically. The color code is located in the lower right corner. The dendrogram for the genes is located just above the color-coded expression values, with one arm connected to every gene in the study. The dendrogram for experiments on the left shows the grouping of the 21 experiments in the study.

Although hierarchical clustering is a commonly employed approach presently, other nonhierarchical (*k*-means) methods are likely to gain popularity in the future with the rapidly growing volume of data and the expanding density of microarray assays. Nonhierarchical approaches provide sufficient clustering without having to create the full distance or similarity matrix or scan the whole dataset excessively.

Data Normalization and Transformation

There are several visualization and statistical techniques (described above) that could be useful for the analysis of microarray data. However, it is important to realize that even with the most powerful statistical methods, the success of the analysis is crucially dependent on the "cleanness" and statistical properties of the data. Essentially, the two questions to ask before starting any analysis are:

1. Does the variation in the data represent true variation in expression values or is it contaminated by experimental variability?
2. Are the data "well behaved" in terms of meeting the underlying assump-

tions of the statistical analysis techniques that are being applied?

It is easy to appreciate the importance of the first point. The significance of the second one derives from the fact that most multivariate analysis techniques are based on underlying assumptions such as normality and homoscedasticity. If these assumptions are not met, at least approximately, then the entire statistical analysis could be distorted, and statistical tests might be invalid. Fortunately, a variety of statistical techniques are available to help us answer "yes" to the above

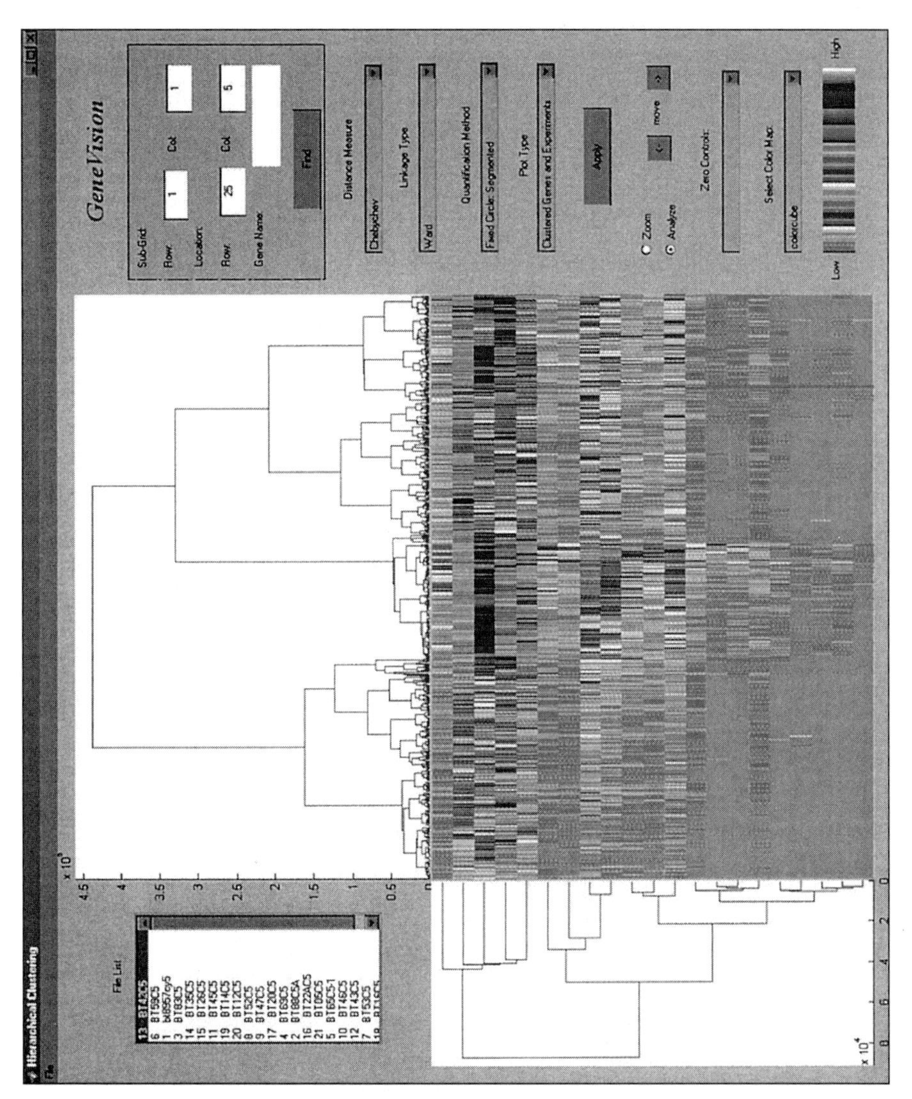

Figure 14. Color-coded gene expression values. Values are given for 600 genes (displayed horizontally) over 21 experiments (displayed vertically). Simultaneous clustering of genes and experiments is visualized by the top and left side dendrograms, respectively. **(See color plate A19.)**

questions. These are normalization (standardization) and transformation.

Normalization, and standardization, as a special form of normalization, can help us separate true variation in expression values from differences due to experimental variability. This step is necessary since it is quite possible that due to the complexity of manufacturing, hybridizing, scanning, and quantifying microarrays, variation originating from the experimental process may contaminate the data. During a typical microarray experiment, many different variables and parameters can affect the measured expression levels. Among these are slide quality, pin quality, amount of DNA spotted, accuracy of arraying device, dye characteristics, scanner quality, and quantification software characteristics, just to name a few. The various methods of normalization aim at removing or minimizing expression differences due to variability in these parameters.

As is discussed later, the various transformation methods all aim at changing the variance and distribution properties of the data in such a way so as to meet the underlying assumptions of the statistical techniques applied to it in the analysis phase. The most common requirements of statistical techniques are for the data to have homologous variance (homoscedasticity) and normal distribution (normality). Several popular ways of normalizing and transforming microarray data are discussed below.

Normalization of the Data

One of the most popular ways to control for spotted DNA quantity and surface chemistry anomalies involves the use of two-color fluorescence. For example, a Cy5 (red)-labeled probe prepared from healthy tissue could be used as control to examine expression profiles in a Cy3 (green)-labeled probe prepared from a tumor tissue. The normalized expression values for every gene would then be calculated as the ratio of experimental and control expression. This method can obviously eliminate much (but not all!) experimental variation by allowing two samples to be compared on the same chip. Even more sophisticated three-color experiments are under way in which one channel serves as a control for the amount of spotted DNA, and channels two and three allow two samples to be compared.

In addition to the local normalization method described above, global methods are also available in the form of "control" spots on the slide. With a set of control spots, it is possible to control for global variation in overall slide quality or scanning differences.

The above procedures describe some array-based measures one can use to normalize microarray data. However, even with multiple color fluorescence and control spots, undesired experimental variation can contaminate expression data. It is also possible that all or some of these physical normalization techniques are missing from the experiment, in which case it is even more important to find additional ways of normalization. Fortunately, additional statistically based normalization methods are available to add additional precision to the data.

For example, expression values for a given set of genes in one experiment may be consistently and significantly different from another experiment due to quality differences among the slides, printing, scanning, or some other factor (Figure 15). It might be very misleading to compare the expression values of the two files plotted in Figure 15 without normalization. The same data plotted after normal-

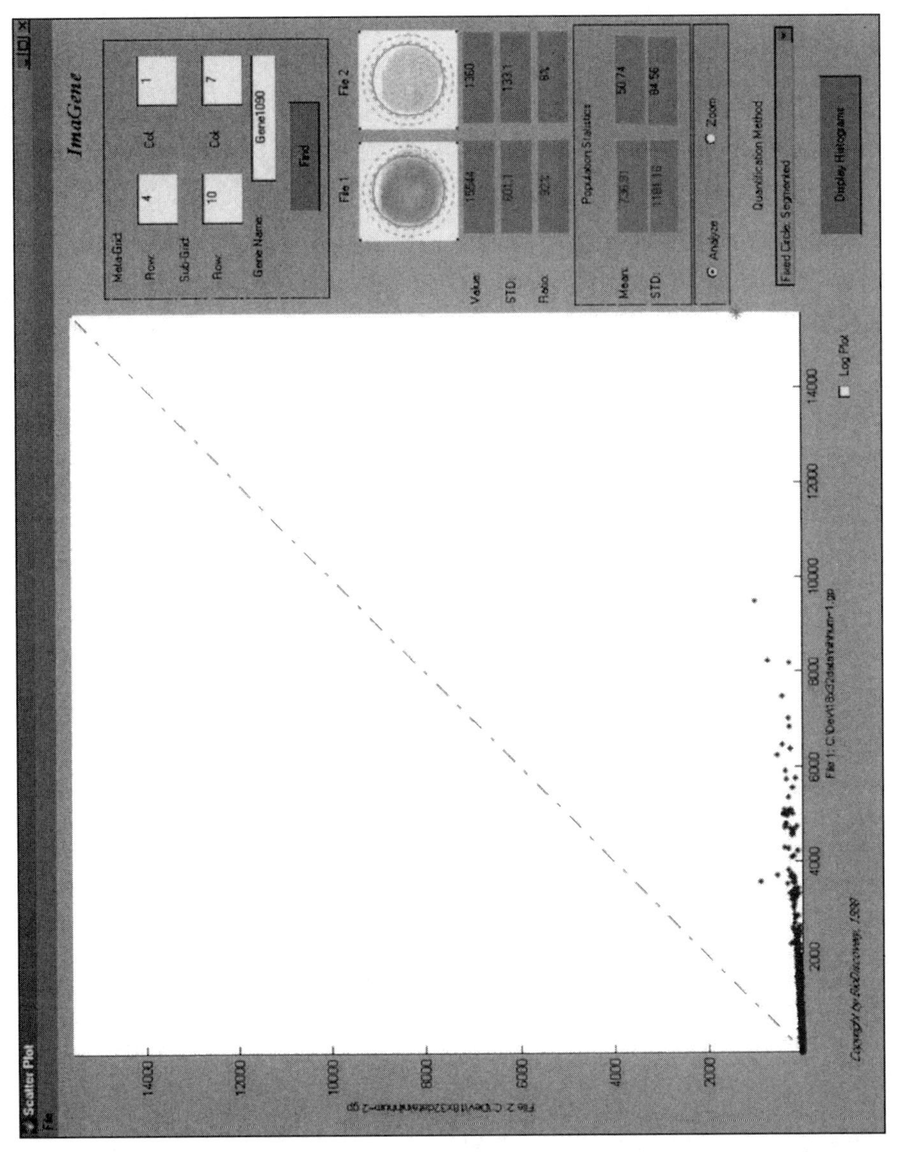

Figure 15. Comparison of fluorescent signals. Scatter plot of two experiments in which the overall fluorescent signal for one file is significantly stronger than the other. (**See color plate A20.**)

ization, however, reveal that most of the genes fall on the identity line, as would be expected for two experiments with the same set of genes (Figure 16).

Transformation of the Data

Although there are many different ways to transform data, the most frequently used procedure in the microarray literature is to take the logarithm of the

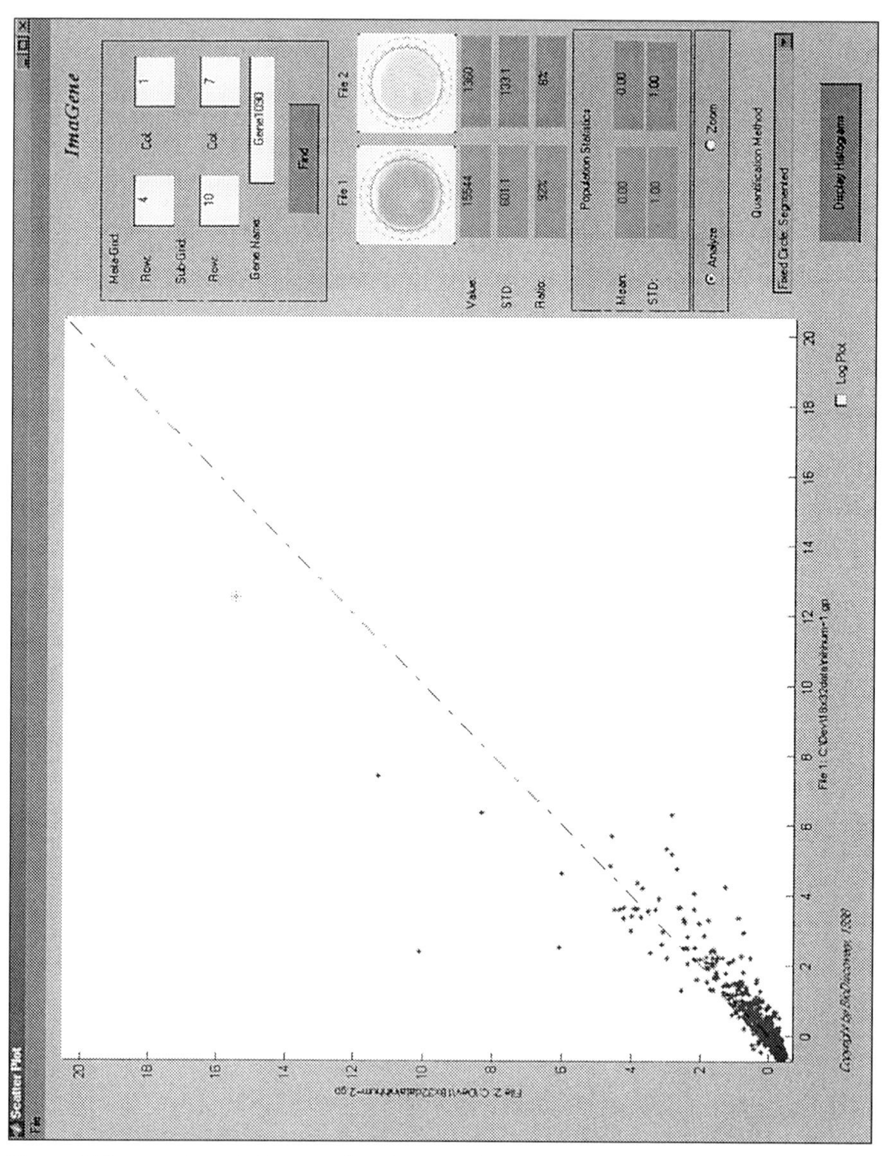

Figure 16. The same data as in Figure 15, but after normalization. (See color plate A21.)

quantified expression values (2,3). An often cited reason for applying such a transform is to equalize variability in the wildly variable raw expression scores. If the expression values were calculated as a ratio of experimental over control, then an additional effect of the log transform would be to equate up- and down-regulation by the same amount in absolute value scores ($\log_{10} 2 = 0.3$ and $\log_{10} 0.5 = -0.3$). Another important consequence of the log transform is bringing the distribution of the data closer to normal. Because the log transform has a good chance of meeting the normality and homoscedasticity assumptions, the use of a variety of parametric statistical analysis methods is also better justified.

As shown in Figure 17 (left panel), without a log transform the data distribution appears as a sharp peak with a very long tail. This distribution is far from normal, which violates the assumptions made by many standard parametric statistical analysis methods. The distribution of the log-transformed data (right panel) is clearly much closer to that of the normal distribution, which could also be verified by a simple normality test. Certainly, other types of transforms could also be explored. In fact, the choice of transformation should be dependent on which one brings the data closest to the requirements of homogeneity of variance and normal distribution (4). For most microarray expression data, the log transform typically provides the best solution.

CONCLUSIONS

Microarray technology has become a powerful tool for genetic research. From an information processing perspective, microarray technology aids the researcher in transforming and supplementing data available on genes and cells into useful information about gene expression, and ultimately, a broader biological understanding. In this chapter, many issues involved in data management and track-

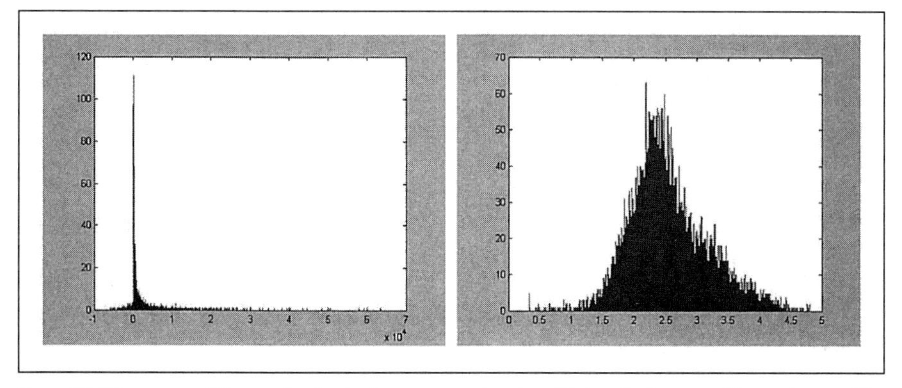

Figure17. Histograms of the previously cited gene expression data. Six hundred genes × 21 experiments. The *x*-axis indicates the expression values, and the *y*-axis shows the number of genes with a particular gene expression level. The left and right panels show the data before and after the log-transform, respectively.

ing, array production, analysis of high-density array images, and manipulation of massive data sets are described. Some methods and tools available for addressing many of these issues are also presented. Many challenges lie ahead as software tools for microarray analysis become increasingly powerful.

REFERENCES

1. **Chen, Y., E.R. Dougherty, and M.L. Bittner.** 1997. Ratio-based decisions and the quantitative analysis of cDNA microarray images. J. Biomed. Optics *2*:364-374.
2. **Eisen, M.B., P.T. Spellman, P.O. Brown, and D. Botstein.** 1998. Cluster analysis and display of genome-wide expression patterns. Proc. Natl. Acad. Sci. USA *95*:14863-14868.
3. **Hilsenbeck, S.G., W.E. Friedrichs, R. Schiff, P. O'Connell, R.K. Hansen, C.K. Osborne, and S.A.W. Fuqua.** 1999. Statistical analysis of array expression data as applied to the problem of tamoxifen resistance. J. Natl. Cancer Inst. *91*:453-459.
4. **Johnson, R.A. and D.W. Wichern.** 1998. Applied Multivariate Statistical Analysis. Prentice Hall, Upper Saddle River.

9 Microarray Tools, Kits, Reagents, and Services

Todd Martinsky[1] and Paul Haje[2]
[1] TeleChem International, Inc. and [2] arrayit.com, Sunnyvale, CA, USA

INTRODUCTION

This chapter describes some of the tools, kits, reagents and services we have developed at TeleChem and arrayit.com (http://arrayit.com), and the science that underlies these products and services. TeleChem's products for the microarray industry are sold under the ArrayIt™ brand name. ArrayIt products are mainly consumables to help scientists make microarrays of high quality and obtain the finest scientific results with their biological "chips." Our microarray products currently fall into three categories: DNA purification systems, microarray manufacturing technology, and surface and hybridization chemistry. Additional custom microarray services are provided by arrayit.com, a new internet-based microarray company offering Flex-Chips™ and eChips™, as well as DiscoverChips™, which contain segments of genes from a diverse set of organisms including yeast, *Drosophila, Caenorhabditis elegans*, mouse, rat and human.

Microarray Experimental Cycle

Microarray analysis is revolutionizing biological research (5,9,10,18–21,23, 26, 29,30,34,39). Each microarray experiment contains five separate steps: biological question, sample preparation, microarray reaction, detection, and data analysis and modeling (33). This methodological life cycle provides the foundation for the microarray industry. Current products and services from TeleChem and arrayit.com fall into four of the five phases of the life cycle: biological question, sample preparation, microarray reaction, and microarray detection phases (Figure 1). These products and services and the underlying science are described below.

Microarray Biochip Technology
Edited by Mark Schena

BIOLOGICAL QUESTION

Forming solid biological questions is very important in microarray research. A question helps focus the experiment and prevents the researcher from getting distracted by the power and "flash" of microarray technology. An example of a question might be "How are the gene expression patterns different between these two healthy individuals?" Being able to ask a good question depends on understanding both the biology and the technology.

For this reason TeleChem is developing a Microarray Resource Center (MRC) at its corporate headquarters (Sunnyvale, CA, USA). The MRC contains a growing and comprehensive collection of books, publications, patents, and product literature relevant to microarray science. The collection contains hundreds of documents and will eventually contain ten times this number.

TeleChem has also developed the world's first microarray demonstration and

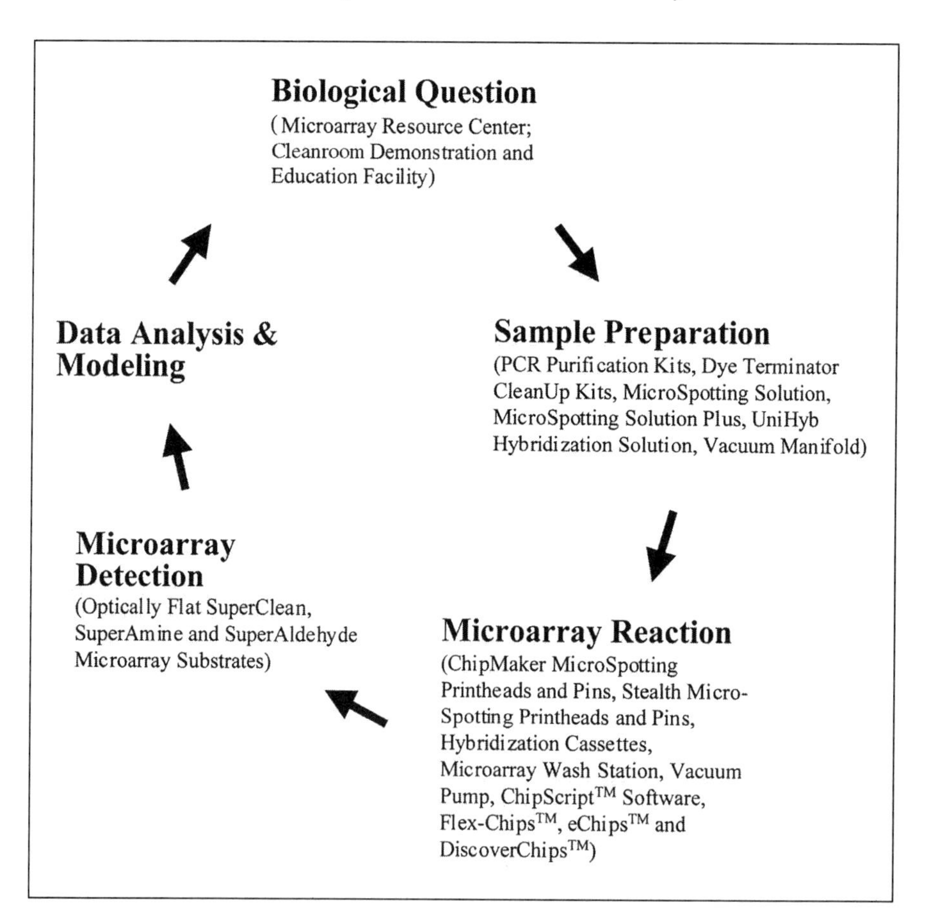

Figure 1. Microarray experimental life cycle. How TeleChem and arrayit.com products fall into the experimental life cycle. The life cycle supports products (TeleChem) and services (arrayit.com).

education facility at its corporate headquarters. The facility is a state-of-the-art cleanroom that houses all the best instruments, tool, kits, and reagents required for complete microarray experimentation. The demonstration facility was developed to help researchers, business leaders, and the media better understand microarray research. Corporate partners and companies with products currently under testing in this effort include Axon Instruments, Cartesian Technologies, Genpak Limited, Genetix, GSI Lumonics, Hewlett-Packard, Jouan, Qiagen, and Savant.

SAMPLE PREPARATION

PCR Purification Kits

The polymerase chain reaction (PCR) is a powerful tool for amplifying DNA (28). Gene segments amplified by PCR provide a common source of genetic material for microarray assays. In addition to the DNA of interest, PCRs contain salts, enzymes, small DNA fragments, and other components that can interfere with microarray experiments. The contaminants can clog pins and ink jets used to make microarrays; they can also attach to the surface of the chip and interfere with hybridization. Contaminants can also elevate nonspecific binding and background fluorescence. These components must be completely removed from PCR products prior to microarray printing. Simple precipitation methods involving ethanol precipitation have proved to be insufficient.

TeleChem has developed a series of PCR purification kits allowing purification in both 96- and 384-well formats (http://arrayit.com/pcr-100/; http://arrayit.com/PCR-384/). All the kits can use either vacuum or centrifugation for purification. The kits are based on the principle that DNA sticks to a membrane when it is mixed with special buffers. This serves as the basis for purification. The PCR products are suspended in binding buffer and pipetted into a filter plate containing a membrane. Either vacuum or centrifugation is used to pass the contaminants through the filter. A dilute wash buffer is then used to remove remaining contaminants, and residual wash buffer is removed from the membrane by centrifugation. The PCR products are then eluted with dH_20 or 0.1× Tris-EDTA and dried down in a microplate. Analysis on gels reveals the high purity of the products (Figure 2). The binding capacity of the membrane is very high, so the capacity of the kits is >100 μg. Purified products are then resuspended in 1× Micro Spotting Solution and printed to form microarrays.

Dye Terminator Clean-Up Kits

Highly accurate DNA sequence information is important in microarray research in two ways: *(i)* spotted cDNA clones need to be sequence verified once they are printed on arrays and analyzed, and *(ii)* highly accurate sequence information is required to synthesize DNA chips either in situ or by delivering synthetic DNAs. One key to obtaining a highly accurate (99.99%) sequence is to make sure the extension products are pure before the sequencing run. Contaminating salts and dyes can obscure the correct sequence reads. These contaminants are particularly disruptive to capillary electrophoresis, which uses fine capillaries that can clog with contaminated samples.

TeleChem has developed both 96- and 384-well versions of a dye terminator clean-up kit (http://arrayit.com/dye-terminator-cleanup). These kits work in a similar way to the PCR purification kits. Extension products from sequencing reactions are mixed with binding buffer and bound to membrane inside a 96- or 384-multiwell filter plate. Binding of the sequencing products to the membrane can be accomplished using either vacuum or centrifugation. Once the products are bound, wash buffer is used to remove contaminants. The highly purified products give superior sequence reads in excess of 500 nucleotides (Figure 3). The kits are recommended by PE Applied Biosystems for use with their 3700 capillary sequencer.

Micro Spotting Solution

Obtaining uniform spots on microarrays is important for several reasons. First, spot uniformity increases the precision of the data, by providing uniform signal intensities at each pixel. Second, spot uniformity means that an equal amount of arrayed material (DNA, RNA, protein, small molecule, extract, cells, etc.) is present across the entire spot, ensuring that the binding to labeled material will have

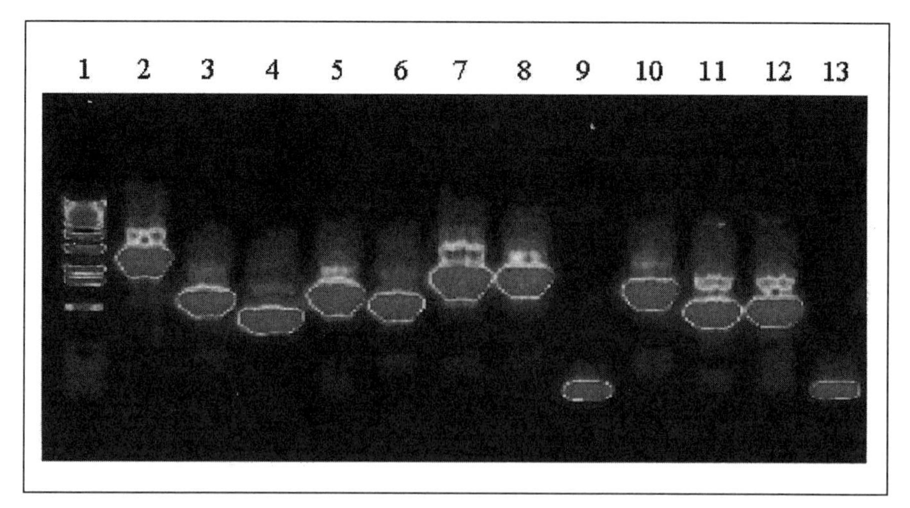

Figure 2. Purified PCR products. A TeleChem PCR Purification Kit was used to purify PCR products. One-tenth of the yield from a 100-µL PCR was analyzed by agarose gel electrophoresis. Lane 1 is a molecular weight marker, and lanes 2 to 13 contain the PCR products. **(See color plate A22.)**

the same rate across the spot. Spot or feature uniformity is difficult, particularly on "oily" substrates such as hydrophobic surfaces containing reactive aldehyde groups. The reactive aldehyde surface tends to cause the hydrophilic samples to dry unevenly, creating spots with an uneven amount of deposited material.

TeleChem's unique printing technology works like an ink stamp. Efficient

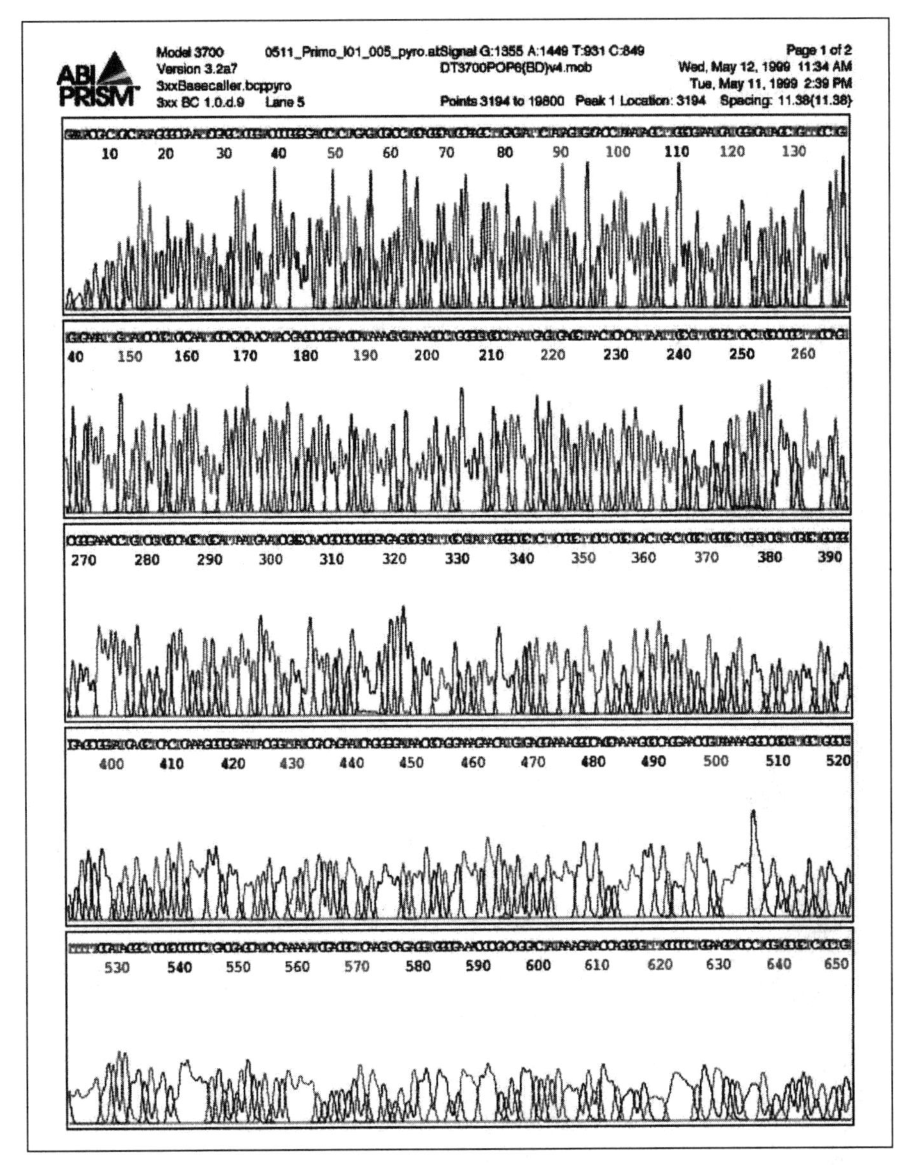

Figure 3. Sequencing purification kits. A TeleChem Dye Terminator Clean-Up Kit was used to purify sequencing products. Sequencing was done on a 3700 machine from PE Applied Biosystems. (Data courtesy of Primo Baybayan's Laboratory.) **(See color plate A23.)**

printing on solid surfaces works because the substrate pulls the sample off the end of the pins and onto the surface of the substrate. Dilute samples that are not very viscous tend to print inefficiently because the hydrophobic surface does not pull very hard on the sample. Also, the printed sample will spread out on the surface if the surface tension of the sample is low. TeleChem has developed Micro Spotting Solution, a general purpose buffer that increases the viscosity and surface tension of sample (http://arrayit.com/micro-spotting-solution/). Micro Spotting Solution provides huge increases in the quality and efficiency of printed arrays, as well as vastly increased uniformity over each spot (Figure 4). Micro Spotting Solution Plus gives similar results to Micro Spotting Solution except that background fluorescence is generally lower, and therefore users who require ultra-low background fluorescence of printed samples prefer the product.

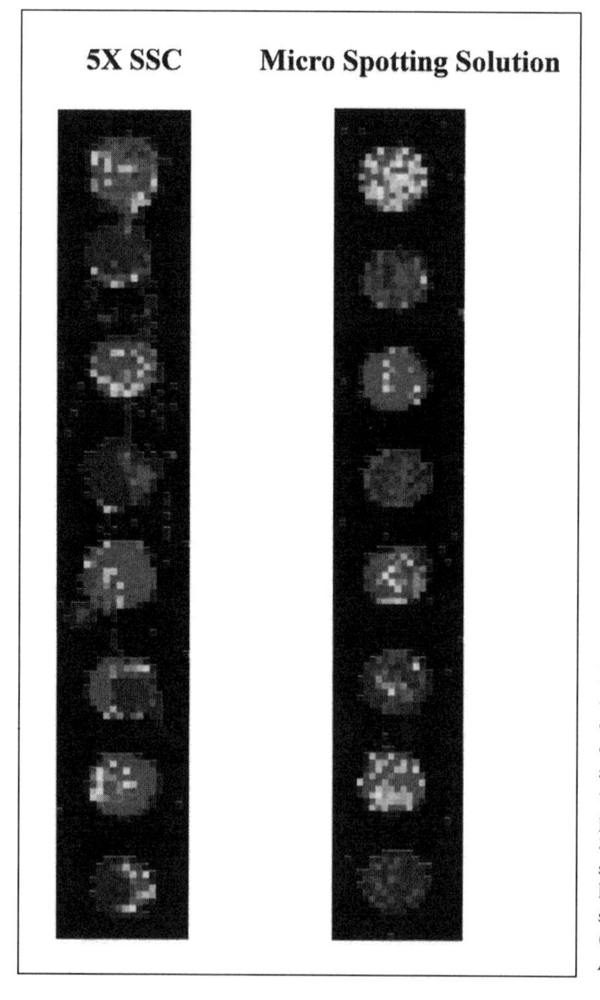

Figure 4. Micro Spotting Solution. Samples print more evenly with Micro Spotting solution, which increases the viscosity and surface tension of the sample and allows uniform drying. Samples were printed with TeleChem's Micro Spotting technology at 150-µm spacing, and hybridized with a Cy3-labeled probe; chips were then scanned with a ScanArray 3000 (GSI Lumonics). (**See color plate A24.**)

UniHyb™ Hybridization Solution

For nucleic acid microarrays, one important consideration is the fact that the G:C base pairs are stronger than the A:T base pairs, because they contain three hydrogen bonds instead of two. The hydrogen bonds between the bases act like a zipper, allowing probe sequences in the sample solution to zipper up or hybridize to targets on the surface of the chip. Nonuniform hybridization can lead to inaccuracies in signal due to differences in the base-pairing strengths of different sequences. For long targets and probes (>200 bases), differences in base pairing strengths are only a minor concern because the total composition of each hybridized region tends to have a similar number of A:T and G:C base pairs for each of the target:probe duplexes on the chip. For short sequences, particularly in the 5- to 75-bp range, however, different numbers of G:C and A:T base pairs can have a large effect on the hybridization signal and thus on how well the chips work for particular applications.

TeleChem has devised a propriety buffer called UniHyb Hybridization Solution (http://arrayit.com/hybridization-solution/) that minimizes the difference between G:C and A:T base pairs. UniHyb (uniform hybridization) solution therefore greatly improves the performance on the microarrays by making sure that signal strength is largely due to differences in the amount each probe species and not differences in the number of G:C and A:T base pairs. Figure 5 shows an experiment with UniHyb compared with a traditional buffer. It is easy to see how UniHyb is superior to traditional buffers. This buffer is useful for many applications involving nucleic acids. UniHyb can be used with many different types of microarrays. It increases signal by accelerating hybridization kinetics, increases sensitivity by reducing background fluorescence background, particularly with aldehyde-containing substrates, and reduces surface tension, providing a uniform hybridization layer. The buffering components stabilize extended reactions. UniHyb arrives premixed and sterile with no preparation required and is very affordable.

MICROARRAY REACTION

Printing Technologies

Good microarray research requires high-quality microarrays. A high-quality microarray manufacturing technology has the following characteristics:

- Uniform spot size
- Regular array pattern
- Same sample amount at each array location
- Durable and long-lasting properties
- Ability to print multiple samples, multiple times on multiple substrates

with one loading of samples

- Cost effectiveness
- Ease of implementation
- Utilization of the smallest amounts of valuable sample possible
- Flexibility to change spot sizes and sample volumes easily.

A number of different manufacturing approaches meet these needs to varying degrees. Some of the approaches that are used include photolithography (14), ink-jetting (2,32), and contact printing (12,22,29,33,35). TeleChem has developed a unique microarray printing technology to help scientists make high-quality microarrays, which is patent pending at this time (http://arrayit.com/biochip4/). The ChipMaker™ and the more recent Stealth technologies work like ink stamps (Figure 6). Direct contact between the pin and the surface is not required for printing, rather, contact between the sample and the substrate is sufficient. Because most printing surfaces and slide holders are not truly flat, most users prefer to touch the pins lightly on the substrate to ensure high printing efficiency on these uneven surfaces. Once the sample is brought in contact with the surface, the pins are moved away from the surface, and a sample is drawn off

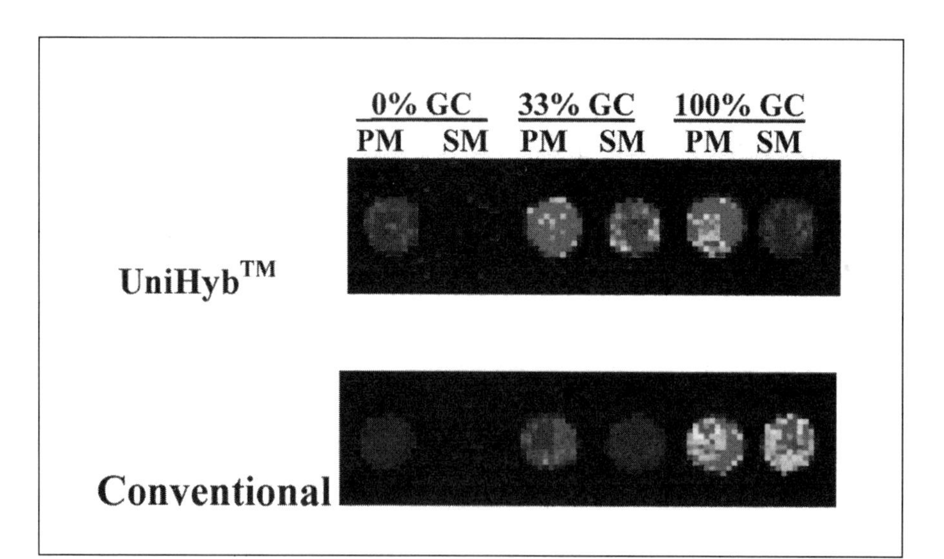

Figure 5. UniHyb Hybridization Solution. Scanned images of oligonucleotide microarrays printed with TeleChem's ChipMaker 3 microspotting device. Spacing is 150 µm center to center on aldehyde slides. The oligonucleotides are three pairs of amino-linked 15-mers with a perfect match (PM) or a single base mismatch (SM) relative to a Cy3-labeled probe. Hybridizations were performed for 4 hours at 42°C with 0.2 pmol/µL probe in Unihyb Hybridization Solution (right) or 5× SSC + 0.2% SDS (bottom). Fluorescent detection was performed using the ScanArray 3000 (GSI Lumonics). Improvements in hybridization signal and specificity are easily observed with Unihyb Hybridization Solution compared with a conventional hybridization buffer. Point mutations are easily identified in sequences spanning the entire range of AT and GC base composition. **(See color plate A25.)**

the end of the pin by surface forces between the substrate and the sample. Spot size and delivery volumes are determined by the diameter of the end of the tip. TeleChem's manufacturing processes allow precise control of the size of the end of the tip. Since each Micro Spotting Pin is manufactured using digital control and each is digitally measured during quality control checks, there are virtually no mechanical differences from pin to pin. This translates into very high printing consistency, provided that sample preparation is good, motion control parameters are set properly, and the printing substrate is homogenous.

Tapping TeleChem pins on the printing slides or substrates is not recommended for printing and greatly reduces the quality of the arrays in several serious ways: *(i)* tapping causes bulk transfer of material out of the Micro Spotting Pins, leading to large spots and spot merging; *(ii)* tapping damages the ends of the pins, leading to larger spots and poor printing quality; and *(iii)* tapping fractures in the surface coating and glass causing doughnut-shaped spots, which lack printed material in the center of the spot.

ChipMaker and Stealth (Figure 7) contain a printhead and slots for 32, 48, or 64 pins. The slots accommodate collars fitted on top of the pins. The combination of the slots designed into the printhead and the collars on the pins performs four important functions: *(i)* keeps the pins from dropping through the holes in the printhead; *(ii)* keeps the pins floating by gravity, only providing the lowest possible force on the ends of the tips; *(iii)* keeps the pins from rotating, thereby increasing spotting accuracy; and *(iv)* ensures that no special tooling is required to

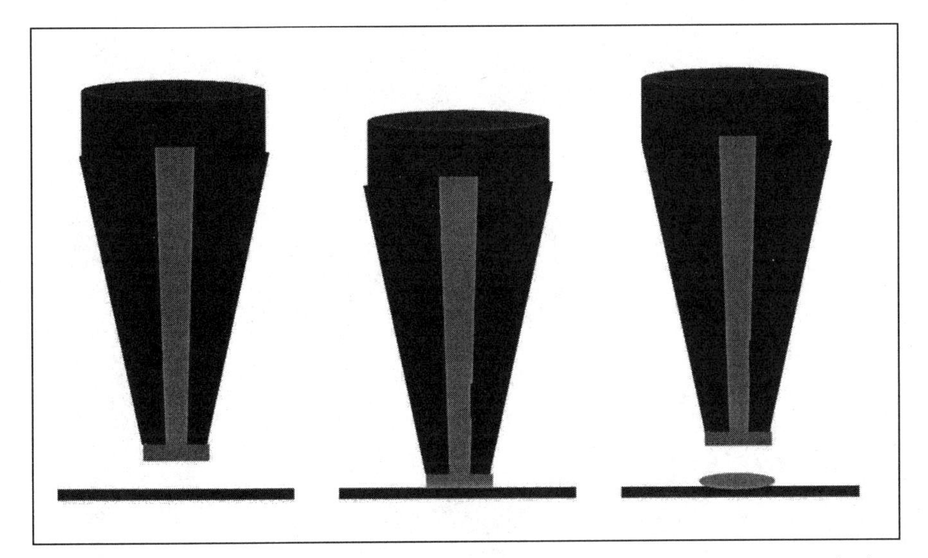

Figure 6. TeleChem's Micro Spotting Technology. TeleChem's novel surface contact printing technology works like an ink stamp. Sample is drawn up into the sample channel, and each pin loads a predefined volume from 0.25 to 0.6 μL. Liquid on the end of the tip is brought in contact with the printing substrate, and the substrate exerts attractive forces on the sample. The pin is moved away from the surface, and a drop of liquid is delivered. Liquid on the end of the tip is replenished by sample in the reservoir for the next round of printing.

change or add pins, which can be done "on the fly." Current designs allow sample uptake from either 96- or 384-well plates, with prototypes capable of handling 1536-well plates. The current printheads can print up to 64 different samples at a time onto as many as 1000 different slides or substrates. The pins are available in 18 different tip diameters covering a range of delivered sample volume from 0.5 to 2.5 nL and spot diameters from 75 to 360 μm. Printing speeds allow 100 microarrays of 10 000 spots to be made in about 6 hours. The technology is capable of putting 129 600 spots on a single 25 × 76-mm printing substrate. Spot sizes keep getting smaller, are pushing the limits of the current scanners, and have surpassed many charge-coupled device (CCD)-based detection systems. A sample microarray printed with TeleChem's Stealth technology is shown in Figure 8.

TeleChem believes that its printing technology is the technology of choice for microarray manufacture. Consistent with this, there are more than 4500 Micro Spotting Pins on robots made by more than 10 manufacturers in the field at the present time. There are many advantages of TeleChem's technology over other microarray technologies. When all the microarray technologies are compared, TeleChem's technology is affordable, high-quality, and easy to implement.

Figure 7. Stealth Micro Spotting Technology. The Stealth-32 printhead and Stealth-3 Micro Spotting Pins provide high-quality microarrays.

Hybridization Cassette

All microarray hybridization reactions require critical parameters including:

- Constant temperature
- Rapid temperature equilibration
- 100% humidity
- Pristine environment
- Support of all 25 × 76-mm (1 × 3-in) formats
- Wide range of reaction temperatures
- Low volumes of hybridization buffers and DNA.

Our microarray Hybridization Cassettes (http://arrayit.com/hybridization/) fulfill these criteria (Figure 9). Each cassette holds one microarray and is made of a chemically resistant, hard-anodized aluminum base and polycarbonate top. Aluminum was selected for the base since it warms up and cools down quickly for accurate control over hybridization temperatures and times. Other manufacturers with lesser success have used other materials such as glass. Glass and other poorly chosen materials can actually insulate the substrate to changes in temperature. TeleChem's cassette design allows easy stacking, which ensures uniform reaction temperatures among multiple cassettes submerged in a water bath. Slots in the cassette base allow easy chip removal, and rugged design ensures durability.

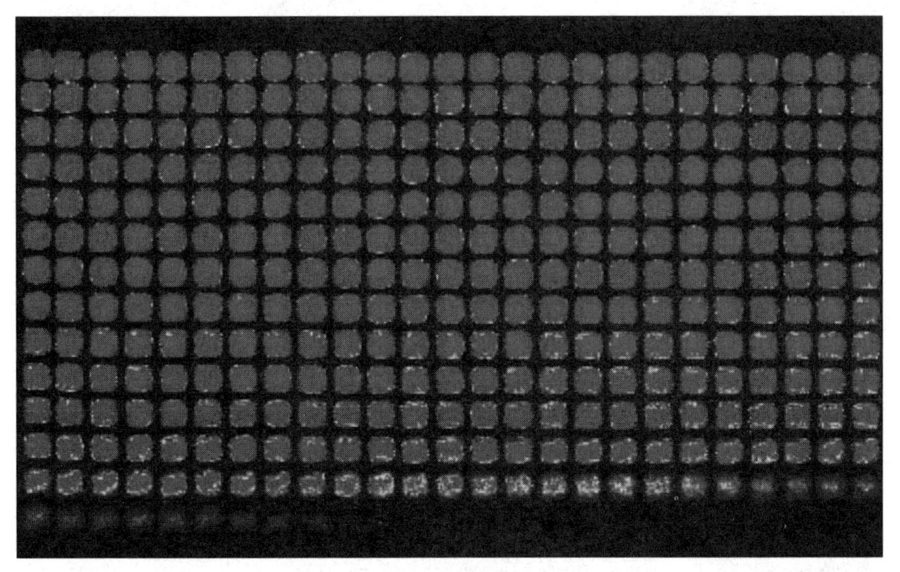

Figure 8. Printed microarray. Stealth 3 pins on a Cartesian PixSys 5500 robot printing at 140-μm spacing. A single loading with a Stealth 3 pin is sufficient to print approximately 250 spots. (**See color plate A26.**)

Inert construction materials provide a pristine reaction environment, and the clear top allows visualization of the microarray during use. They are easy to assemble and disassemble by hand, with no tools required.

Microarray Wash Station

After the microarray sample is added and reacted to the microarray, excess material that is unbound needs to be washed away. This can be accomplished using TeleChem's Microarray Wash Station (Figure 10). The slotted configuration of the Wash Station prevents damage to the substrate surface when it is inserted into the Wash Station. The Wash Station holds six slides or substrates and fits nicely in a 600-mL beaker. A spin bar is used to move wash buffers vigorously around the arrays for very complete and consistent washing, while limiting the handling of the arrays.

Vacuum Pump

Most microarray printing robots use a vacuum to hold the substrates down on the printing surface. A vacuum source is also commonly used in the dry station to dry pins and ink jets between printed samples. TeleChem provides an afford-

Figure 9. Hybridization cassette.

able, durable, and quiet-running vacuum pump that provides high air flow (Figure 11). The Microarray Vacuum Pump is oil free and therefore ideal for cleanroom manufacturing (http://arrayit.com/vacuum_pump/).

ChipScript Software

Microarray manufacturing requires user-friendly software for the end-user. Also, the software needs to contain the right parameters in terms of dwell times and the like for high-quality printing. TeleChem has created a series of software programs called ChipScript (Figure 12) that work with the Aspirate/Dispense software for the PixSys 5500 gridding robot (Cartesian Technologies). It is fully customized to the PixSys 5500 and to TeleChem's ChipMaker Micro Spotting Technology. There are 24 individual routines that allow the user to select any number of pins (4, 8, 16, or 32), any number of 384-well plates (1–75), any number of slides (10, 20, 30, 40, or 50), and single, double, and triple spotting capabilities. Spacing of 180 μm with ChipMaker 2 pins has subgrids of 24 × 24, and spacing of 140 μm with ChipMaker 3 pins has subgrids of 30 × 30. A total of 18 432 features (24 × 24 × 32) or 28 800 features (30 × 30 × 32) are printed in a microarray area of 1.8 × 3.6 cm. The software supports TeleChem's Bubble Pin technology, all 25- × 76- × 1-mm substrates, and all fluorescent detection devices

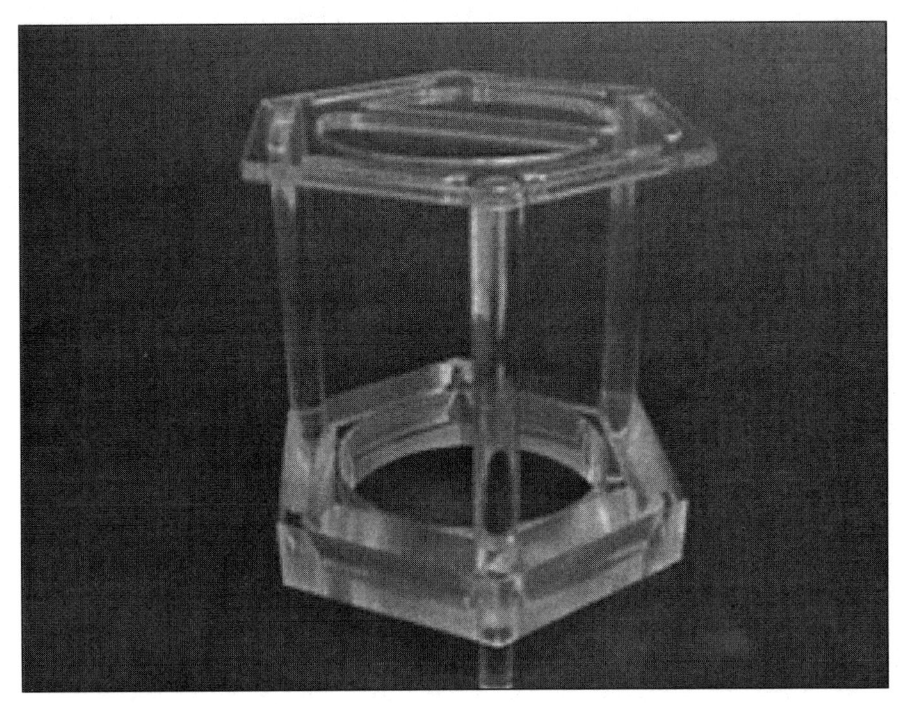

Figure 10. Microarray Wash Station.

that accept the microscope slide format. Intuitive coding architecture also allows for easy user modification (http://arrayit.com/chipscript/).

Flex-Chips, eChips, and DiscoverChips

arrayit.com provides a series of custom microarray services (http://arrayit.com/services). These services provide unprecedented access to microarray technology at an affordable price. The services unite the power of microarray technology with the universality of the Internet, which makes microarrays efficient and easy to make and use. All microarrays are manufactured in a state-of-the-art cleanroom facility (Figure 13), with air filtered down to 0.1 μm. This manufacturing environment provides microarrays of a quality that is similar to that of chips manufactured in the semiconductor industry.

Custom Microarray Services of arrayit.com are of three types: Flex-Chips, eChips, and DiscoverChips (Figure 14). With the Flex-Chips service, the customer provides the samples for up to 25 384-well microplates that can contain a wide range of materials including nucleic acids, proteins, small molecules, extracts, cells, and so forth. Because arrayit.com does not know the identity of the samples, the customer is guaranteed confidential results. This can provide an advantage in both academic and commercial research. With the eChips service, the customer provides the sequences of up to 384 different oligomers. The oligomers can be any synthetic polymer that is available commercially including DNA,

Figure 11. Microarray Vacuum Pump.

RNA, peptide, and the like. The customer then receives printed chips with the synthetic sequences on them. DiscoverChips are the third service area and contain 384 select genes from human, mouse, fly, worm, plant, and yeast. The genes represent a wide spectrum of interesting sequences from widely studied genes. DiscoverChips have a spectrum of uses including mapping, resequencing, evolutionary analysis, expression monitoring, and sequencing by hybridization. For all the services, arrayit.com does not endorse the misuse of its products. There are a number of issued and pending patents in the microarray field (6,11,15–17,27,36,37), and users should make sure they are licensed properly for any applicable patents.

A number of aspects of the arrayit.com services are appealing: affordable pricing, the use of pristine, optically flat glass printing substrates (see below), proprietary and covalent coupling chemistries that are stable to boiling, 25 × 76 × 0.96-mm substrates that fit all commercial microarray detection systems, a substrate corner chamfer that allows easy orientation, bar coding that allows automated chip identification, a wide range of sequences and applications, triple spotting for reliable data, substrate fiducials that mark array regions, and low reaction volumes (2–30 μL) that speed up assays. Custom Microarray Services from arrayit.com use TeleChem's advanced Stealth printing technology for the highest microarray quality. Total confidentiality of all results, the freedom to develop new assays, and publish, discover and patent results, electronic design and ordering, and reasonable turnaround times are added advantages.

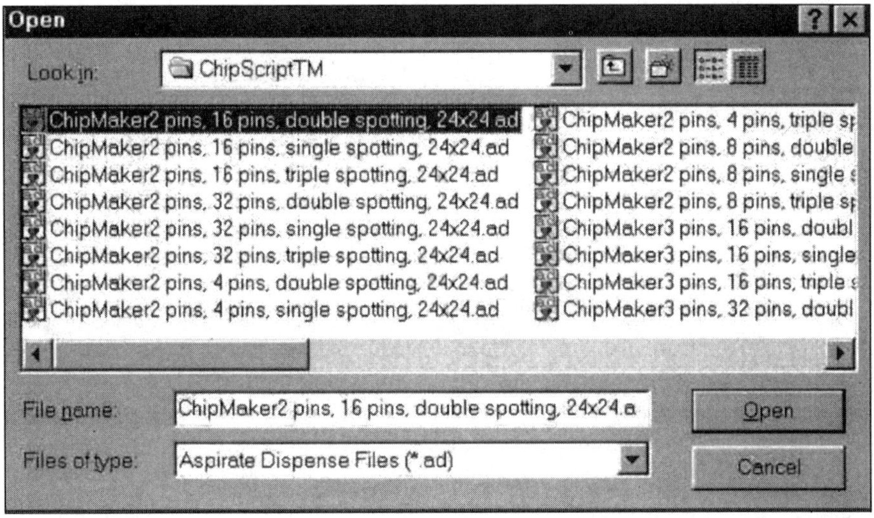

Figure 12. ChipScript software. A software package for the Cartesian PixSys 5500 gridding robot that runs "on top" of Aspirate/Dispense.

DETECTION

Fluorescence detection is a key step in the microarray life cycle. Currently two types of detection instruments are available. The first type uses some form of scanning technology and laser excitation with photomultiplier tube detectors (34). The second type uses imaging with a CCD and a continuous light source for excitation. In all these technologies, an optically flat surface that has low

Figure 13. arrayit.com cleanroom facility.

background noise provides the ideal substrate for fluorescence detection. Optical flatness takes the guesswork out of imaging by ensuring that the distance between the detector optics and the molecules on the surface do not vary because of unevenness in the surface.

TeleChem has developed a product line called Super Microarray Substrates (http://arrayit.com/microarray-substrates/) to address the need for optically flat printing with robust attachment surfaces (Figure 15) and low background. Background noise determines the detectivity of the microarray assay (34). Some other unique features of these products include manufacturing in a class 1000 cleanroom, organo-amine and organo-aldehyde reactive groups, 5×10^{12} reactive groups/cm^2, compatible with contact printing and ink-jetting and with all commercial microarray detection systems, proprietary glass cut to precision dimensions ($25 \times 76 \times 0.96$ mm), chamfer at upper right corner allowing easy substrate orientation, 60/40 scratch-dig surface quality, flat to <0.0005 in over the 76-mm span, refractive index of 1.52 (400–700 nm), transmission of >91% (380–700 nm), and thermal strain point of 490°C. These substrates allow attachment of nucleic acids, proteins, small molecules, extracts, and cells at room temperature and neutral pH. Attached molecules are stable to boiling, and substrates have a 1-year shelf life.

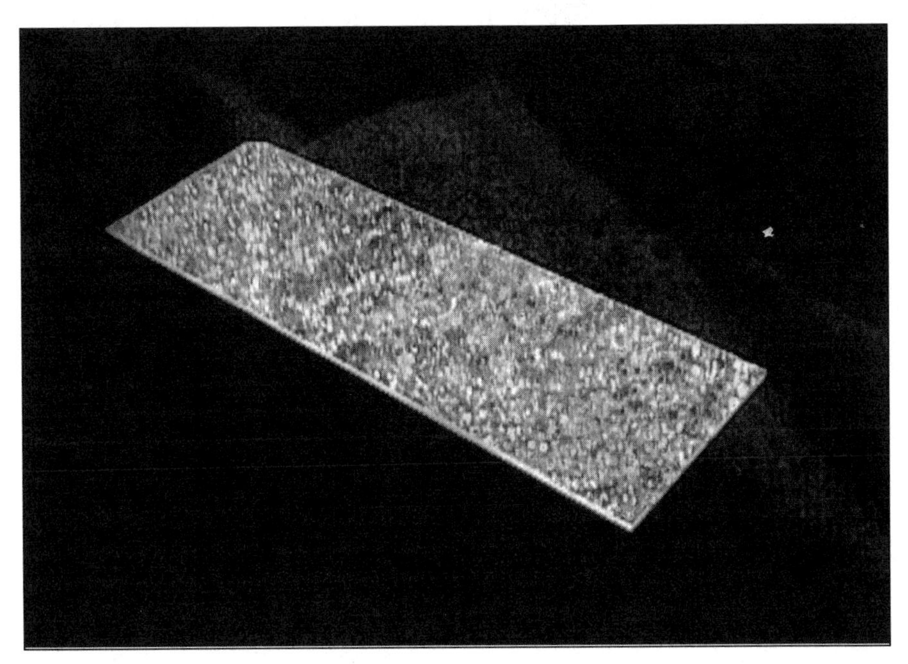

Figure 14. Custom microarray services. Artist's rendition of a custom microarray. **(See color plate A26.)**

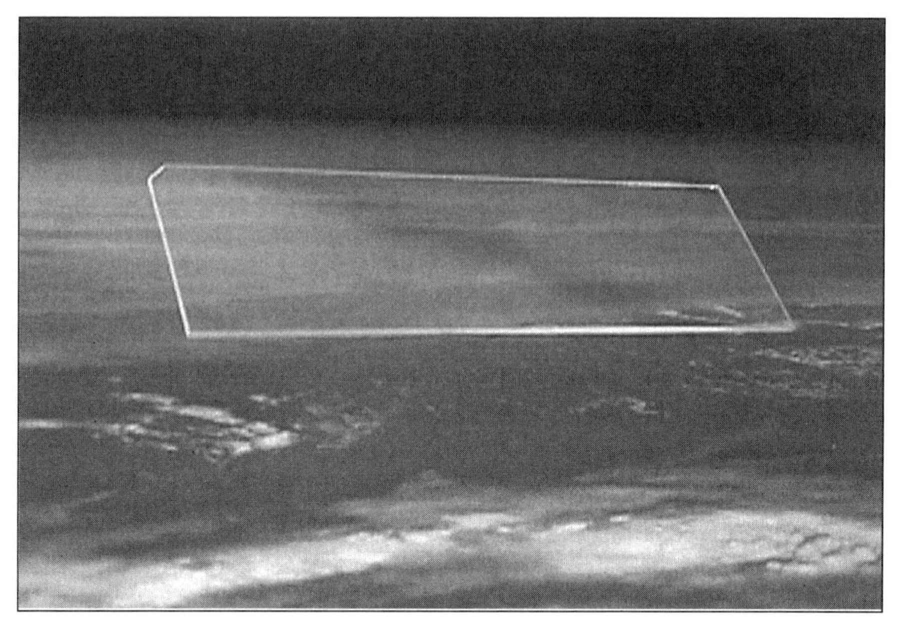

Figure 15. Super Microarray Substrates. SuperClean Microarray Substrate with corner chamfer.

CONCLUSIONS

Good microarray science requires powerful tools and services that are affordable. TeleChem and arrayit.com provide tools, kits, reagents, and services to scientists worldwide at affordable prices. Using microarrays and internet resources, scientists will be able to unravel the functional content of genomes from organisms as diverse as bacteria and humans (1,3,4,7,8,13,24,25,31,38). Speeding up our understanding of the molecular basis of life will usher in a new era in medicine and agriculture.

REFERENCES

1. **Bassett, D.E., M.B. Eisen, and M.S. Boguski.** 1999. Gene expression informatics—it's all in your mine. Nat. Genet. Suppl. *21*:51-55.
2. **Blanchard, A.P., R.J. Kaiser, and L.E. Hood.** 1996. High-density oligonucleotide arrays. Biosen. Bioelectron. *11*:687-690.
3. **Bowtell, D.D.L. 1999.** Options available—from start to finish—for obtaining expression data by microarray. Nat. Genet. Suppl. *21*:25-32.
4. **Brown, P.O. and D. Botstein.** 1999. Exploring the new world of the genome with DNA microarrays. Nat. Genet. Suppl. *21*:33-37.
5. **Chee, M., R. Yang, E. Hubbell, A. Berno, X.C. Huang, D. Stern, J. Winkler, D.J. Lockhart, M.S. Morris, and S.P.A Fodor.** 1996. Accessing genetic information with high-density DNA arrays. Science *274*:610-614.
6. **Chee, M., M.T. Cronin, S.P.A. Fodor, X.X. Huang, E.A. Hubbell, R.J. Lipshutz, P.E. Lobban, M.S.**

Morris, S. MacDonald, and E.L. Sheldon. 1998. Arrays of nucleic acid probes on biological chips. U.S. Patent 5,837,832.
7. **Cheung, V.G., M. Morley, F. Aguilar, A. Massimi, R. Kucherlapati, and G. Childs.** 1999. Making and reading microarrays. Nat. Genet. Suppl. *21*:15-19.
8. **Debouck, C. and P.N. Goodfellow.** 1999. DNA microarrays in drug discovery and development. Nat. Genet. Suppl. *21*:48-50.
9. **DeRisi, J.L., L. Penland, P.O. Brown, M.L. Bittner, P.S. Meltzer, M. Ray, Y. Chen, Y.A. Su, and J.M. Trent.** 1996. Use of a cDNA microarray to analyze gene expression patterns in human cancer. Nat. Genet. *14*:457-460.
10. **DeRisi, J.L, V.R. Iyer, and P.O. Brown.** 1997. Exploring the metabolic and genetic control of gene expression on a genomic scale. Science *278*:680-686.
11. **Drmanac, R.T. and R.B. Crkvenjakov.** 1993. Method of sequencing of genomes by hybridization of oligonucleotide probes. U.S. Patent 5,202,231.
12. **Drmanac, R.T., and R.B. Crkvenjakov.** 1996. Method of sequencing by hybridization of oligonucleotide probes. U.S. Patent 5,525,464.
13. **Duggan, D.J., M. Bittner, Y. Chen, P. Meltzer, and J.M. Trent.** 1999. Expression profiling using cDNA microarrays. Nat. Genet. Suppl. *21*:10-14.
14. **Fodor, S.A., J.L. Read, M.C. Pirrung, L. Stryer, A. Tsai Lu, and D. Solas.** 1991. Light-directed, spatially addressable parallel chemical synthesis. Science *251*:767-773.
15. **Fodor, S.P.A., L. Stryer, J.L. Read, and M.C. Pirrung.** 1995. Array of oligonucleotides on a solid substrate. U.S. Patent 5,445,934.
16. **Fodor, S.P.A., D.W. Solas, and W.J. Dower.** 1998. Method of detecting nucleic acids. U.S. Patent 5,800,992.
17. **Fodor, S.P.A., D.W. Solas, and W.J. Dower.** 1999. Methods for nucleic acid analysis. U.S. Patent 5,871,928.
18. **Hacia, J.G., L.C. Brody, M.S. Chee, S.P.A. Fodor, and F.S. Collins.** 1996. Detection of heterozygous mutations in *BRCA1* using high density oligonucleotide arrays and two-colour fluorescence analysis. Nat. Genet. *14*:441-447.
19. **Heller, R.A., M. Schena, A. Chai, D. Shalon, T. Bedilion, J. Gilmore, D. E. Woolley, and R.W. Davis.** 1997. Discovery and analysis of inflammatory disease-related genes using cDNA microarrays. Proc. Natl. Acad. Sci. USA *94*:2150-2155.
20. **Iyer, V.R., M.B. Eisen, D.T. Ross, G. Schuler, T. Moore, J.C.F. Lee, J.M. Trent, L.M. Staudt, J. Hudson Jr., M.S. Boguski, et al.** 1999. The transcriptional program in the response of human fibrolasts to serum. Science *283*:83-87.
21. **Kozal, M.J., N. Shah, N. Shen, R. Yang, R. Fucini, T.C. Merigan, D. D. Richman, D. Morris, E. Hubbell, M. Chee, and T.R. Gingeras.** 1996. Extensive polymorphisms observed in HIV-1 clade B protease gene using high-density oligonucleotide arrays. Nat. Med. *2*:793-799.
22. **Khrapko, K.R., A.A. Khorlin, I.B. Ivanov, B.K. Chernov, Iu.P. Lysov, S.K. Vasilenko, V.L. Florent'ev, and A.D. Mirzabekov.** 1991. Hybridization of DNA with oligonucleotides immobilized in gel: a convenient method for detecting single base substitutions. Mol. Biol. *25*:581-591.
23. **Lashkari, D.A., J.L. DeRisi, J.H. McCusker, A.F. Namath, C. Gentile, S.Y. Hwang, P.O. Brown, and R.W. Davis.** 1997. Yeast microarrays for genome wide parallel genetic and gene expression analysis. Proc. Natl. Acad. Sci. USA *94*:13057-13062.
24. **Lemieux, B., A. Aharoni, and M. Schena.** 1998. Overview of DNA chip technology. Mol. Breeding *4*:277-289.
25. **Lipshutz, R.J., S.P.A. Fodor, T.R. Gingeras, and D.J. Lockhart.** 1999. High density synthetic oligonucleotide arrays. Nat. Genet. Suppl. *21*:20-24.
26. **Lockhart, D.J., H. Dong, M.C. Byrne, M.T. Follettie, M.V. Gallo, M.S. Chee, M. Mittman, C. Wang, M. Kobayashi, H. Horton, and E.L. Brown.** 1996. Expression monitoring by hybridization to high-density oligonucleotide arrays. Nat. Biotechnol. *14*:1675-1680.
27. **Matson, R.S. , P.J. Coassin, J.B. Rampal, and E.M. Southern.** 1995. Method and apparatus for creating biopolymer arrays on a solid support surface. U.S. Patent 5,429,807.
28. **Mullis, K.B.** 1987. Process for amplifying nucleic acid sequences. U.S. Patent 4,683,202.
29. **Schena, M., D. Shalon, R.W. Davis, and P.O. Brown.** 1995. Quantitative monitoring of gene expression patterns with a complementary DNA microarray. Science *270*:467-470.
30. **Schena, M., D. Shalon, R. Heller, A. Chai, P.O. Brown, and R.W. Davis.** 1996. Parallel human

genome analysis:microarray-based expression monitoring of 1000 genes. Proc. Natl. Acad. Sci USA *93*:10614-10619.
31. **Schena, M.** 1996. Genome analysis with gene expression microarrays. Bioessays *18*:427-431.
32. **Schena, M., R.A. Heller, T.P. Theriault, K. Konrad, E. Lachenmeier, and R.W. Davis.** 1998. Microarrays: Biotech's discovery platform for functional genomics. Trends Biotechnol. *16*:301-306.
33. **Schena, M. and R.W. Davis.** 1998. Parallel Analysis with Biological Chips: PCR Methods Manual. Academic Press, San Diego, p. 445-456.
34. **Schermer, M.J.** 1999. Confocal Scanning Microscopy in Microarray Detection in DNA Microarrays: A Practical Approach. Oxford University Press, Oxford .
35. **Shalon, D., S.J. Smith, and P.O. Brown.** 1996. A DNA microarray system for analyzing complex DNA samples using two-color fluorescent probe hybridization. Genome Res. *6*:639-645.
36. **Southern, E.M. and U. Maskos.** 1995. Support-bound oligonucleotides. U.S. Patent 5,436,327.
37. **Southern, E.** 1997. Apparatus and method for analyzing polynucleotide sequences and method of generating oligonucleotide arrays. U.S. Patent 5,700,637.
38. **Southern, E., K. Mir, and M. Shchepinov.** 1999. Molecular interactions on microarrays. Nat. Genet. Suppl. *21*:5-9.
39. **Wodicka, L., H. Dong, M. Mittman, M-H. Ho, and D. J. Lockhart.** 1997. Genome-wide expression monitoring in *Saccharomyces cerevisiae.* Nat. Biotechnol. *15*:1359-1367.

10 MICROMAX™: A Highly Sensitive System for Differential Gene Expression on Microarrays

Karl Adler[1], Jeffrey Broadbent[1], Russell Garlick[1], Richard Joseph[1], Anis Khimani[1], Alvydas Mikulskis[1], Peter Rapiejko[2], and Jeffrey Killian[1]
[1]NEN Life Science Products, Inc., Boston, MA; [2]AlphaGene, Inc., Woburn, MA, USA

INTRODUCTION

Microarray technology for studying gene networks among various nucleic acid samples is gaining widespread acceptance. Naturally, this impending sense of permanence has spawned great interest in accomplishing more with the technique. Microarray surface chemistry, quality of arrayed sequences, assay chemistry, specialized detection instruments, and bioinformatics are among the many areas in which tremendous improvement efforts are under way. MICROMAX™ cDNA Microarray Products, from NEN Life Science Products (Boston, MA, USA), feature complete cDNA microarray and assay systems that have been optimized to deliver high-sensitivity differential gene expression results. Perhaps the the most appealing aspect of the MICROMAX technology is the novel integrated chemistry, which culminates in massive signal amplification and greatly enhanced genetic detection.

Much of the glass-based cDNA microarray technology to date has been based on using fluorescent labeled cDNA, generated from differentially expressed messenger RNA (mRNA) pools. Specifically, most current protocols include a cDNA synthesis step that incorporates cyanine 3- and cyanine 5-labeled dNTP analogs, such that the two mRNA samples are labeled with one of these analogs in the presence of nonlabeled nucleotides and reverse transcriptase. Sensitive detection of low-abundance mRNA molecules and/or the ability to use very small amounts of nucleic acid source material (e.g., extracted from tissue samples) are ultimately limited with direct labeling strategies.

Microarray Biochip Technology
Edited by Mark Schena

The MICROMAX microarray strategy differs from direct labeling in that cyanine fluorescence is introduced into the procedure *after* hybridization, by way of an enzymatic reaction called Tyramide Signal Amplification or TSA™ (Figure 1). Using this procedure, TSA has been experimentally demonstrated to increase sensitivity by 10- to 50-fold over direct cDNA labeling cDNA methods. This can be exploited to allow gene expression monitoring of rare transcripts or samples prepared from small numbers of cells.

MICROMAX microarrays are manufactured by AlphaGene, Inc. (Woburn, MA, USA), for exclusive distribution by NEN Life Science Products, Inc. Using its patented Full Length Expressed Gene FLEX™ technology (2), AlphaGene generates double-stranded cDNA clone libraries that predominate in full-length cDNAs, representing complete expressed sequences. After the cDNAs are isolated from the library, the inserts are characterized by sequencing, amplified by the polymerase chain reaction (see Notes), and printed onto the microarray. The microarray is processed further in order to attach and denature the cDNAs, as well as minimize autofluorescence and other background effects that could interfere with data interpretation.

MICROMAX EXPERIMENTAL PROCEDURE

1. A researcher prepares total RNA from cell culture or tissue samples. It is imper-

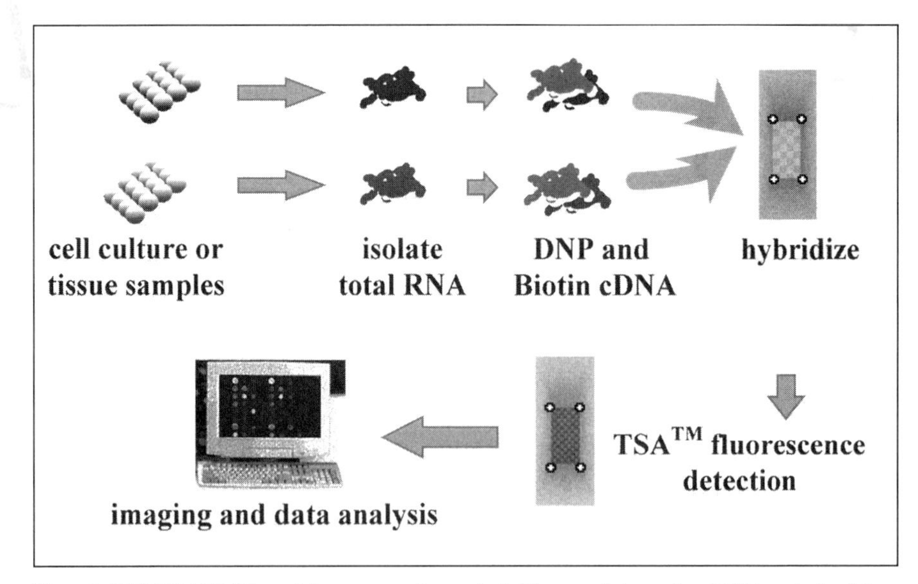

Figure 1. MICROMAX differential gene expression analysis. Two populations of total RNA are isolated from cells or tissues and labeled by incorporating dinitrophenol (DNP) or biotin, respectively, into cDNA. The cDNA is hybridized to a microarray and stained with Tyramide Signal Amplification (TSA) technology. Cyanine 3 and cyanine 5 signals are acquired for the two samples, and the data are analyzed in software. (**See color plate A27.**)

ative that RNA be analytically pure as determined by both spectrophotometric measurements at 260 and 280 nm and agarose gel analysis. A high-quality control RNA should be used for the gel analysis, because 260:280 measurements can be misleading: intact and degraded RNA give similar readings.

2. Total RNA is used as the substrate in a cDNA synthesis reaction whereby anchored oligo-dTTP primers (4) and a specialized nucleotide mix are enzymatically incorporated to generate cDNA by reverse transcriptase. Included in the nucleotide mix are either dinitrophenyl (DNP) or biotin-dCTP analogs. Incorporation of these analogs facilitates cDNA detection, using the appropriate labeling scheme, in latter procedural steps. The anchored oligo-dTTP primers are used in the mix so that primer attachment and subsequent 5′ to 3′ elongation is limited to the coding sequence of the genes comprising the mRNA population of total RNA (Figure 2). Theoretically, this method should enhance the concentration of specific gene representation relative to the DNP or biotin cDNA pools.
3. In addition to the total RNA samples used in the procedure, equivalent aliquots of a control RNA mixture are spiked into each reaction to function as controls for the entire microarray experimental process. This control RNA mixture represents three plant genes, which should be reverse transcribed with the same efficiency as the total RNA under investigation. Because there are complementary sequences for these genes on the microarray, their respective fluorescence signals can be equalized by adjusting the laser power of the detection instrument used for analysis.
4. Once cDNA synthesis is completed, reactions can be purified by cartridge (e.g., molecular weight sieve media or membranes) or simple ethanol precipitation. It should be noted that, with ethanol precipitation, the recovered cDNA pellet is often large and sticky, possibly due to coprecipitation of

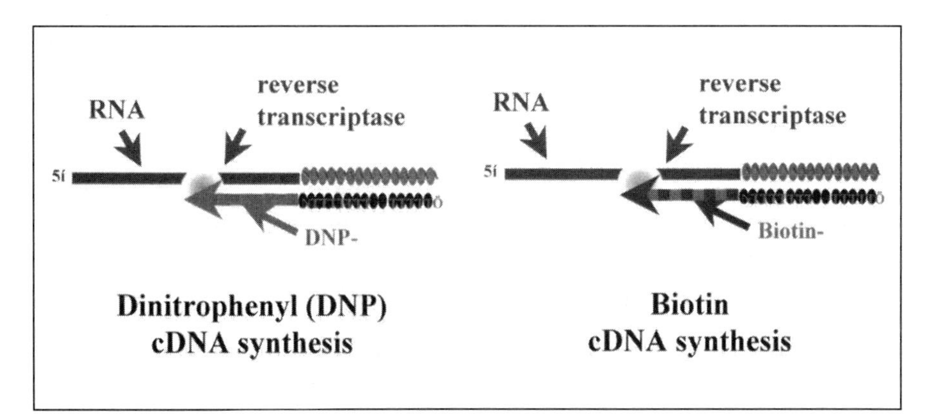

Figure 2. MICROMAX cDNA synthesis. Two samples of total RNA are primed, and reverse transcriptase is used to incorporate dinitrophenol (DNP)- and biotin-modified nucleotides, respectively, into two cDNA samples. (**See color plate A27.**)

reaction buffer salts. Therefore, extra care must be taken to ensure that resuspension is as complete as possible. After this, any remaining insoluble material can be removed from solution by centrifugation.

5. Each reaction should be compared with a control DNP or biotin cDNA population using a chemiluminescence-based membrane detection assay that proceeds at room temperature. This procedure indicates the success of the DNP or biotin cDNA synthesis reaction. Briefly, the labeled DNP or biotin control cDNA is serially diluted and spotted onto a nylon membrane. Adjacent to the spotted dilution series on the membrane, dilutions of the synthesized sample cDNA are spotted. The bound cDNA is cross-linked to the membrane. The membrane is then blocked with a protein mixture, incubated with the appropriate detector conjugate [e.g., streptavidin-horseradish peroxidase (HRP) for biotin-labeled cDNA], washed, and incubated with enhanced luminol (8) chemiluminescence substrate followed by exposure to X-ray film or a chemiluminescent imager. HRP catalyzes the production of chemiluminescence. Collection of a chemiluminescent signal will enable the researcher to compare the control and reaction spot series. Both sets of spots should emit a signal. If the sample reaction spots do not emit signal, the researcher should not proceed with the microarray hybridization, and the total RNA sample should be reexamined.
6. An overnight microarray hybridization is performed at 65°C in a closed, hydrated vessel (e.g., 50-mL sealed conical tube with wet tissue or a hybridization cassette) using the labeled cDNA samples in the presence of a specialized buffer. MICROMAX hybridization buffer contains sodium citrate-sodium chloride (SSC), some detergents to increase stringency, and various nucleic acids intended to block hybridization to nonspecific and highly redundant sequences.
7. After hybridization, the microarray is subjected to stringent washes (i.e., copious volumes of buffers with progressively lower salt concentrations) to eliminate excess and nonspecifically bound cDNA. These washes are performed at room temperature in a closed vessel (e.g., sealed 50-mL conical tube or a microarray wash station).
8. TSA is a posthybridization process that ultimately results in the sequential, enzymatic deposition of cyanine 3 and cyanine 5 fluorescence at locations containing DNP- or biotin-labeled cDNA, respectively. The procedure generally takes approximately 1 hour at room temperature and requires no special instrumentation. A detailed description of TSA principles and application to MICROMAX Microarray Systems are provided below.
9. The microarray is scanned for fluorescence emission using dedicated, commercially available instrumentation. A high-resolution (e.g., <25 μm/pixel) instrument, such as a confocal laser scanner, is favorable. Currently, cyanine 3 and cyanine 5 fluorescence is the convention for differential expression analysis using a single microarray. The fluorescence signal information for each feature on the array must be processed by series software programs to extract the data and interprete the experimental results.

PRINCIPLES OF TYRAMIDE SIGNAL AMPLIFICATION

A fundamental principle of differential gene expression analysis is the requirement for sensitive and accurate detection of hybridized cDNA products on the microarray surface. Many important cellular genes are expressed at relatively low levels, so there is a need for a highly sensitive, robust method that detects mRNA transcripts while maintaining the high spatial resolution required for microarray analysis. TSA enables the MICROMAX system to meet the critical demands of sensitivity and spatial resolution.

TSA works on the principle of catalyzed reporter deposition (3,5). In the initial step, the enzyme HRP becomes bound to the cDNA microarray surface by way of conjugation to an antibody or through streptavidin bound to DNP- or biotin-labeled cDNA (Figure 3). In the presence of hydrogen peroxide, HRP oxidizes the phenolic ring of tyramide conjugates (in this case a cyanine-tyramide reporter molecule) to produce highly reactive, free radical intermediates. These activated substrates subsequently form covalent bonds with tyrosine residues of nearby protein molecules used to block the microarray surface. In a short period, multiple depositions are possible through HRP-catalyzed substrate conversion. Typically, a 50- to 100-fold increase in sensitivity over direct fluorescence hybridization and detection techniques is found with MICROMAX using TSA.

Spatial resolution on the microarray surface is maintained because the activated tyramide reacts either with a surface-linked protein molecule or, alternatively, with another cyanine-tyramide molecule in solution. In the latter instance, the dimer product is easily washed from the surface prior to detection and does not result in any significant background fluorescence. As required for all TSA applications, the MICROMAX system is optimized for maximal deposition by way of enzyme, substrate, and surface chemistry. Two-color detection is accomplished using matched hapten/cyanine-tyramide pairs and sequential detection (DNP cDNA/cyanine 3-tyramide or biotin cDNA/cyanine 5-tyramide) as described.

TSA has been used in many other formats requiring sensitive chromogenic, fluorescent, or colorimetric detection. Maintaining optical resolution (0.2 μm) using TSA was demonstrated with single-copy detection of human papillomavirus type 16 in SiHa cells using fluorescent in situ hybridization (1).

MICROMAX TSA PERFORMANCE

Sensitivity

Fluorescence reporter deposition with the MICROMAX (NEN Catalog #MPS101) TSA process occurs after hybridization, whereas direct incorporation of cyanine nucleotide analogs into cDNA makes it possible to hybridize fluorescent-labeled material directly onto the microarray. The latter method is less complex than the TSA method; however, the limitations of sensitivity using the direct fluorescence method has made it necessary to investigate amplification strategies. Whereas techniques such as the polymerase chain reaction and in vitro RNA transcription allow amplification of nucleic acid samples, these schemes may skew the ratios of individual molecules in the complex mRNA mixture. By contrast, TSA is an endpoint signal amplification process. After the appropriate cDNA synthesis and microarray hybridization, fluorescence is applied in an ordered series of TSA steps that require approximately 1 hour to perform. Experiments indicate that the TSA process, when compared with the direct fluorescence method, allows the use of 10- to 50-fold less RNA sample for cDNA synthesis and hybridization.

MICROMAX TSA and direct fluorescence methods were compared by measuring cyanine 3 and cyanine 5 fluorescence resulting from each detection strategy. Total RNA was isolated from LNCaP and PC3 cell lines. Both cell lines are associated with prostate cancer: LNCaP is androgen sensitive, whereas PC3 is androgen insensitive and more malignant than LNCaP (9). For the MICROMAX system, 2 μg of total RNA from each cell line was used. For the direct method, 120 μg of total RNA from each of the same RNA pools were used for cDNA synthesis in the presence of cyanine 3 and cyanine 5 nucleotide analogs.

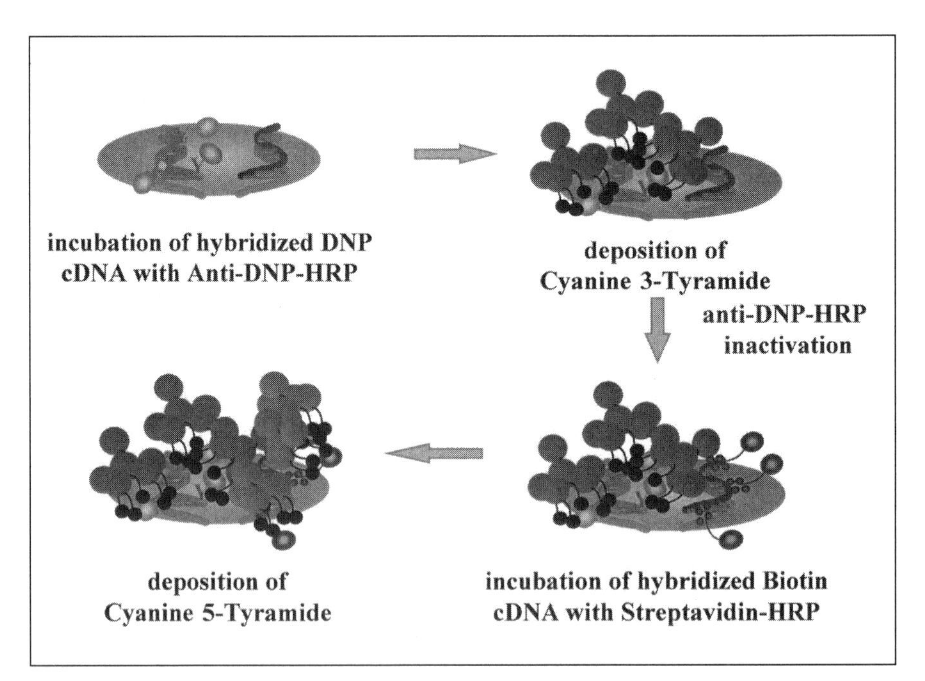

Figure 3. Fluorescence deposition via sequential TSA detection steps. The hybridized microarray is reacted with an α-DNP/HRP conjugate and mixed with a cyanine 3-tyramide reagent. In a second step, the microarray is reacted with a streptavidin/HRP conjugate and mixed with a cyanine 5-tyramide reagent. The cyanine 3 and cyanine 5 dyes are then detected by fluorescence emission. **(See color plate A28.)**

The observed signal with the TSA slide was considerably enhanced relative to the slide that used direct fluorescence slide (Figure 4). TSA makes it possible to obtain greater gene representation (i.e., more spot signals) with a microarray experiment using 60-fold less RNA sample.

In another experiment, 15-fold differential detection was observed for the immediate early expression of the transcription factor ETR101 when comparing RNA from unstimulated and phorbol myristate acetate (PMA)-stimulated cultured Jurkat T cells. According to the time course depicted in Figure 5, there is significant induction of ETR101 within the first 2 hours after cellular stimulation by PMA. This observation is consistent with previous data that reported gene induction by phorbol esters measured by Northern blot analysis (7) or receptor-mediated activation of B- and T-cell line expression patterns profiled by nylon membrane hybridization (6).

Linearity and Dynamic Range

If the same template RNA is used for cyanine 3 and cyanine 5 detection in the two-color MICROMAX labeling scheme, the theoretical signal for each spot on the microarray should be identical in both channels. Complexities in the hybridization, labeling, and detection steps leads to small signal variation, for some spots, in the two detection channels. We have observed that all signal variation is less than twofold with the direct method, whereas 85% of the spots detected with the TSA method are within a twofold range, and 95% of the spots vary by less than three-fold variation (Figure 6). It can be inferred, from the log plot in

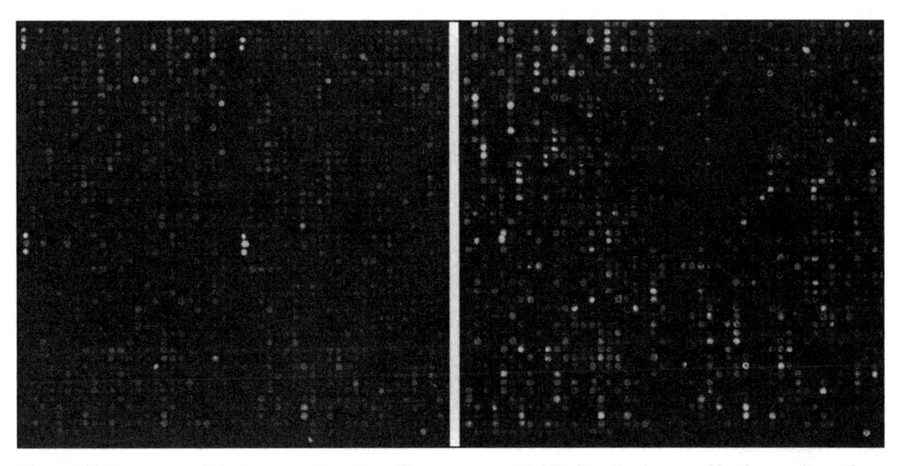

Figure 4. Microarray analysis comparing direct fluorescent and TSA. Overlay bitmap files from a direct fluorescence method with 120 µg total LNCaP and PC3 RNA after cDNA synthesis and hybridization (left), versus the TSA fluorescence method with 2 µg of total RNA input (right). Color overlays are composite images of cyanine 3 and cyanine 5 fluorescence signals: red represents cyanine 3 > cyanine 5, green represents cyanine 5 > cyanine 3, and yellow represents cyanine 3 = cyanine 5. **(See color plate A28.)**

Figure 6, that the observable dynamic range is nearly 1000-fold. This has been our experience with both the direct and TSA fluorescence detection methods.

The MICROMAX TSA system has more nonspecific signal or "noise" between the two channels than the direct method. Currently, there is no empirical evidence pinpointing which step(s) generates the extra noise, although the TSA steps are likely candidates given the nature of enzymatic fluorescence deposition. It is important to remember that there is a series of enzymatic reactions occurring at each individual spot on the microarray and although the overall TSA process is greatly controlled, there is still the potential for a small amount of nonlinearity.

CONCLUSIONS

The main advantage of MICROMAX cDNA microarray chemistry is signal amplification by sequential TSA steps. Starting with a qualified total RNA population, first-strand cDNA synthesis with DNP-dCTP or biotin-dCTP is the only required manipulation of the sample nucleic acid. A large number of differentially expressed genes can be examined with a small amount of input sample RNA. This is highly beneficial, particularly when working with scarce tissue specimens with limited RNA.

The signal output is semiquantitative because it can measure the relative fluorescence of each sample under investigation; however, it cannot report an absolute value of gene expression for an individual sample. This is because, with enzymatic signal amplification, cDNA fluorescence is not measured directly.

Other hapten nucleotide analogs can be used for cDNA synthesis with suitable HRP conjugates. Tyramines can be conjugated to a variety of fluorophores. The MICROMAX detection strategy described above is adaptable, as long as the HRP, tyrosine, and tyramine interactions are conserved in the system. This is another

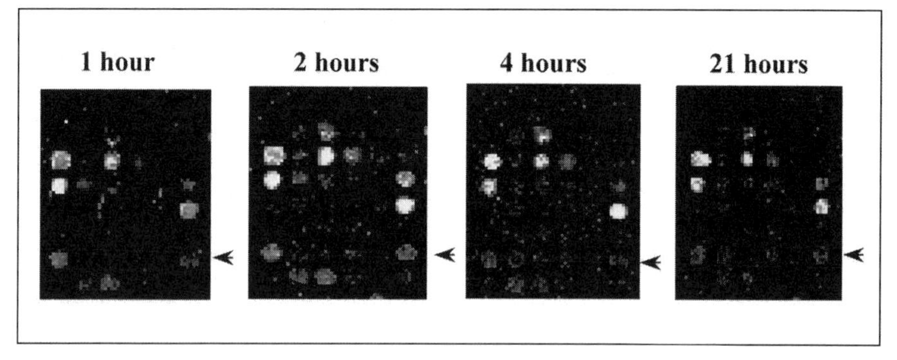

Figure 5. Gene expression time course obtained with MICROMAX. Transcription factor ETR101 induction (arrowhead) is measured in a time course experiment at various times (1 to 21 hours) after stimulation of Jurkat cells with phorbol myristate acetate. (**See color plate A29.**)

key advantage, particularly when considering different experimental formats such as instruments that allow detection of fluors other than cyanine 3 and cyanine 5.

The need for high-sensitivity systems featuring integrated chemical components, like MICROMAX, will expand as cDNA microarray techniques continue to become the standard for high-throughput gene analyses. If these systems are reliable and easy to use, gene characterization efforts can be greatly accelerated compared with conventional methods that examine one gene at a time.

NOTES

MICROMAX and TSA are trademarks of NEN Life Science Products, Inc. MICROMAX Microarray Systems are jointly developed by AlphaGene, Inc. and NEN Life Science Products, Inc. and are distributed exclusively by NEN Life Science Products, Inc. The full-length expressed gene technology is covered by patents owned by AlphaGene, Inc. (U.S. patents 5,162,209 and 5,643,766) and pending patent applications. TSA and the use of TSA in microarray applications are protected by NEN U.S. patents (5,731,158; 5,583,001; and 5,196,306), foreign equivalents, and patents pending.

MICROMAX Microarray Systems are made using the Polymerase Chain Reaction ("PCR") Process, which is covered by patents owned by Roche Molecular Systems, Inc. and F. Hoffman-La Roche Ltd ("Roche"). No license to use the

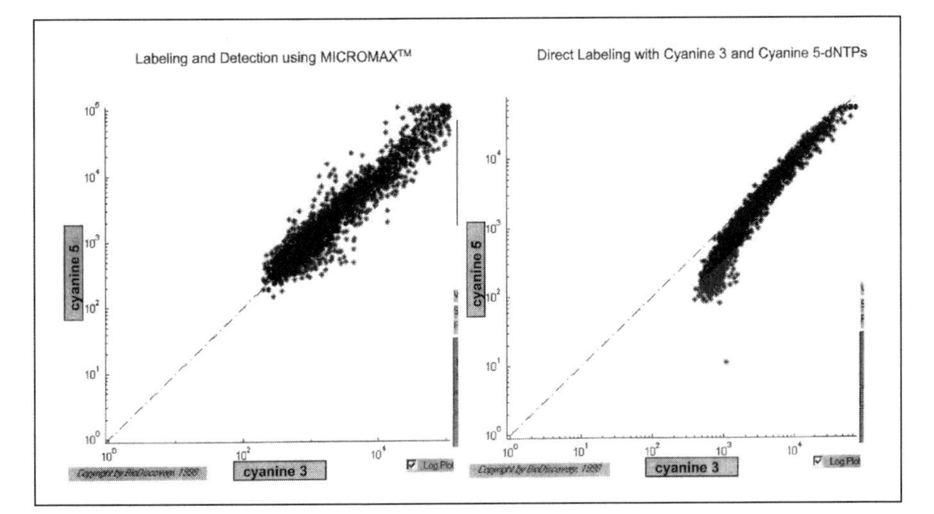

Figure 6. Scatter plots comparing TSA and direct labeling. Log versus log plots of a single source of RNA compared with two different double-labeling methods. A single sample of mRNA from Jurkat cells was divided into four equal aliquots and labeled with cyanine 3 (*x*-axis) and cyanine 5 (*y*-axis) using TSA fluorescence (left) or direct fluorescence (right). Blue dots represent gene sequences that have fluorescent signals within twofold and pink dots represent gene sequences that show more than twofold difference. (**See color plate A29.**)

PCR Process is conveyed expressly or by implication to the purchaser by the purchase of these products. Information on purchasing licenses to practice the PCR Process may be obtained by contacting the Director of Licensing at Roche Molecular Systems, Inc., 1145 Atlantic Avenue, Almeda, California 94501.

REFERENCES

1. **Adler K., T. Erickson, and M.N. Bobrow.** 1997. High sensitivity detection of HPV-16 in SiHa and CaSki Cells using FISH enhanced by TSA. Histochem. Cell Biol. *108*:321-324.
2. **AlphaGene, Inc.** U.S. Patents 5,162,209 (1992) and 5,643,766 (1997).
3. **Bobrow M.N., T.D. Harris, K.J. Shaughnessy, and G.J. Litt.** 1989. Catalyzed reporter deposition, a novel method of signal amplification. Application to immunoassays. J. Immunol. Methods *125*:279-285.
4. **Liang P. and A. Pardee.** 1993. Distribution and cloning of eukaryotic mRNAs by means of differential display: refinements and optimization. Nucleic Acids Res. *21*:3269-3275.
5. **NEN Life Science Products, Inc.** U.S. Patents 5,731,158 (1998); 5,583,001 (1996); 5,196,306 (1993) and foreign equivalents.
6. **Ollola, J. and M. Vihinen.** 1998. Stimulation of B and T cells activates expression of transcription and differentiation factors. Biochem. Biophys. Res. Commun. *249*:475-480.
7. **Shimizu, N., M. Ohta, C. Fujiwara, J. Sagara, N. Mochizuki, T. Oda, and H. Utiyama.** 1991. Expression of a novel immediate early gene during 12-*O*-tetradecanoylphorbol-13-acetate-induced macrophagic differentiation of HL-60 cells. J. Biol. Chem. *266*:12157-12161.
8. **Thorpe, G.H.G., L.J. Kricka, S.B. Mosely, and T.P. Whitehead.** 1985. Phenols as enhancers of the chemiluminescent horseradish peroxidase-luminol-hydrogen peroxide reaction: application in luminescence-monitored enzyme immunoassays. Clin. Chem. *31*:1335-1341.
9. **Yang, M., M. Loda, and A.J. Sytkowski.** 1998. Identification of genes expressed differentially by LNCaP or PC3 prostate cancer cell lines. Cancer Res. *58*:3732-3735.

11 Production of Microarrays on Porous Substrates Using Noncontact Piezoelectric Dispensing

David Englert
Packard Instrument Company, Meriden, CT, USA

INTRODUCTION

Wide acceptance of microarray technology will require sensitive and reproducible systems that are accessible to most scientists. Versatility in type of substrates, molecules that can be immobilized, and immobilization chemistry is desirable to realize the potentially broad range of applications for microarray technology. Packard Instrument Company is developing microarray tools based on piezoelectric, noncontact dispensing of presynthesized probe molecules. This technology will afford a broad range of biologists great versatility, for example, to create microarrays for expression analysis with either oligonucleotide or cDNA probes, or even both on the same array. Piezoelectric noncontact dispensing can provide great precision and speed, and it is well suited for dispensing onto porous substrates, which may provide enhanced sensitivity and reproducibility compared with planar substrates.

Porous substrates (membranes and polymer matrices) can permit immobilization of large amounts of probe molecules and can provide a three-dimensional hydrophilic environment similar to free solution for biomolecular interactions to occur (1,12). Some porous substrates are fragile and are likely to be damaged by devices that make contact with them. Furthermore, capillary forces in these substrates cause wicking of liquid from a contact dispenser, resulting in undesirable spread of probe molecules in the substrate and poorly controlled dispense volumes. Piezoelectric noncontact dispensing, however, delivers a small, precisely controlled volume that is imbibed into the matrix in a consistent manner.

Microarray Biochip Technology
Edited by Mark Schena

PIEZOELECTRIC DISPENSING TECHNOLOGY

The BioChip Arrayer™ developed by Packard Instrument Company is based on glass capillary dispensers (8) that are able to aspirate probe solutions from source wells and dispense many droplets to destinations in microarray formats. The construction of a piezoelectric dispenser is shown in Figure 1. It consists of a glass capillary that has an orifice of approximately 75 μm at one end and a connection to a precision syringe pump at the other end. The syringe pump applies vacuum to aspirate solutions through the dispenser tip. A piezoelectric transducer around the center of the capillary exerts pressure on the capillary when activated by an electronic pulse. The electronic pulse creates a pressure wave in the capillary that ejects a droplet of about 350 pL from the orifice. As droplets are ejected, the dispenser end of the capillary refills from the system reservoir fluid by capillary flow. A pressure sensor automatically detects the resulting decrease in pressure, and the syringe pump is activated to restore the volume as droplets are dispensed. This proprietary pressure management system is essential for maintaining consistent operating conditions in the capillary.

The deck of the BioChip Arrayer accommodates six microplate-size source plates or destination blocks. It is typically configured with one 384-well source plate and five destination blocks, each of which can hold five microscope slides. A typical deck configuration thus contains one 384-well plate and 25 microscope slides. The dispense head consists of four piezoelectric dispensers at 9-mm spacing. A precision X-Y-Z stage moves the head to the source microplate, where solutions of probes are aspirated, and then over to array substrates on the destination blocks, whereby many droplets are dispensed from a single loading to make the arrays. A camera mounted on the dispense head allows positioning of discrete targets on the microarray substrates (e.g., gel pads) to be located (12).

The tips travel close to the microarray substrates (approximately 0.5 mm) but do not make contact, so the volume dispensed is completely independent of the surface characteristics of the substrate. One or more droplets (approximately 350 pL each) can be dispensed at each array address. Because of the dominant influence of surface energy in the small droplets, there is no splashing of the droplets and no aerosol production. The greatest efficiency is achieved when many arrays are produced with each aspirated sample. Approximately 100 different samples can be dispensed onto 25 arrays in 1 hour with the four-tip BioChip Arrayer. The maximum number of arrays that can be prepared in one batch is the maximum number of slides (i.e., 25) times the number of arrays on each slide.

Both the source plates and the destination blocks can be cooled. Cooling of source plates is very important to minimize evaporation. It may be desirable to cool the destination slides in some cases to slow evaporation of the droplets after they are dispensed. Droplets will dry within seconds at ambient temperature. Cooling can maintain hydration to facilitate chemical reactions required for some immobilization chemistries.

Continuous measurements of the pressure in the capillaries confirms the delivery of the expected volumes from each dispenser. If the system software detects the failure of a tip to dispense the expected volume during a predispense verification test, the instrument can automatically aspirate the sample again. Dispense failures sometimes occur when small gas bubbles are aspirated; a second aspiration attempt usually corrects the problem and fills the capillary dispenser prop-

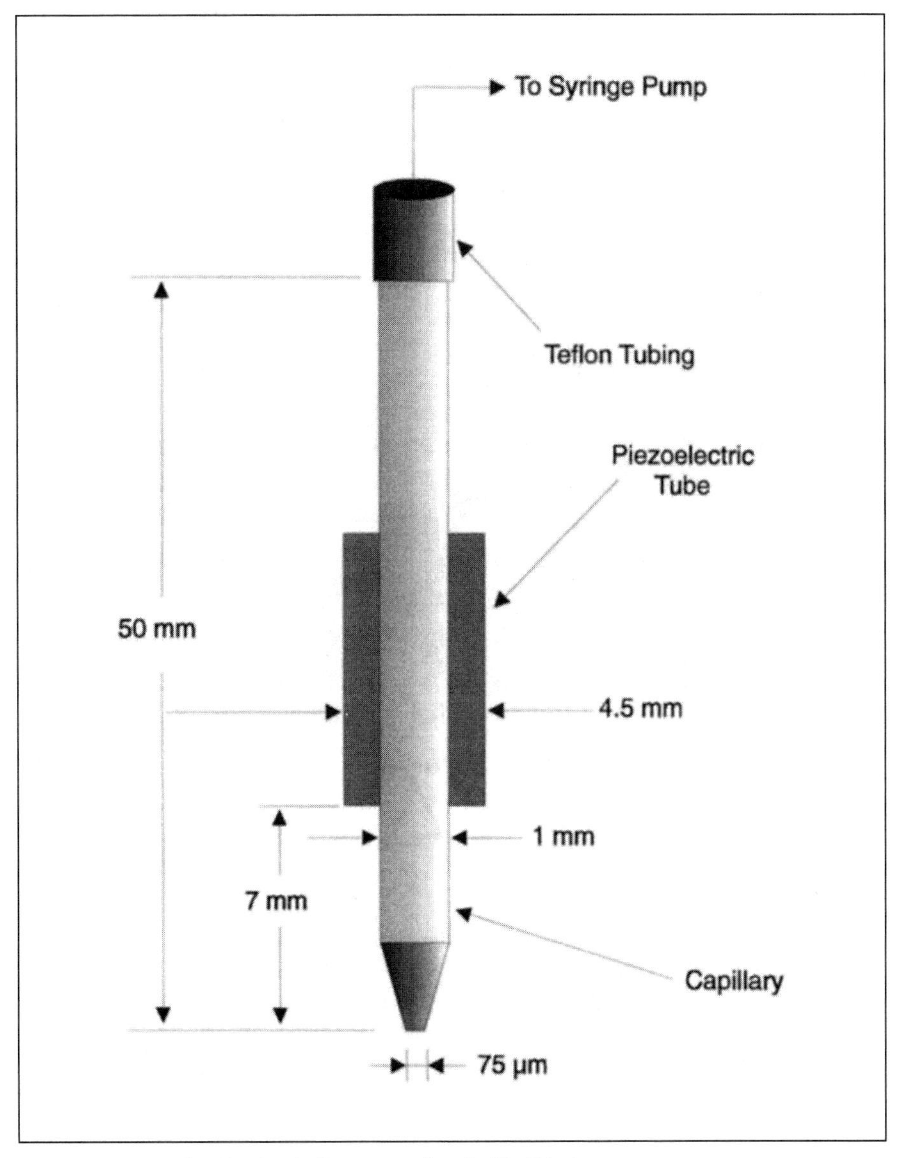

Figure 1. Diagram of a piezoelectric dispenser used on the BioChip Arrayer.

erly. We have developed operating protocols for sample degassing, system fluid degassing, and tip cleaning, which have made the operation of the BioChip Arrayer highly reliable (approximately 98% dispense success rate). If probe molecules are dispensed onto the arrays with twofold redundancy, the probability that a given probe will not be present on an array is extremely small.

With careful management of the pressure in the capillaries, the droplet volume delivered by the piezoelectric dispensers is extremely uniform (8). This is illustrated in Figure 2, which presents the coefficients of variation (CVs) for many columns of spots of a ^{33}P-labeled oligonucleotide dispensed serially with the BioChip Arrayer. Except for the first column of spots, the CVs average about

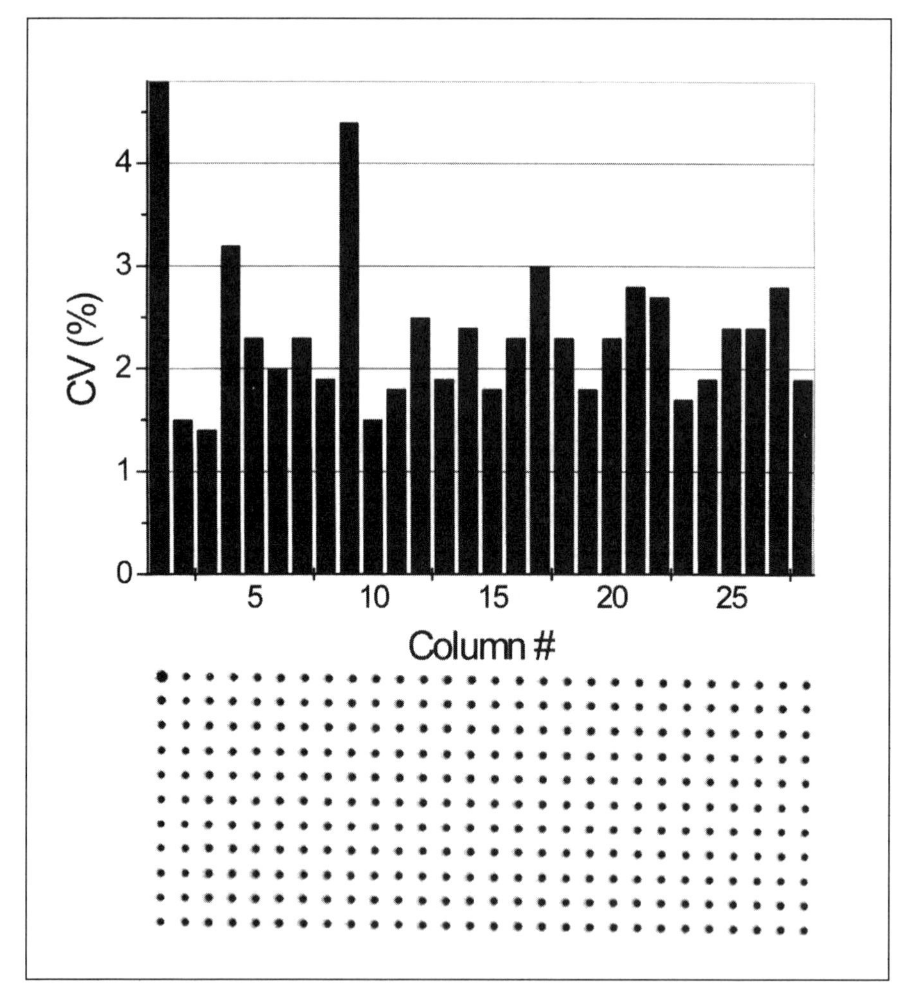

Figure 2. Consistency of dispense volume of piezoelectric dispenser. A ^{33}P-labeled oligonucleotide was dispensed onto a glass slide, and the glass slide was then imaged by storage phosphor imaging. The coefficient of variation (CV) was calculated for each column of the array.

2%, which represents the combined error of the dispenser and the detector (a storage phosphor screen scanned with Cyclone™). The variability in the first column is due to the "first droplet effect," which can be avoided by appropriate aspirate/dispense protocols. It is not possible to differentiate the relative errors of the dispenser and the detector, since we have found no way to dispense a test source more uniformly than 2% and no way to measure a label with greater precision than 2%. In our experience, detection of fluorescence with a laser scanner or a microscope/CCD imager is generally more variable than storage phosphor imaging with the Packard Cyclone confocal scanner.

When droplets of a sample are dispensed onto many destinations distributed over the deck of the BioChip Arrayer, the variability is somewhat greater, presumably due to small variations in pressure within the capillaries as the tips move over the work surface. The CVs of the amounts of a ^{33}P-labeled oligonucleotide dispensed onto 10 microscope slides distributed over extreme positions on the BioChip Arrayer deck are shown in Table 1. For all four dispensers (each assessed by quantifying four separate spots on the arrays), the CVs are approximately 5% to 10%.

After each sample is dispensed onto the arrays, the dispense head moves to a cleaning station. The syringe pumps force system fluid through the tips to clean the insides of the capillaries, and a peristaltic pump simultaneously passes fluid over the outsides of the tips. Cleaning is facilitated ultrasonically, by oscillation of the piezoelectric transducers on the capillary dispensers and by activation of an ultrasonic transducer in the cleaning bath. Efficient cleaning maintains the surface of the capillary tips for consistent operation and minimizes carryover between samples. We have assessed the carryover between samples by first dispensing a ^{33}P-labeled oligonucleotide in a series of spots, and then dispensing an unlabeled oligonucleotide in second series of spots. The radioactive counts in the second series were just barely visible above background and were difficult to quantify, but the carryover was clearly less than 1 part/10 000 (Figure 3).

POROUS MICROARRAY SUBSTRATES

Microarrays can be printed with the BioChip Arrayer on most substrates, including glass slides, membranes such as nylon, nitrocellulose, and Anopore™ (Whatman, Maidstone, UK), and polyacrylamide gels. Spot sizes when single droplets are dispensed on a planar glass surfaces depend on surface properties, but they are generally 150 to 180 µm when dispensed on glass surfaces prepared as described by Guo et al. (2). We have found that the distribution of oligonucleotide is often nonuniform on glass supports. The solution dries on the surface nonuniformly, and the oligonucleotide becomes immobilized in the pattern in which it dries. Because probe density is likely to affect the accessibility of probe molecules and their interaction with targets (9), such nonuniformity may be a

Table 1. BioChip Arrayer Precision

Tip number (%)	Spot number (%)			
	1	2	3	4
1	5.4	6.0	6.8	7.4
2	5.9	6.1	7.0	6.5
3	7.3	7.8	8.0	6.1
4	10.1	9.6	8.1	9.0

Coefficients of variation (CVs) of ^{33}P activity measured from spots dispensed with the BioChip Arrayer on a set of 10 microscope slides in extreme positions on the instrument deck. A group of four droplets was dispensed onto separate positions on each slide with all four tips of the BioChip Arrayer. The mean and standard deviation ($N = 10$) of the activity was calculated for each spot on the 10 slides.

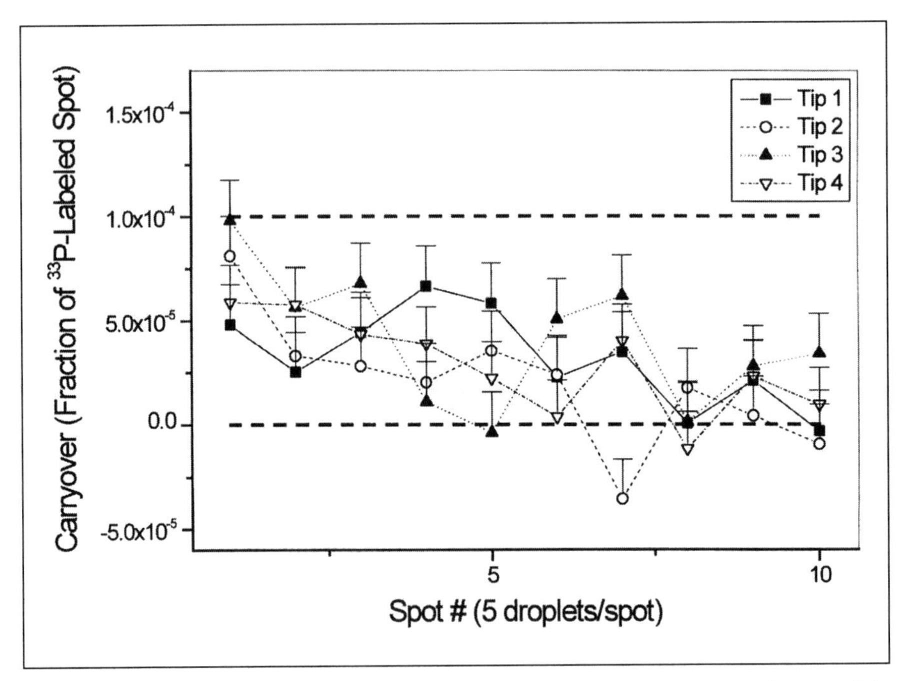

Figure 3. Assessment of the carryover of a ^{33}P-labeled oligonucleotide to other spots. All four tips of the BioChip Arrayer were used to dispense one droplet of a 100-µM solution of labeled oligonucleotide as a series of spots on a microscope slide, and then five droplets of a 100-µM solution of an unlabeled oligonucleotide were dispensed on another series of addresses. The amount of ^{33}P present in the second series of spots was measured by storage phosphor imaging using Cyclone and calculated as a fraction of the activity in the first series of spots containing the labeled oligonucleotide; with values normalizing for the number of droplets dispensed per spot. The error bars indicate the standard deviation of 20 background regions identical to those used to quantify the radioactivity. Zero carryover and 0.01% carryover are indicated on the graph by dashed lines.

source of experimental variability. Furthermore, steric crowding limits the useful density of probes on planar surfaces (2,9), and this in turn limits sensitivity. Porous substrates can overcome some of the limitations of planar surfaces.

MICROARRAYS ON MEMBRANES

Compared with nonporous surfaces, conventional nitrocellulose or nylon membranes have a much larger area available for surface interactions per unit of macroscopic area. Furthermore, a liquid dispensed onto these membranes will immediately distribute into the membrane by capillary flow, which may result in relatively uniform distribution. The BioChip Arrayer can be used to dispense oligonucleotide probes on these membranes, if the membranes are mounted on microscope slides (Figure 4). Nylon and nitrocellulose membranes mounted on microscope slides are now commercially available from Schleicher and Schuell (Keene, NH, USA). We have found that single droplets of a fluorescently labeled oligonucleotide form uniform spots of <200 μm on nylon membranes. However, the background fluorescence of nylon and nitrocellulose membranes is very high, so radioactive labels are preferred over common fluorophors. Spots of ^{33}P-labeled oligonucleotides on nylon membranes are uniform in storage phosphor images, with about 90% of the activity contained within 0.25 mm of the center of the spot. The profiles of the spots in the storage phosphor image (Figure 4) were about the same whether one, two, or four droplets were dispensed per spot.

Anopore is an inorganic microporous membrane with a highly controlled, uniform capillary pore structure. It is 60 μm thick and is available with 200-nm capillary pores, which afford the membrane a large surface area for the immobilization of probe molecules per unit of macroscopic area. Since it is composed of inorganic material (Al_2O_3), its background fluorescence is low compared with nylon or nitrocellulose membranes, and fluorescence detection is relatively sensitive. High flow rates are possible through the membrane, and solutions of target molecules can be passed through the membrane repeatedly to effect fast, efficient capture. Washing of the capillary pores can be performed efficiently by pressure or vacuum filtration.

Anopore is quite fragile and would probably be damaged by contact dispensers. However, the BioChip Arrayer can be used to dispense onto Anopore, and the spot sizes obtained are relatively small since the 350-pL droplets dispensed by the BioChip Arrayer are immediately drawn into the capillary pores of the membrane rather than spreading laterally. The spot size produced when single droplets of a Cy3-labeled nucleotide are dispensed onto Anopore is about 125 μm (Figure 5). Preliminary hybridization experiments with Anopore suggest that it will be a good substrate on which to make sensitive and reproducible measurements, and that it will be possible to make arrays on Anopore at a higher density than with most other substrates.

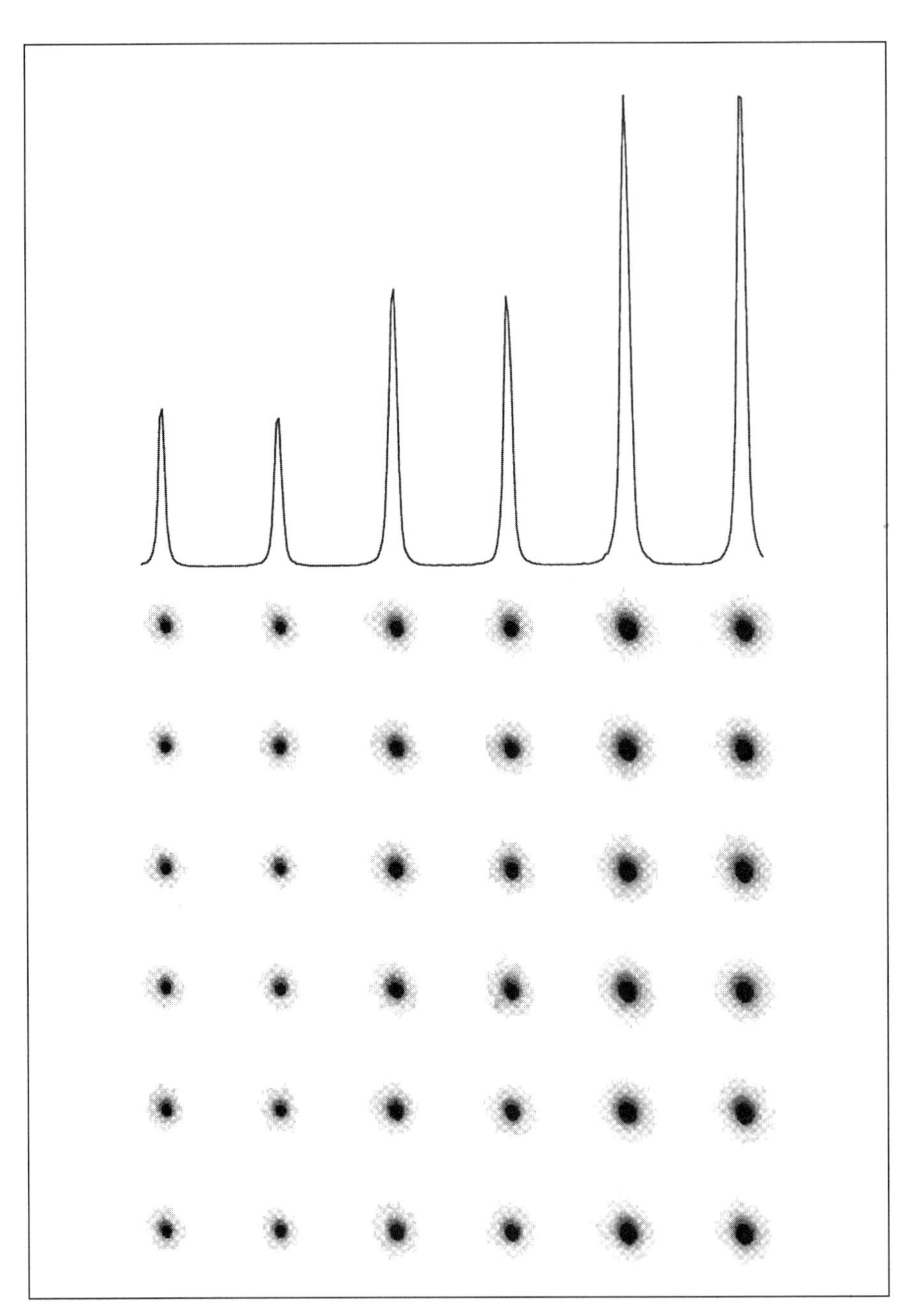

Figure 4. Storage phosphor image of a ^{33}P-labeled oligonucleotide dispensed onto a nylon membrane with the BioChip Arrayer. A 100-μM solution of a ^{33}P-labeled oligonucleotide was dispensed onto a charged nylon membrane (GeneScreen Plus®; NEN Life Sciences, Boston, MA, USA), which was taped onto a microscope slide, and the membrane was imaged with an SR storage phosphor screen and Cyclone. One, two, or four droplets per spot were dispensed in pairs of columns every 2 mm left to right. A profile through one of the rows of spots is shown above the image to show the shape of the spots.

MICROARRAYS ON POLYACRYLAMIDE GELS

The polyacrylamide gel pad technology developed in the laboratories of Andrei Mirzabekov (10,12) provides a microarray format that permits the immobilization of probe molecules within a three-dimensional hydrophilic matrix. Relatively large amounts of probe per unit area of the supporting substrate can be achieved (12), while avoiding the steric crowding that occurs on planar, two-dimensional surfaces (9). Compared with a planar surface, the polyacrylamide matrix more closely approximates solution hybridization conditions, because the tethered probes are not hindered on a solid surface. On the other hand, there are restrictions on the size of molecules that can diffuse into the gel, and the kinetics of hybridization reactions are influenced by the gel (9,12). Probes and targets

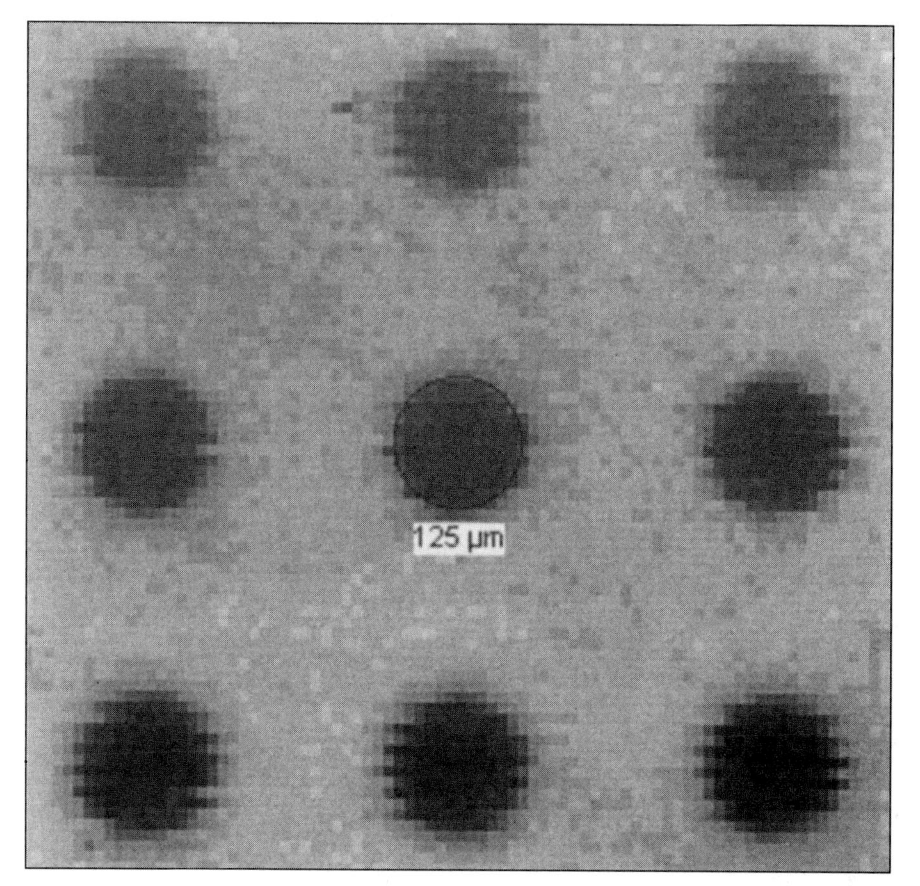

Figure 5. Spots produced on Anopore membrane by dispensing single droplets of Cy3-labeled oligonucleotide with the BioChip Arrayer. Spots were imaged with a laser scanner. The rows of spots contained 1.8, 3.6, and 7.2 amol of Cy3 from top to bottom. The circle in the center has a diameter of 125 µm.

must be of a suitably small size or must be fragmented (7,9). Fragmentation is also performed when using planar arrays, to enhance hybridization efficiency and specificity and to minimize secondary structure in target molecules (9,11).

Deposition of probes on gel pads has been effected previously with a special pin tool robot (12). We have dispensed probes onto gel pads by noncontact dispensing using the BioChip Arrayer. This has required the implementation of a camera to visualize the gel pads and software routines to establish registration between the piezoelectric tips and these discrete target sites. Although these protocols have been established and are used routinely in our laboratory, the requirement to dispense onto discrete target sites complicates array production somewhat, and we have found it more convenient to dispense droplets onto continuous films of polyacrylamide. Because the BioChip Arrayer dispenses droplets that are >1000 times smaller than the volumes originally used by Khrapko et al. on continuous polyacrylamide films (3,4), we are able to create much smaller spots on polyacrylamide films than the 1- to 2-mm spots obtained by the earlier authors. We have optimized the protocol for immobilizing aminated oligonucleotides to avoid diffusion of the probes in the polyacrylamide matrix before fixation and hence to obtain small, discrete spot sizes.

We have assessed both discrete polyacrylamide gel pads and continuous polyacrylamide films in DNA or RNA hybridizations and in minisequencing reactions (6) for single-nucleotide polymorphism (SNP) scoring. Hybridizations in polyacrylamide gels have been documented (4,12), and we have found in experiments that will be reported elsewhere that <100 amol of target can be detected, making possible the detection of rare transcripts in gene expression analysis (5). The spot size observed when DNA hybridization is performed with single droplets dispensed onto a continuous film is shown in Figure 6. Note the homogeneity of the spots. Similar spots are observed with minisequencing.

Minisequencing involves the capture of target DNA by hybridization to oligonucleotides immobilized by their 5′ ends and template-directed enzymatic extension of the 3′ ends of the oligonucleotide primers with dideoxy nucleotide terminators (6). SNPs in the targets immediately adjacent to the immobilized oligonucleotides are scored by noting which type of DNA base or bases are incorporated. Multiplex analysis of many targets can be performed on microarrays if the hybridization and base extension reactions are highly specific. Presented here are some results illustrating how we have optimized minisequencing reactions in polyacrylamide gels to obtain the high specificity required for multiplex analysis. Details of these investigations will be presented in subsequent publications.

Gel pad arrays in which eight different minisequencing primers were immobilized are shown in Figure 7. Each row contained a primer to a different SNP target at three different concentrations in triplicate. The microarrays were incubated with a polymerase chain reaction amplicon containing a sequence complementary to the primer (FMAH) in the third row. The two images at the top of the figure were obtained from the same array incubated at 60°C, and the two images at the bottom of the figure are from another array incubated at 85°C. Images at the left were scanned with a green laser that revealed only the fluorescent label on the ddGTP nucleotide analog and hence indicates the incorporation of ddGTP, whereas the images at the right were scanned with a red laser that revealed the incorporation of the ddTTP nucleotide analog. We expect only ddGTP to be incorporated into the gel pads in the third row (FMAH primer). However, at 60°C we also observed significant incorporation of ddTTP in the third row. Furthermore, there was also significant incorporation of ddGTP into the gel pads at 60°C in the fourth row of the array, which contained an oligonucleotide complementary to a different SNP locus (F8C primer). These artifacts were largely eliminated when the incubation was performed at 85°C, although the signals are generally weaker.

Quantification of the fluorescent signals as a function of the oligonucleotide primer concentration is shown in Figure 8. It can be seen that the greatest signal from the incorporation of the expected oligonucleotide (ddGTP on the FMAH primer) is observed with intermediate primer concentrations, whereas the artifactual signals (ddTTP incorporation on the FMAH primer and ddGTP incorporation on the F8C primer) are minimized with low primer concentrations. The magnitude of the signal with the expected nucleotide, relative to the artifactual signals, is maximized at intermediate primer concentrations and high temperature.

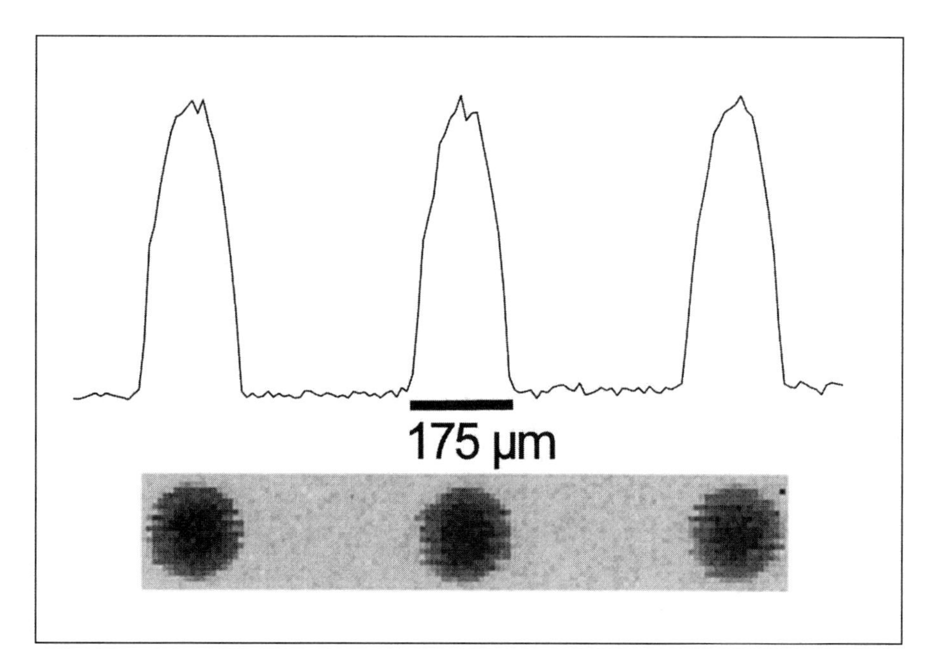

Figure 6. Image of spots on a continuous polyacrylamide film resulting from a DNA hybridization experiment with a Cy5-labeled target. Single 350-pL droplets of a hybridization probe were dispensed with the BioChip Arrayer. The array was imaged with a laser scanner. The profile above the image reveals the intensity and uniformity of the spots.

We have observed the same phenomena with other SNP targets, whether the experiments are performed on gel pads or continuous polyacrylamide films. We dispense no more than approximately 20 to 50 fmol of aminated minisequencing primer per spot for good discrimination. A minisequencing experiment performed on a continuous polyacrylamide film is shown in Figure 9. The alleles are clearly determined to be G for FMAH and T for CFTR. The spots sizes are approximately 175 μm. We have found the minisequencing signals on continuous polyacrylamide films to be 20 to 30 times greater than the signals from activated glass slides. Minisequencing is quite reproducible on the continuous films, as shown in Table 2, which gives the average signals and CVs at different primer concentrations for groups of 12 primer spots across an array on a continuous film. The signal does not increase linearly with primer concentration.

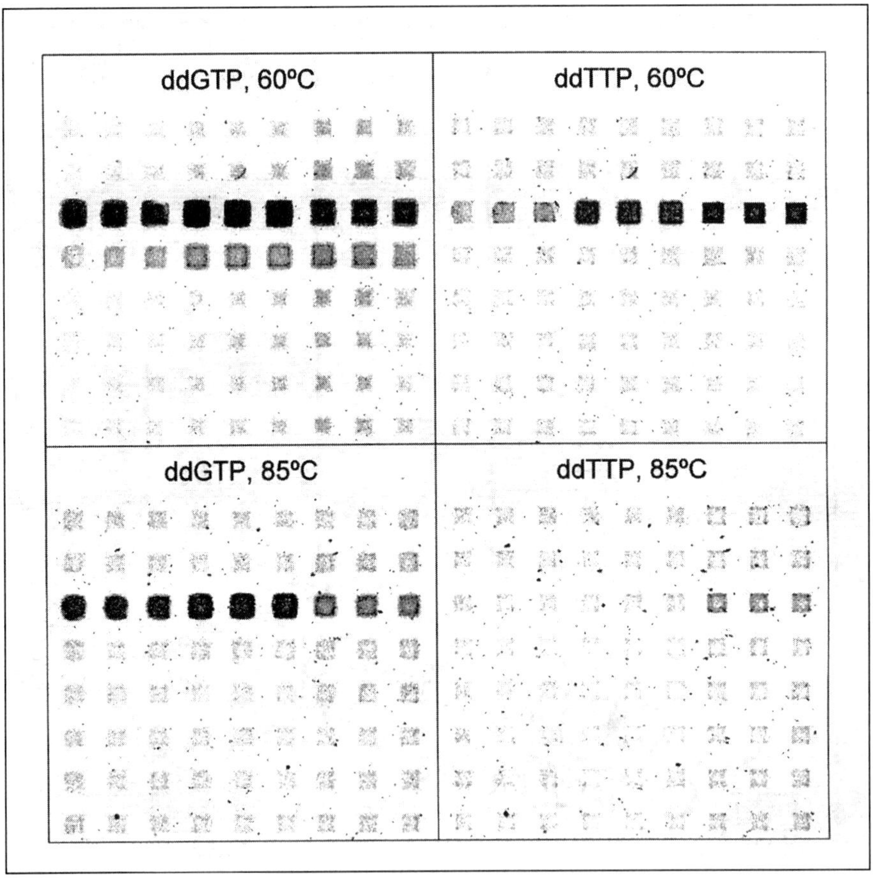

Figure 7. Minisequencing on gel pad arrays. Each row of nine pads contained a different minisequencing primer at three different concentrations in triplicate, increasing from left to right. The arrays were incubated in a minisequencing reaction with a target (FMAH) complementary to the primer in row 3. The arrays were imaged with a laser scanner. See text for further details.

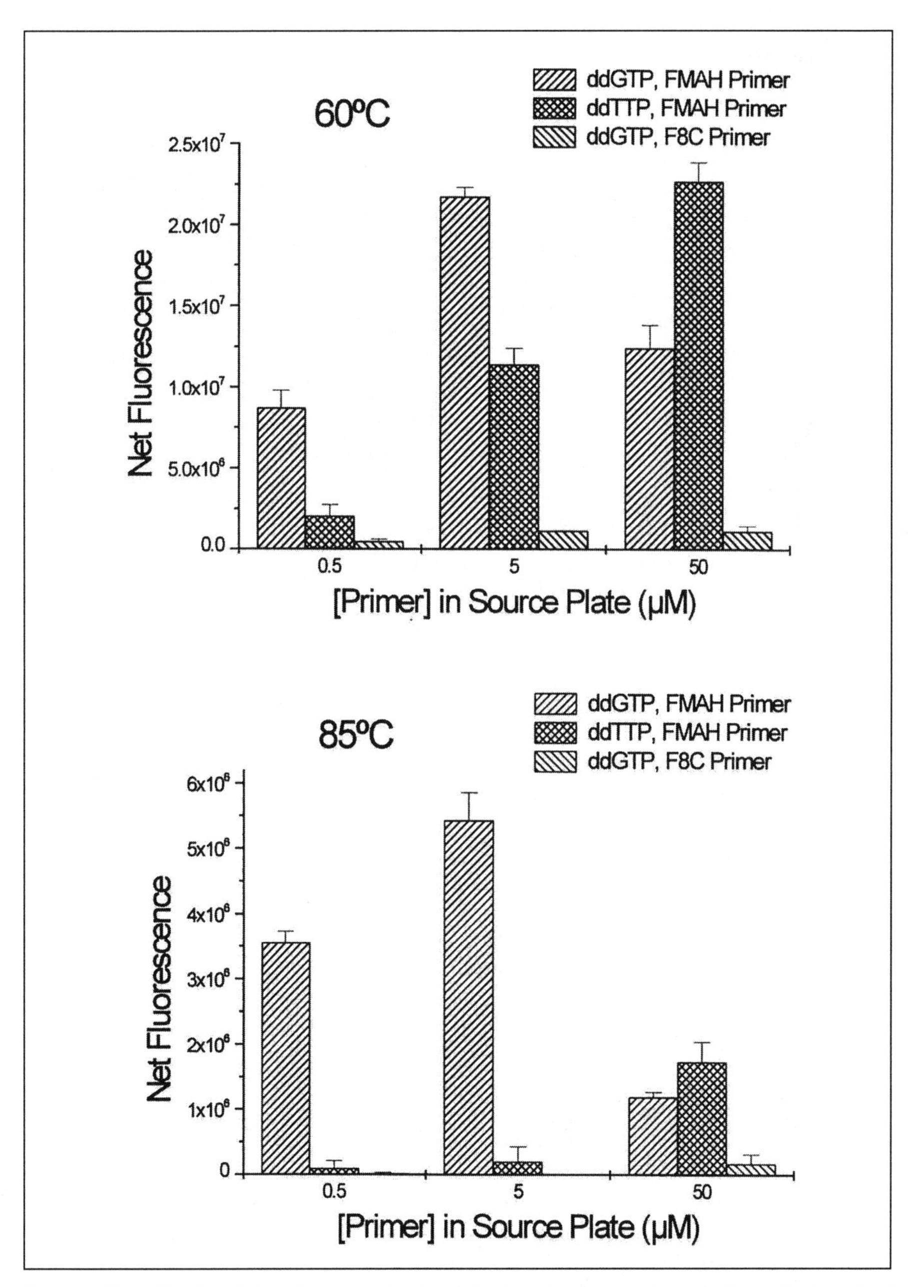

Figure 8. Quantification of the minisequencing signals in the gel pad arrays shown in Figure 7. For each of the arrays (one incubated at 60°C and the other at 85°C) the net fluorescence in channel 1 (incorporation of ddGTP) and in channel 2 (incorporation of ddTTP) in row 3 (FMAH primer) is shown. The net fluorescence in row 4 of channel 1 (ddGTP) and for row 4 (F8C primer) is also shown. Net fluorescence is shown as a function of the primer concentration in the source plate (increasing from left to right in the rows in Figure 7). Ten 350-pL droplets were dispensed onto each gel pad. Means and standard deviations are shown for triplicates of each primer concentration.

CONCLUSIONS

Piezoelectric dispensing with the BioChip Arrayer can be used to create microarrays on a variety of substrates. The BioChip Arrayer dispenses consistent quantities of probe and is well suited for dispensing onto porous substrates. Arrays of spots with diameters <200 μm result when single 350-pL droplets are dispensed onto activated glass slides and continuous polyacrylamide films. Spot sizes of about 125 μm are obtained on Anopore membranes. Sensitive hybridizations and/or enzymatic reactions can be performed with microarrays on these

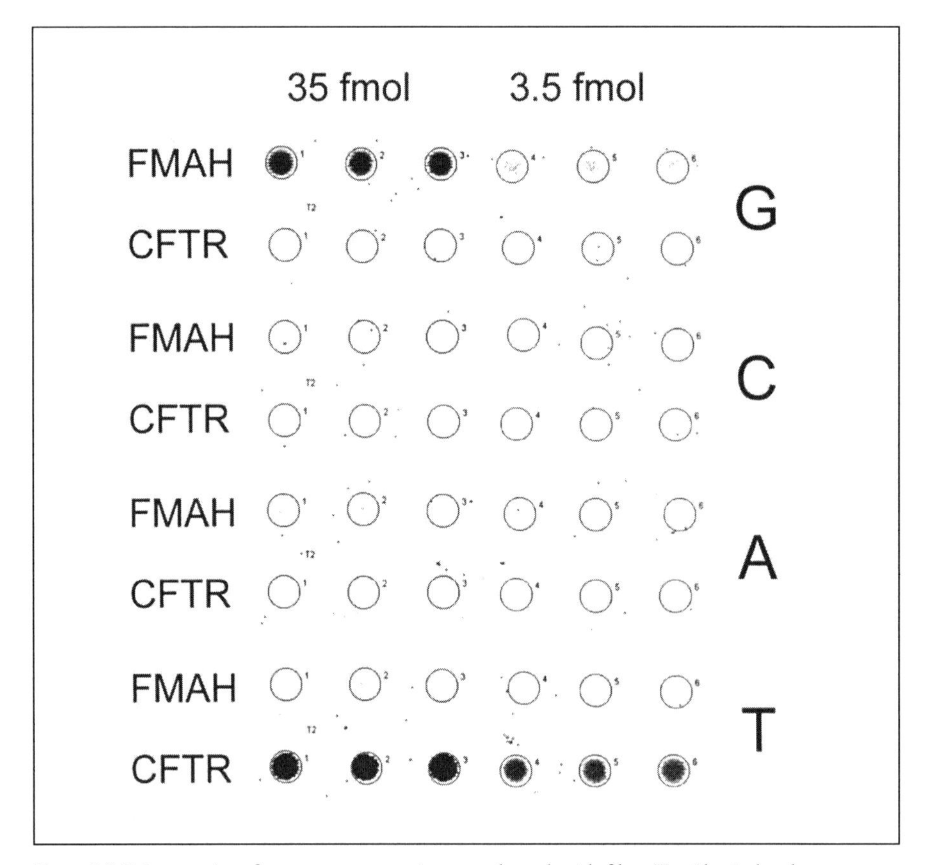

Figure 9. Minisequencing of two targets on continuous polyacrylamide films. Two identical probe arrays were created on polyacrylamide-coated microscope slides by dispensing single droplets of aminated oligonucleotide at discrete addresses. Each array contained two different quantities of minisequencing primers arrayed in triplicate for two SNPs from FMAH and CFTR, respectively. The 35- and 3.5-fmol quantities correspond to the amount of oligonucleotide dispensed onto the polyacrylamide layer. Minisequencing reactions were performed with 50 fmol of the two synthetic targets on each array using two fluorescently labeled dideoxy nucleotide terminators (TMR-ddGTP and Cy5-ddTTP or TMR-ddCTP and Cy5-ddATP). Both arrays were scanned in two channels (TMR and Cy5) with the laser scanner to obtain the four images shown, each of which reveals the incorporation of one of the four nucleotides (G, C, A, T). The circular regions around the spots on the images are 200 μm in diameter.

Table 2. Minisequencing on Polyacrylamide Films

Primer (fmol)	Average	CV (%)
35	944200	1.7
3.5	227030	11.4
0.35	68210	8.8
0.035	14100	8.3

Average signals and coefficients of variation (CVs) for minisequencing reactions performed in groups of 12 spots across a continuous polyacrylamide film. The indicated amount of aminated oligonucleotide primer was dispensed onto each spot as a single droplet.

porous substrates. For minisequencing, signal levels are up to 20 to 30 times greater on polyacrylamide films than on activated glass slides.

ACKNOWLEDGMENTS

I thank my colleagues at Packard Instrument Company, including Kevin Croker, Eric Hoffman, Alicia Hulbert, Lori Losh, Roeland Papen, Dee Shen, Shane Weber, and Karen Woodward, whose work I have described here. I thank Staf van Cauter at Packard for reviewing the manuscript.

REFERENCES

1. **Dubiley, S., E. Kirillov, Y. Lysov, and A. Mirzabekov.** 1997. Fractionation, phosphorylation and ligation on oligonucleotide microchips to enhance sequencing by hybridization. Nucleic Acids Res. *25*:2259-2265.
2. **Guo, Z., R. Guilfoyle, A. Thiel, R. Wang, and L. Smith.** 1994. Direct fluorescence analysis of genetic polymorphisms by hybridization with oligonucleotide arrays on glass supports. Nucleic Acids Res. *22*:5456-5465.
3. **Khrapko, K., Y. Lysov, A. Khorlyn, V. Shick, V. Florentiev, and A. Mirzabekov.** 1989. A oligonucleotide hybridization appproach to DNA sequencing. FEBS Lett. *256*:118-122.
4. **Khrapko, K., Y. Lysov, A. Khorlin, I. Ivanov, G. Yershov, S. Vasilenko, V. Florentiev, and A. Mirzabekov.** 1991. A method for DNA sequencing by hybridization with oligonucleotide matrix. DNA Sequence - J. DNA Sequencing Mapping *1*:375-388.
5. **Lockhart, D., H. Dong, M.C. Byrne, M. Follettie, M. Gallo, M. Chee, M. Mittmann, C. Wang, M. Kobayashi, H. Horton, and E. Brown.** 1996. Expression monitoring by hybridization to high density oligonucleotide arrays. Nat. Biotechnol. *14*:1675-1680.
6. **Pastinen, T., J. Partanen, and A.-C. Syvanen.** 1996. Multiplex, fluorescent, solid-phase minisequencing for efficient screening of DNA sequence variation. Clin. Chem. *42*:1391-1397.
7. **Proudnikov, D. and A. Mirzabekov.** 1996. Chemical methods of DNA and RNA fluorescent labeling. Nucleic Acids Res. *24*:4535-4542.
8. **Schober, A., R. Günther, A. Schwienhorst, M. Döring, and B. Lindemann.** 1993. Accurate high-

speed liquid handling of very small biological samples. BioTechniques *15*:324-329.
9. **Southern, E., K. Mir, and M. Shchepinov.** 1999. Molecular interactions on microarrays. Nat. Genet. Suppl. *21*:5-9.
10. **Timofeev, E., S. Kochetkova, A. Mirzabekov, and V. Florentiev.** 1996. Regioselective immobilization of short oligonucleotides to acrylic copolymer gels. Nucleic Acids Res. *24*:3142-3148.
11. **Wodicka, L., H. Dong, M. Mittmann, M.-H. Ho, and D. Lockhart.** 1997. Genome-wide expression monitoring in *Saccharomyces cerevisiae.* Nat. Biotechnol. *15*:1359-1366.
12. **Yershov, G., V. Barsky, A. Belgovskiy, E. Kirillov, E. Kreindlin, I. Ivanov, S. Parinov, D. Guschin, A. Drobishev, S. Dubiley, and A. Mirzabekov.** 1996. DNA analysis and diagnostics on oligonucleotide microchips. Proc. Natl. Acad. Sci. USA *93*:4319-4918.

12 Arrayed Primer Extension on the DNA Chip: Method and Applications

Neeme Tõnisson[1,2], Ants Kurg[1], Elin Lõhmussaar[1], and Andres Metspalu[1,2]
[1]Institute of Molecular and Cell Biology, University of Tartu, Estonian Biocentre, Tartu, Estonia; [2]Asper Ltd., Tartu, Estonia

INTRODUCTION

The first draft of the complete human genome will be available during the year 2000. Its impact on medicine will be the same as Mendeleev's periodic table on chemistry (i.e., chemistry was differentiated from alchemy). The rapidly accumulating human genetic databases predict an equally rapid increase in the number of genetic tests for both basic biomedical research and medical genetics. Structural (polymorphisms, mutations) and functional data of gene expression patterns are needed to obtain better knowledge about the genetic basis of multifactorial diseases, to allow effective drug discovery, therapy, and personalized medicine. New, cost-effective technologies with the potential of automation and parallelism, together with affordable costs, are needed to satisfy these demands. The rapidly developing DNA array technologies will fulfill a major proportion of these needs.

DNA ARRAY TECHNOLOGIES

DNA Array Formats

A number of reviews on DNA arrays have been published recently; the most comprehensive and recent are collected in the "Chipping Forecast" supplement to *Nature Genetics* (18). An overview of some of the early developments can be

Microarray Biochip Technology
Edited by Mark Schena

found in *Microsystem Technology: A Powerful Tool for Biomolecular Studies* (14). DNA microarrays are devices that contain specific oligonucleotides or longer DNA fragments attached in a discrete order on activated solid surfaces. Three main types of DNA microarrays are used currently: oligonucleotide arrays (15–25 mers), gene expression arrays containing complementary DNAs (cDNAs) and expressed sequence tags (ESTs), and genomic arrays of bacterial artificial chromosomes (BACs) and yeast artificial chromosomes (YAC clones). Oligonucleotides on arrays can be either prefabricated and delivered to the surface or synthesized in situ (4,12,22).

Applications of DNA Microarrays

The two main applications for use of DNA arrays are DNA resequencing and comparative gene expression analysis. Oligonucleotide arrays are used for both purposes (2,7,11,22). Differential gene expression patterns can be characterized by cDNA microarrays, when amplified polymerase chain reaction (PCR) products are bound to a solid surface and hybridized simultaneously to mRNA-derived probes from two different sources. The two mRNA-derived samples are labeled with two different fluors, and relative expression levels from two different sources can be obtained by measuring fluorescence intensities; moreover, relative abundance of individual mRNAs can be visualized by comparing the ratios of labels (20). Genomic arrays are used similarly for comparative genomic hybridization (CGH), but are 2 orders of magnitude higher in resolution than the conventional CGH (23).

In this chapter, we focus on the use of oligonucleotide arrays for DNA sequence analysis. This includes resequencing and mutation detection of known genes and single-nucleotide polymorphism (SNP) testing. Although mutation detection is the first obvious application of oligonucleotide arrays, SNP testing to find disease genes underlying complex diseases appears to be the most important one. Moreover, the introduction of personalized drug therapy and early genetic risk assessment will require DNA array techniques to be specific, robust, and cost-effective.

Allelic Discrimination with Enzymatic Reactions

Hybridization is an essential step in all DNA microarray platforms. However, signal-to-noise ratio for single nucleotide discrimination is quite low if the reaction is based only on the DNA hybridization. Single nucleotide discrimination is not overly critical in gene expression monitoring but is extremely important in the detection of heterozygous mutations or polymorphisms and somatic mutations in cancer. This fact has led several groups to look for methods with better allelic discrimination capabilities. A promising alternative lies in the use of enzymes, either polymerase (16,22), ligase (5,11), or cleavase (1), as an additional recognition mechanism when coupled to hybridization.

ARRAYED PRIMER EXTENSION

Arrayed primer extension (APEX) is similar to minisequencing (16,25) and genetic bit analysis (GBA) (15). The APEX method is based on incorporation of four dye terminators into oligonucleotide primers with a DNA polymerase (Figure 1). This can be viewed as similar to Sanger dideoxy sequencing technology.

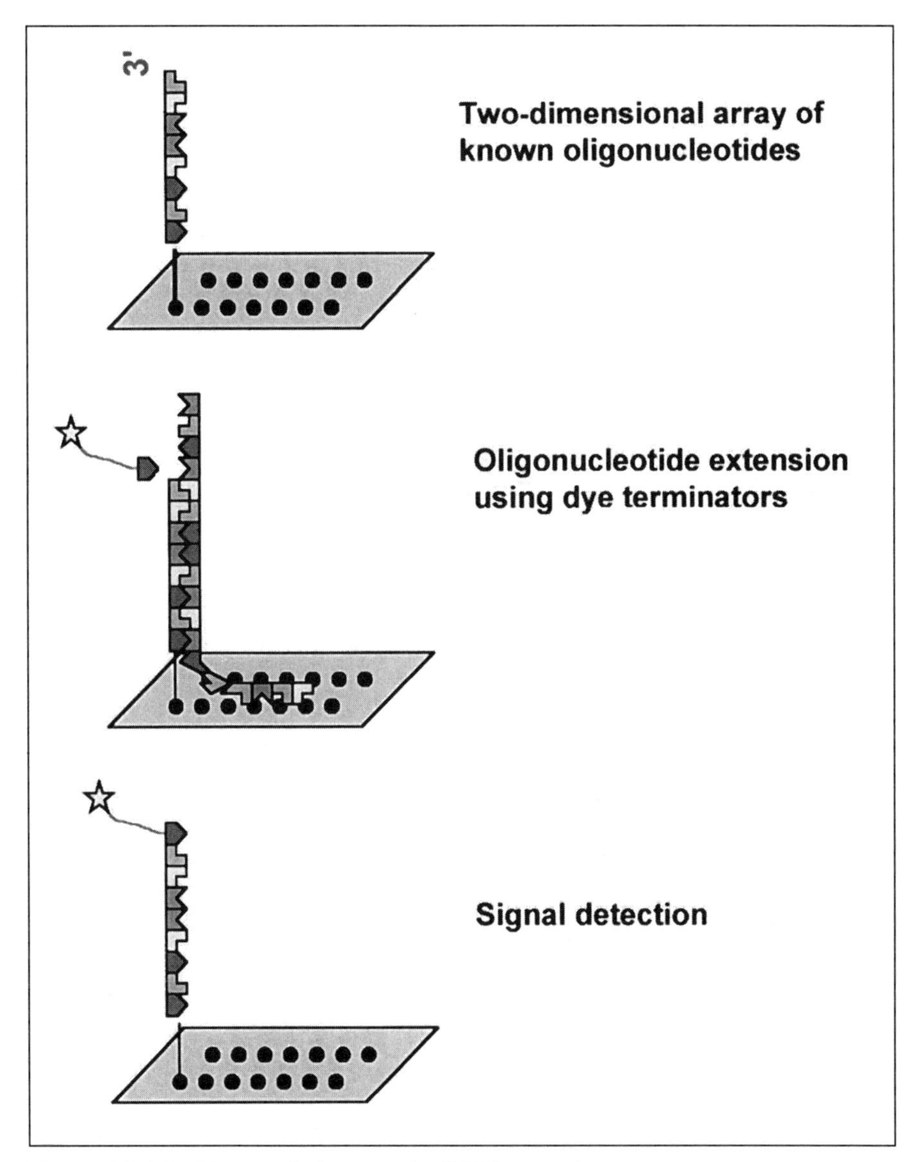

Figure 1. Principle of the arrayed primer extension (APEX) approach.

Instead of one primer being elongated into many fragments and fragment lengths used for their separation, APEX arrays use many primers positioned on the array that are elongated by only one base at each array site.

OLIGONUCLEOTIDE LIGATION REACTION

The oligonucleotide ligation reaction utilizes either one (padlock probe type) or two oligonucleotides with a ligation site located at the nucleotide to be identified (5,11). Both types of ligation assays can be adapted to a microarray platform.

CLEAVASE-BASED ASSAYS

A number of assays have been developed by Third Wave Technologies (http://www.twt.com) that utilize a family of structure-specific enzymes termed Cleavase™. Cleavase cuts junctions between single- and double-stranded regions of DNA (1). The assays are based on the observation that single strands of DNA form distinct higher order structures by folding on themselves. Although they are highly allele specific, robust, and reproducible, the assays are difficult to transform into a microarray format.

APEX ASSAY

The APEX reaction combines the high information content of oligonucleotide microarrays with the specificity of molecular recognition by DNA polymerase. APEX is capable of identifying mutations and polymorphisms (Figure 2), and can be used for resequencing (13,14). For testing known mutations, primers are selected to allow discrimination between wild-type and mutant alleles. For unknown mutations, the only option is to use the DNA resequencing application on the chip. This assay may be one of the most used genetic tests for mutation detection in the future.

Hybridization platforms like Affymetrix require many oligonucleotides and complicated analysis algorithms to analyze signal intensity and make comparisons across multiple features on the chip to obtain valid data. APEX, on the other hand, utilizes only two oligos based on a wild-type sequence to analyze one base pair. The analysis is similar to that of an automated ABI type DNA sequencer that uses normalized intensity comparisons from different fluorescent labels on the same band on a gel.

Oligonucleotide Selection

APEX technology can be tailored to different applications simply by changing the arrayed oligonucleotides. Oligonucleotides are synthesized according to

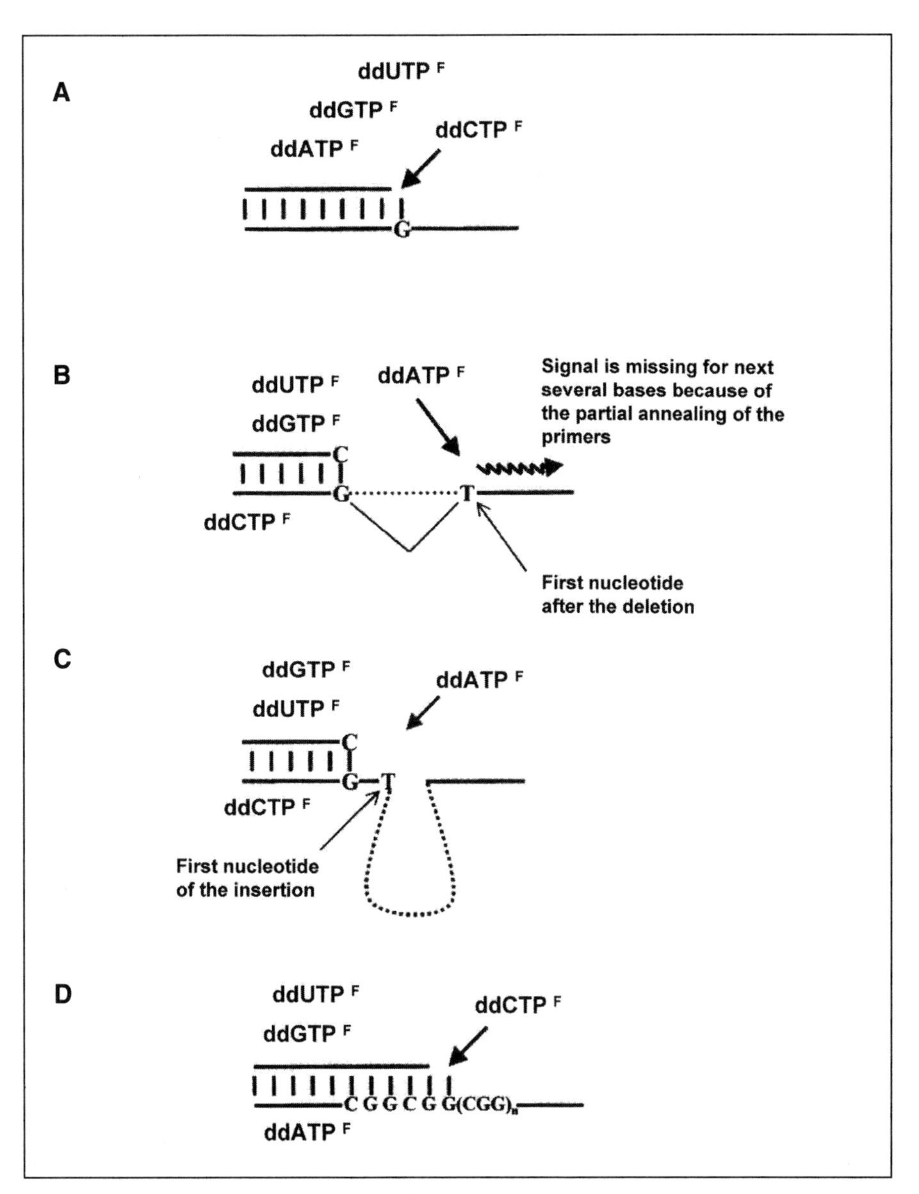

Figure 2. DNA sequence alterations analyzed by APEX. (A) Single-nucleotide polymorphism (SNP) and point mutations. Since all four dye terminators (ddNTPs) are present in the primer extension reaction, oligonucleotides are extended by a complementary nucleotide. (B) A deletion analysis primer immediately upstream of the deletion is extended by a dye terminator complementary to the base after the deletion. This allows detection of the borders of the deletion. By analyzing both strands, nucleotide sequence upstream and downstream of the deletion can be obtained. The length of the detection equals the "footprint" (region with no signal) on the resequencing chip. (C) Insertions. By analyzing both strands, one can detect the first and the last nucleotide of the inserted sequence. Mono- or dinucleotide insertions are identified completely. (D) Repeats. Di-, tri-, tetra (etc.) nucleotide repeats can be analyzed if the repeat length is shorter than the oligonucleotide primer on the oligonucleotide microarray. Long repeats (as in the fragile X case) cannot be analyzed at the present time using this version of the APEX assay.

the wild-type sequence, with an amino linking group at the 5′ end. The amino link is connected to a spacer arm of 12 to 18 carbon atoms and ensures covalent immobilization of the oligonucleotides via their 5′ end. The 3′ end of the oligonucleotides is free for enzymatic extension by DNA polymerase. The length of the spacer arm has a significant impact on oligonucleotide accessibility by the template DNA strand and consequently on hybridization efficiency (24). The 3′ end of each oligonucleotide is positioned one nucleotide upstream of the nucleotide to be analyzed. For resequencing with the APEX assay, oligonucleotides are designed with a one-nucleotide shift in their sequence. For example, 1000 oligonucleotides are needed to identify 1000 bases in the sample DNA. As APEX allows analysis of both strands in the same reaction, two oligos are arrayed per base pair. We have used 20- and 25-mer oligonucleotides in the APEX assay. In our experience, signals with 25-mers are 10% to 50% stronger than with 20-mers. Unfortunately, 25-mers also have more stable secondary structures (hairpins or oligonucleotide dimers), resulting in a loss of specificity due to self-extension. Approximately 10% to 15% of oligonucleotides do not give satisfactory signals on the first attempt and need to be replaced with new oligonucleotides. There is currently no reliable method to predict the overall hybridization and self-extension behavior of arrayed primers prior to running the APEX assay.

Surface Chemistry Used for Oligonucleotide Arrays

A variety of solid supports such as glass, plastic, nylon, and polyacrylamide have been used in DNA chip manufacturing. An ideal substrate for APEX would have certain technical and chemical properties. These would include mechanical and chemical stability, ease of chemical modification and derivatization, absence of autofluorescence, and low cost. We and many other laboratories have chosen glass microscope slides as the substrate material, due to low cost and ready availability. Glass provides a nonporous, relatively homogenous chemical surface that can be used in conjunction with silane chemistries. However, low-cost glass may be heterogeneous, and this variability may cause problems during surface treatment. High-quality glass substrates are now available (http://arrayit.com and http://www.cmt.corning.com), providing greater reliability during surface modification.

The attachment chemistry must also meet several criteria. The accessibility and functionality of the surface-bound DNA for hybridization, the density of attached oligonucleotides, and the reproducibility of the attachment chemistry are the most critical parameters. The chemical linkage must be stable, exhibit minimal nonspecific binding to the reaction components, and provide sufficient clearance from the solid surface to minimize or eliminate steric hindrance.

The two most commonly used silane-modified surfaces for DNA array preparation include epoxysilane-modified (10) and aminosilane-modified surfaces (6). Oligonucleotide primers are attached to epoxy-activated glass surfaces by secondary amine formation between the epoxy-derivatized glass surface and 5′

amino linkers. Amino-linked primers are diluted in sodium hydroxide and spotted onto the activated surface (22).

For amino derivatization, the slides are silanized with 3-aminopropyltrimethoxysilane solution. To convert the amino groups to amino-reactive phenylisothiocyanate groups, the slides are treated with a solution of 1,4-phenylene diisothiocyanate. Amino-modified oligonucleotide primers are diluted in sodium carbonate/bicarbonate buffer (pH 9.0) and spotted onto the activated surface as described above. The slides are later blocked with NH_4OH, washed with water, dried with nitrogen flow, and stored at room temperature. In our experience, they are stable at least for 6 months from the time of preparation.

Template Preparation for APEX

Preparation of template DNA for APEX starts from (multiplex) amplification by the PCR. Both single- and double-stranded target DNA can be used in APEX reactions. Single-stranded DNA has the advantage of lacking a complementary strand, thereby eliminating competing hybridization with arrayed primers. The main disadvantage is that preparing sufficient amounts of single-stranded target is laborious and expensive, and one loses 50% of the genetic information in the assay. The use of a double-stranded target permits single base changes to be identified from both strands in a single reaction, thereby increasing the reliability of the assay.

One important issue in the APEX assay is the length of the template. We have found that the maximum length of a double-stranded PCR product that can be used is approximately 200 bp, but 100 bp is optimal. Longer PCR products require fragmentation before use in the APEX reaction. Fragmentation is performed by replacing a fraction of the dTTPs with dUTPs in the amplification mix, followed by the use of uracil *N*-glycosylase (UNG) to fragment the template (3). UNG is highly specific to uracil bases in DNA, and therefore the reaction can be controlled by dUTP incorporation during PCR. By changing the dUTP concentration from 15% to 30% of dTTP, one can vary the mean length of the template DNA fragments. Several other possibilities exist for DNA target fragmentation, including DNase I treatment, restriction enzyme digestion, and mechanical shearing. However, none of these worked reproducibly in our hands.

The APEX principle for the single-nucleotide extension reaction works only if no deoxyribonucleotide triphosphates are carried over from the amplification reaction. A simple and quick way to inactivate the dNTP leftover from PCR is enzymatic digestion with shrimp alkaline phosphatase (SAP). This is performed in a single-tube reaction together with UNG treatment, and both enzymes are thermally inactivated prior to the APEX reaction (Figure 3).

APEX Reaction

APEX is a complex, single-step reaction consisting of target annealing to the oligonucleotide array and an enzymatic primer extension reaction with fluorescent dideoxy nucleotides. Engineered DNA polymerases (26) are able to incorporate dye terminators quite efficiently. We use Thermo Sequenase (Amersham Pharmacia Biotech, Piscataway, NJ, USA) and 20 to 50 pmol of each fluorescent terminator per APEX reaction. Other commercially available polymerases are also capable of incorporating labeled dideoxy nucleotides, including DynaSeq

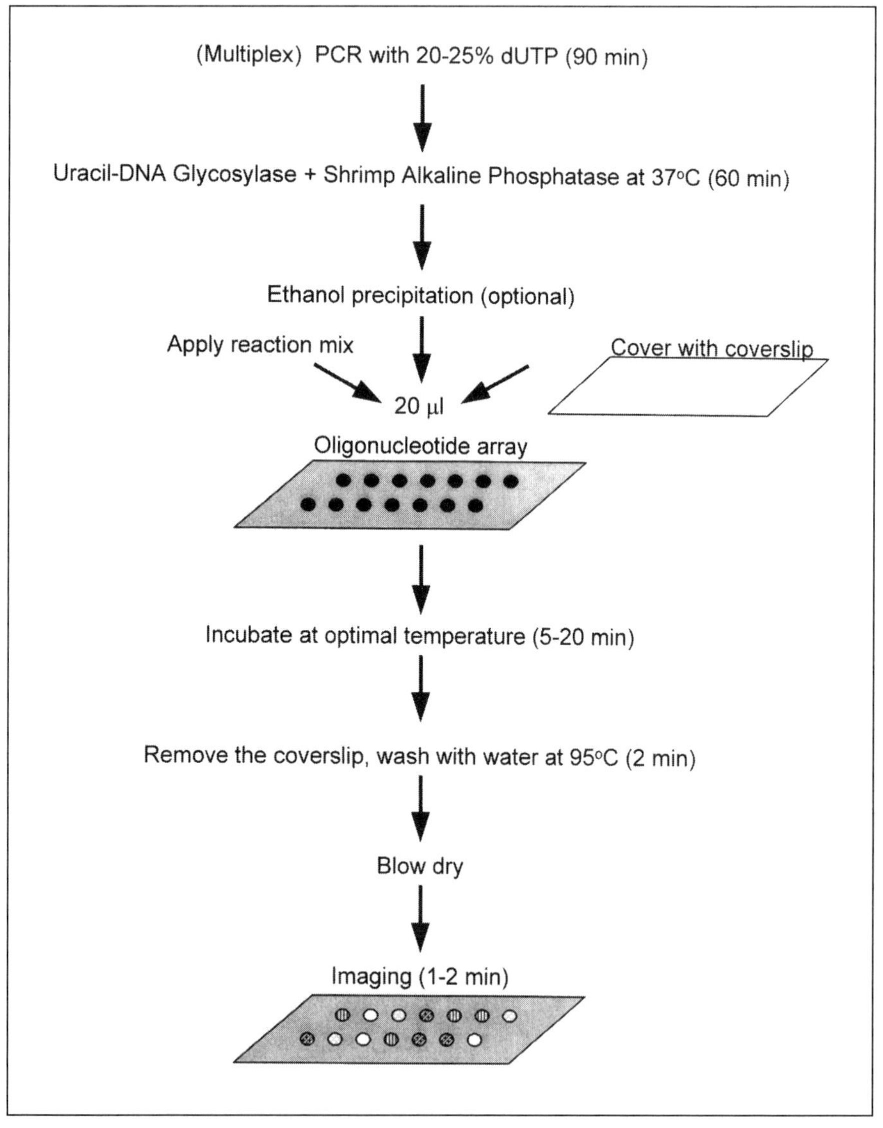

Figure 3. One-tube, one-chip reaction scheme of the APEX reaction. It takes about 3 hours for the entire test, including template amplification by PCR.

(Finnzymes OY, Espoo, Finland), AmpliTaq® FS (Roche Molecular Systems, Branchburg, NJ, USA), and others. Template hybridization to the microarray and the primer extension reaction are different in terms of their optimal temperatures. Hybridization is more effective at lower temperatures, whereas polymerase works better at higher temperatures. For this reason, some primer extension protocols are designed as two separate steps, whereby hybridization is performed first and primer extension with an appropriate DNA polymerase second (16). Unbound target is removed by washing between these two steps. Our goal is to make the APEX assay as robust as possible, and therefore we have used single-step reactions. Hybridization and target-dependent extension of arrayed primers can be performed simultaneously. The signal-to-noise ratio of APEX is currently 30:1 to 100:1. The primer extension reactions are routinely performed at a constant temperature at 48°C for 20 minutes (9). Incorporation of labeled terminators is a rapid reaction, but hybridization is an equilibrium process that requires longer reaction times to obtain strong signals.

The choice of dye labels is determined by two factors: (*i*) the labels must be well separated spectrally, and (*ii*) they must be efficiently incorporated by the DNA polymerase used. Currently, we use the following labels listed in increasing wavelength of emitted light: fluorescein, Cy3, Texas Red, and Cy5. All but fluorescein are very stable to photobleaching. Because of the photosensitivity of fluorescein, we use the antifading reagent SlowFade® Light (Molecular Probes, Eugene, OR, USA) for the imaging step.

Arraying Equipment

Microarrays containing a small number of spots (e.g., 1–2-mm center-to-center spacing) can be made with standard laboratory pipetting robots. Denser arrays require advanced arraying equipment for preparation. Current robotic systems are capable of placing up to 10 000 DNA spots/cm^2 (21). Arraying is done either by mechanical microspotting or ink-jet-type piezoelectric dispensing. A number of arrayers are commercially available from a number of different vendors, and most use different variations of pin spotting (18). An interesting technology developed by Genetic MicroSystems (Woburn, MA, USA; http://www.geneticmicro.com) combines pin spotting and sample loading in a ring. In the Pin and Ring technology, a small-diameter ring is first immersed into the sample solution and then the pin is driven through the ring to deliver nanoliter volumes of the solution to the array surface. One drawback of the Pin and Ring is that the ring holds a rather large volume of sample, requiring a large sample volume for ring loading. Regular spot geometry and uniform density are both very critical parameters for microarray manufacture.

A number of different companies are also offering custom-made microarrays and silanized glass slides, which are probably the best options for small laboratories requiring only a small number but high-quality arrays. Custom microarray services (see, for example, http://arrayit.com) obviate the need for expensive hardware, sample storage, surface chemistry, labor, and many other issues.

Imaging Equipment

A high-quality imaging system must allow at least 10 × 10 pixels per arrayed element. This means that the minimal pixel diameter of the detector must be at least 10 μm for a 100-μm center-to-center distance array. Some arrayers operate in the 200-μm center-to-center distance, allowing acceptable detector readout with 20-μm pixel size. Commercially available detectors mainly use confocal scanning for array imaging (18). Confocal scanning has a high sensitivity and dynamic range but is significantly slower than total internal reflection (TIR)-based imaging.

We have developed an automatic TIR-based system for fast analysis (FD-003) of APEX reactions (Figure 4). Four lasers of 3- to 10-mW power, with light-emitting wavelengths of 488, 532, 594, and 635 nm, are used to excite the fluorescent labels on the APEX microarrays. The optical system consists of light reflectors and cylindrical lenses that transform a beam from each laser into a homogeneously illuminated stripe that is directed into the interior of the glass slide. Due to TIR of the slide surface, the incoming beam spreads only inside the slide and therefore the intensity of light remains uniform along the entire length of the substrate. The evanescent light field excites the bound fluorophores residing near the surface of the slide, and a gated charge-coupled device (CCD) camera cooled to -25°C records the emitted fluorescence. The respective narrowband interference filters fixed on the revolving wheel in front of the CCD camera depress noise radiation from scattered laser light on the slide surface. Using one, two, or four dye markers, four images corresponding to four laser wavelengths are obtained, one for each dye-labeled ddNTP.

Custom software controls all the parameters of the detection procedure including switching on and off the shutters of the lasers, choosing respective interference filters and setting the recording time at each wavelength. Because of precise spectral separation of exciting and emitting wavelengths, the system demonstrates a high signal-to-noise ratio. The time required for one fluorescence channel is 10 to 60 seconds, enabling a theoretical throughput of 60 samples per hour. More information on the TIR-based fluorescence detector can be found at http://www.asper.ee. Additional software was developed to convert the fluorescence signals into DNA sequence information.

APEX APPLICATIONS

Detecting Common Mutations in the Human β-Globin Gene

β-Thalassemia is a common autosomal recessive disorder caused by mutations in the human β-globin gene, resulting in imbalanced globin β-chain production.

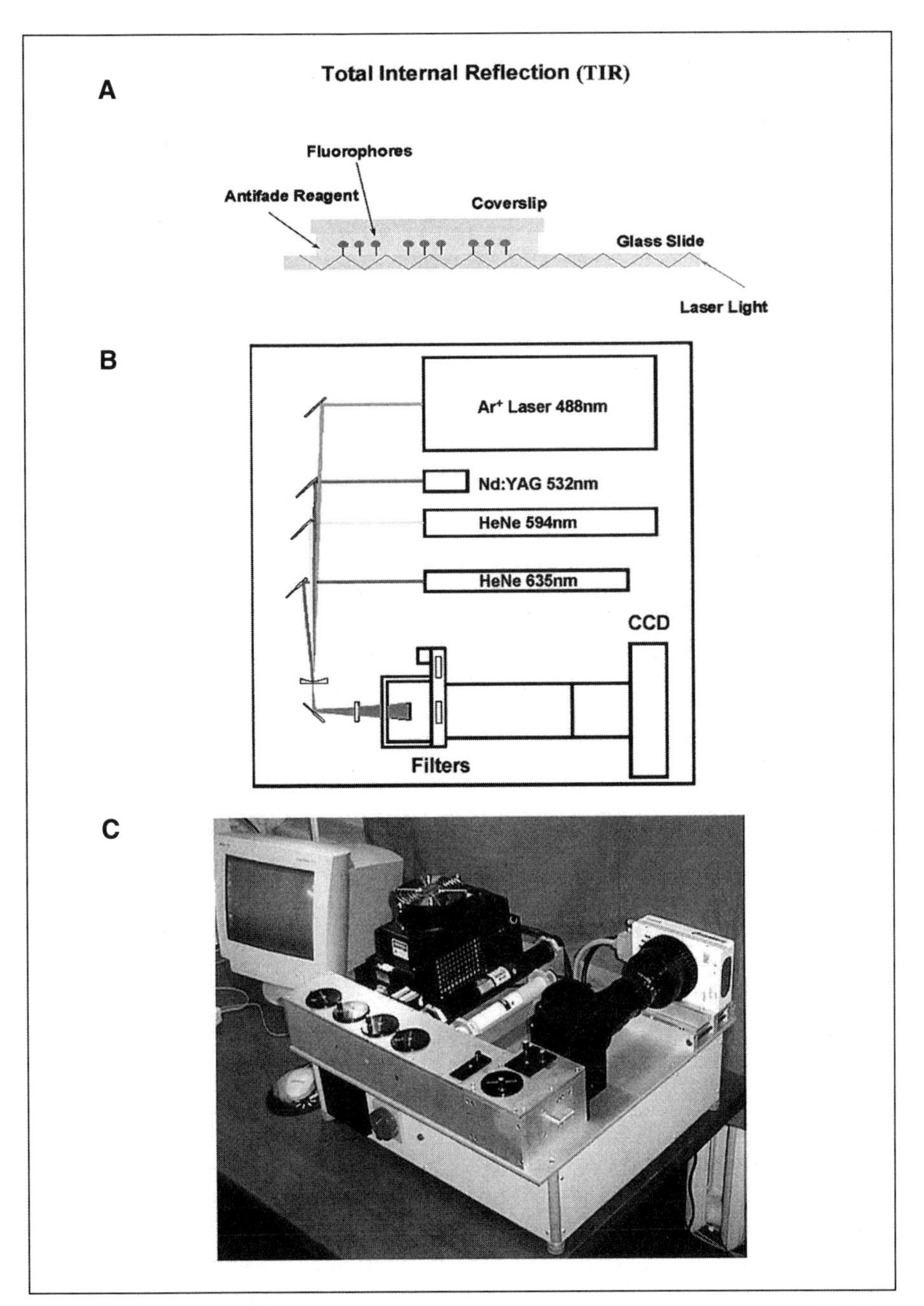

Figure 4. Four-color fluorescence detector based on total internal reflection (TIR). (A) Principle of the fluorophore excitation mechanism based on TIR. The glass slide acts as a waveguide, and fluorophores residing near glass surface are excited by the evanescent light field. (B) Layout of main detector components. Light from four lasers is collected into a common path, transformed into a light stripe, and introduced into the glass slide. Emission is filtered and collected into a cooled CCD camera. Four images are taken per APEX reaction, one for each fluorophore. (C) Photo of the fluorescence detector.

By World Health Organization estimates, approximately 240 million people worldwide are heterozygous for β-thalassemia, and annually at least 200 000 affected individuals are born. Current screening methods are performed via chromatography, protein electrophoresis, and evaluation of the hematological indices. As a model platform for high-throughput genetic testing of β-thalassemia, we used APEX technology to identify 10 β-globin gene mutations, accounting for >95% of all cases found in the Mediterranean region.

To perform the test, a 1.4-kb fragment was amplified from genomic DNA. In the reaction, 25% of the dTTP was replaced with dUTP in the amplification mix. The amplification products were purified and concentrated by ethanol precipitation. Fragmentation of the target and cleanup of the leftover dNTPs was done simultaneously with UNG and SAP. A 200-ng aliquot of the amplified target was used for each APEX reaction. A readout of the β-thalassemia APEX analysis is shown in Figure 5. Nine patient samples containing either homozygous or heterozygous mutations in the β-globin gene were correctly identified using the approach described above (9).

Population Genetics: Finding Risk Factors for Myocardial Infarction

Because it was efficient and fast, a minisequencing method similar to the APEX platform described was used by a Helsinki group to identify genetic risk factors of myocardial infarction in the Finnish population (17). A total of 152 individuals with a history of myocardial infarction and an identical number of control individuals, matched for age, sex, and geography were simultaneously studied for 12 polymorphisms in eight genes known to play roles in platelet adhesion, coagulation, or fibrinolysis. Two alleles were found to be associated with an increased risk of myocardial infarction. Exemplifying the advantage of multiplex genetic analysis, carrier status for both of these polymorphisms conferred additive risk. This is the first study in which minisequencing on oligonucleotide arrays was applied to a large number of individuals. In total approximately 4000 genotypes were identified in this study, greatly exceeding the scope of the model experiment.

SNP Testing

SNPs are the most common type of DNA sequence variation. SNPs are estimated to occur once every 500 to 1000 bp when any two chromosomes are compared (19,27). Recent population modeling demonstrates that one has to map as many as 500 000 SNPs from an individual to perform linkage disequilibrium (LD) analysis (8). SNPs offer a number of advantages with respect to population-based analysis of the human genome. They are very frequent and allow binary assays to be developed, thereby permiting automation and digitalization of the data, which are prerequisites for storing and analyzing the huge amount of genetic information acquired from DNA chip experiments.

A

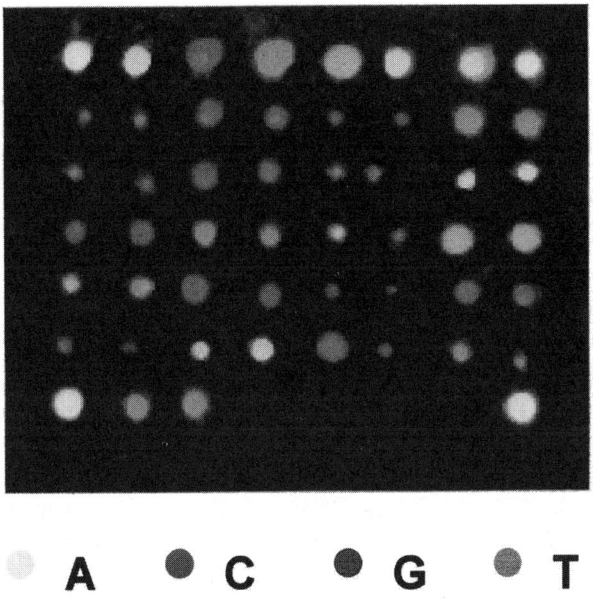

A C G T

B

	S 25mer	S 20mer	As 25mer	As 20mer	S 25mer	S 20mer	As 25mer	As 20mer
1	N	A	C	G	T	AC	AG	N
2	IVS-I- 110 G to A	IVS-I- 110 G to A	IVS-I- 110 C to T	IVS-I- 110 C to T	Codon 39 C to T	Codon 39 C to T	Codon 39 G to A	Codon 39 G to A
3	IVS-I- 1 G to A	IVS-I- 1 G to A	IVS-I- 1 C to T	IVS-I- 1 C to T	IVS-I- 6 T to C	IVS-I- 6 T to C	IVS-I- 6 A to G	IVS-I- 6 A to G
4	IVS-II 745 C to G	IVS-II 745 C to G	IVS-II 745 G to C	IVS-II 745 G to C	Codon 6 A to G	Codon 6 A to G	Codon 6 T to C	Codon 6 T to C
5	IVS-II 1 G to A	IVS-II 1 G to A	IVS-II 1 C to T	IVS-II 1 C to T	IVS-I- 5 G to A	IVS-I- 5 G to A	IVS-I- 5 C to T	IVS-I- 5 C to T
6	Codon 5 C to T	Codon 5 C to T	Codon 5 A to G	Codon 5 A to G	-87 C to G	-87 C to G	-87 G to C	-87 G to C
7	N	Codon 5 G	Codon 5 G					N

Figure 5. β-Thalassemia APEX with a wild-type DNA sample. (A) Color code of the incorporated nucleotides is shown below the image. (B) Key code of the array. Seven rows and eight columns of oligonucleotides were arrayed onto the slide. Two oligonucleotides (20- and 25-mer) for the sense strand (s) and two for the antisense strand were used for each mutation site. Upper row and four corners are self-elongating marker oligonucleotides. Wild-type signals and signals expected from mutations are indicated at the bottom of each cell. **(See color plate A30.)**

We are working on APEX-based DNA arrays to analyze genome-wide SNP markers—one from each chromosome. We have tested the array with individually amplified SNP-specific templates and provide a complete experimental protocol below. Quantitative results of the SNP tests are shown in Figure 6. Although all the SNPs were tested using specifically amplified templates, we fully understand that it would be impractical to amplify 10 000 to 500 000 fragments for SNP mapping in LD studies. To solve the problem, we are working in two directions: first, we are using whole-genome amplification to generate enough DNA for APEX, and second, we are increasing the sensitivity of the APEX system including optimization of surface chemistry, enzyme reactions, and imaging so that 100 000 to 500 000 copies of the genome obtained from 10 to 100 μL of whole blood would be sufficient for the genetic test.

Template DNA preparation for SNP assay. PCR primers were obtained from Life Technologies (Gaithersburg, MD, USA). The PCR mixture (50 μL) contained the following: 5 μL of 10× PCR buffer [200 mM Tris-HCl (pH 8.4), 500 mM KCl; Life Technologies], 2.5 mM $MgCl_2$, 0.2 mM of each deoxynucleotide triphosphate (dATP, dCTP, dGTP), 0.16 mM dTTP, 0.04 mM dUTP (Amersham Pharmacia Biotech), 40 pmol of each primer, 50 ng of template DNA, and 1 unit of Platinum™ Taq DNA Polymerase (Life Technologies). The amplification reactions were performed in a PTC-200 thermocycler (MJ Research, Watertown,

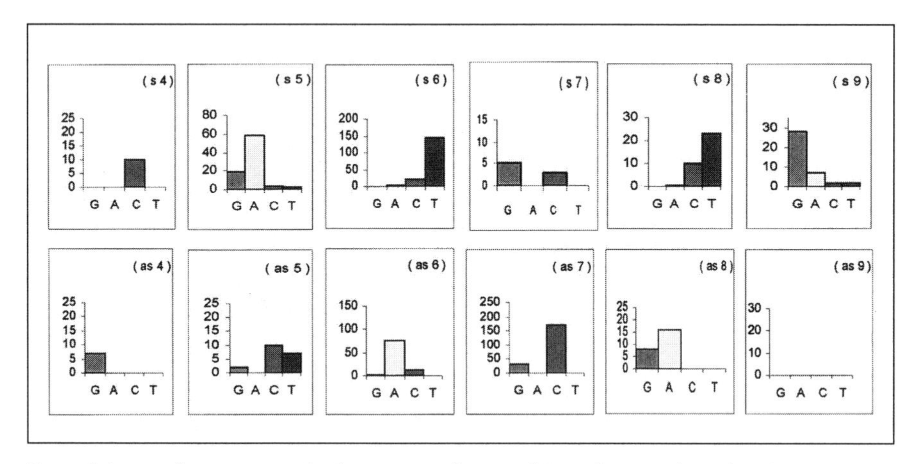

Figure 6. Average fluorescent signals of six sequentially arrayed SNPs from six identical APEX experiments with the same DNA sample. Both sense (s) and antisense (as) strands were analyzed on the same microarray. Positions 5, 7, 8, and 9 are heterozygous. The rest of positions (4 and 6) are homozygous. The signals from correctly incorporated nucleotides are clearly different from other signals. Position 4, for example, is homozygous with a C/G base pair. Position 5 is heterozygous with G plus A signal from the sense strand and corresponding C plus T signal from the antisense strand. Although these signals result only from correct nucleotide incorporation, incorrect nucleotides are sometimes incorporated also, for example, G signal in the antisense position of SNP 5. This misincorporation may result because of oligonucleotide self-extension or low stringency in the hybridization reactions. In some cases, as in SNP 9, the signal was obtained only from the sense strand of the DNA analyzed. The difficulty may arise from poor quality of the oligonucleotides or stable oligonucleotide secondary structures, which require substantially more arrayed oligonucleotide or more template in the reaction mix.

MA, USA). An initial incubation was performed at 95°C for 3 minutes, followed by 35 amplification cycles consisting of denaturation at 95°C for 30 seconds; primer annealing at 56°C for 30 seconds; extension at 72°C for 25 seconds; and the final extension at 72°C for 5 minutes. The amplification products were initially concentrated and purified by ethanol precipitation in the presence of ammonium acetate. DNA fragmentation and functional inactivation of unincorporated dNTPs was achieved in a one-step reaction by addition of 0.5 U of SAP (Amersham Pharmacia Biotech) and 0.5 U of thermolabile uracil *N*-glycosylase (Epicentre Technologies, Madison, WI, USA) per amplification. The reaction mixture was incubated at 37°C for 1 hour and used directly in primer extension reactions.

SNP model array. Oligonucleotides for primer extension were designed according to the wild-type sequence for each SNP in both the sense and antisense orientations. The 25-mer oligonucleotides with 12 carbon amino linkers at their 5′ end were obtained from Genset (Paris, France). Oligonucleotide primers were attached to an epoxy-activated glass surface via amino linkers at their 5′ ends (14). The 24 × 60-mm glass microscope coverslips (Fisherfinest Premium Cover Glasses, Fisher Scientific, Pittsburgh, PA, USA) were sonicated in acetone and 100 mM NaOH for 5 minutes each; rinsed in dH_2O (MilliQ) water; and finally sonicated for 2 minutes in 2% 3-glycidoxypropyltrimethoxysilane (Gelest, Tullytown, PA, USA) in 95% ethanol. Unbound silane and residual water were removed by a brief rinsing of the coverslip in 100% ethanol. Primers were diluted to 50 μM in 0.1 M NaOH and spotted onto the activated surface with a custom-manufactured 25-gauge, 96-channel spotter. The slides were stored in a dust-free environment at room temperature until needed and washed twice in 95°C dH_2O (MilliQ) water prior to the APEX reactions.

APEX reactions for SNP testing. The 20-μL primer extension reactions consisted of 10 μL fragmented DNA (200 ng of each amplified product), 6 U of Thermo Sequenase DNA polymerase (Amersham Pharmacia Biotech), 4 μL Thermo Sequenase reaction buffer concentrate [260 mM Tris-HCl (pH 9.5), 65 mM $MgCl_2$; Amersham Pharmacia Biotech] and 12 μM of each fluorescently labeled ddNTP (FITC-G, Cy3-C, Texas Red-A, and Cy5-U; NEN, Boston, MA, USA, and Amersham Pharmacia Biotech). The DNA in the buffer was denatured at 95°C for 10 minutes. The enzyme and dye terminators were immediately added to other components, and the whole mix was applied to washed and prewarmed slides at 48°C. The reactions were allowed to proceed for 5 to 25 minutes under coverslips and stopped by washing three times at 95°C for 90 seconds in dH_2O (MilliQ) water. A droplet of SlowFade® Light Antifade Reagent (Molecular Probes, Eugene, OR, USA) was applied to the chips before imaging. The signals were acquired with a custom-built TIR-based detector with a CCD camera FD-003 (Asper Ltd., Tartu, Estonia).

Analysis of data. Images were processed with Image Pro Plus™ software (Media Cybernetics, Silver Springs, MD, USA) and custom-written software to obtain the allele data from the analyzed SNPs.

CONCLUSIONS AND PERSPECTIVES

At the end of the first decade of DNA microarray technology development, no doubt remains that the DNA chips are useful and effective in many applications. Although large genomics and pharmaceutical companies have solved many of the initial questions concerning DNA microarrays, some problems remain. We need more standardization, better internal controls, and uniform data formats, all of which will make the results obtained by different groups comparable.

The arrayed primer extension technology (APEX), described in detail here, can be viewed as an advance in hybridization-based microarray technology with an additional allele-discriminating mechanism. We have developed a fully integrated genetic testing system consisting of template preparation, multiplex primer extension on the microarray, fluorescence imaging, and data analysis. One developmental cycle for our laboratory is complete. The next step will be to demonstrate the versatility and usefulness of the APEX technology in daily experimentation.

ACKNOWLEDGMENTS

The authors thank Heidi Saulep, Viljo Soo, and Jana Zernant for technical assistance, Evgeny Berik (Estla Ltd., Tartu, Estonia) for constructing the TIR-based imaging system, Mrs. Krista Liiv for the artwork, and Dr. Hiljar Sibul for critical reading of the manuscript. This work was supported by research grants from the European Community (IC15-CT98-0309), the Estonian Science Foundation (2492), and the Estonian Ministry of Education (core grant 0180518s98).

REFERENCES

1. **Brow, M.A., M.C. Oldenburg, V. Lyamichev, L.M. Heisler, N. Lyamicheva, J.G. Hall, N.J. Eagan, D.M. Olive, L.M. Smith, L. Fors, and J.E. Dahlsberg.** 1996. Differentiation of bacterial 16S rRNA genes and intergenic regions and Mycobacterium tuberculosis katG genes by structure-specific endonuclease cleavage. J. Clin. Microbiol. *34*:3129-3137.
2. **Chee, M., R. Yang, E. Hubbell, A. Berno, X.C. Huang, D. Stern, J. Winkler, D.J. Lockhart, M.S. Morris, and S.P.A. Fodor.** 1996. Accessing genetic information with high-density DNA arrays. Science *274*:610-614.
3. **Cronin, M.T., R.V. Fucini, S.M. Kim, R.S. Masino, R.M. Wespi, and C.G. Miyada.** 1996. Cystic fibrosis mutation detection by hybridization to light generated DNA probe arrays. Hum. Mutat. *7*:244-255.
4. **Fodor, S.P.A., J.L. Read, M.C. Pirrung, L. Stryer, A.T. Lu, and D. Solas.** 1991. Light-directed, spatially adressable parallel chemical synthesis. Science *251*:767-773.
5. **Gunderson, K.L., X.C. Huang, M.S. Morris, R.J. Lipshutz, D.J. Lockhart, and M.S. Chee.** 1998. Mutation detection by ligation to complete *n*-mer DNA arrays. Genome Res. *8*:1142-1153.
6. **Guo, Z., R.A. Guilfoyle, A.J. Thiel, R.Wang, and L.M. Smith.** 1994. Direct fluorescence analysis of genetic polymorphisms by hybridization with oligonucleotide arrays on glass supports. Nucleic Acids Res. *22*:5456-5465.
7. **Hacia, J.G., L.C. Brody, M.S. Chee, S.P.A. Fodor, and F.C. Collins.** 1996. Detection of heterozy-

gous mutations in BRCA1 using high density oligonucleotide arrays and two-colour fluorescence analysis. Nat. Genet. *14*:441-447.

8.**Kruglyak, L.** 1999. Prospects for whole-genome linkage disequilibrium mapping of common disease genes. Nat. Genet. *22*:139-144.

9.**Kurg, A., N. Tõnisson, I. Georgiou, J. Shumaker, J. Tollett, and A. Metspalu.** Arrayed primer extension: solid phase four-color DNA resequencing and mutation detection technology. Genet. Test. (In press).

10.**Lamture, J.B., K.L. Beattie, B.E. Burke, M.D. Eggers, D.J. Ehrlich, R. Fowler, M.A. Hollis, B.B. Kosicki, R.K. Reich, S.R. Smith, et al.** 1994. Direct detection of nucleic acid hybridization on the surface of a charge coupled device. Nucleic Acids Res. *22*:2121-2125.

11.**Landegren, U., R. Kaiser, J. Sanders, and L. Hood.** 1988. A ligase-mediated gene detection technique. Science *241*:1077-1080.

12.**Lockhart, D.J., H. Dong, M.C. Byrne, M.T. Follettie, M.V. Gallo, M.S. Chee, M. Mittmann, C. Wang, M. Kobayashi, H. Horton, and E.L. Brown.** 1996. Expression monitoring by hybridization to high-density oligonucleotide arrays. Nat. Biotechnol. *14*:1675-1680.

13.**Metspalu, A., H. Saulep, A. Kurg, and N. Tõnisson.** 1998. Primer extension from two-dimensional oligonucleotide grids for DNA sequence analysis. *In* G.K. Dixon, L.G. Copping and D. Livingstone (Eds.), Genomics Commercial Opportunities from a Scientific Revolution. BIOS Scientific Publishers, Oxford.

14.**Metspalu, A. and J.M. Shumaker.** 1999. DNA resequencing, mutation detection and gene expression analysis by oligonucleotide microchips, p. 371-397. *In* J.M. Köhler, T. Mejevaia and H.P. Saluz (Eds.), Microsystem Technology: A Powerful Tool for Biomedical Studies. Birkhäuser Verlag, Basel.

15.**Nikiforov, T.T., R.B. Rendle, P. Goelet, Y.-H. Rogers, M.L. Kotewicz, S. Anderson, G.L. Trainor, and M.R. Knapp.** 1994. Genetic bit analysis: a solid phase method for typing single nucleotide polymorphisms. Nucleic Acids Res. *22*:4167-4175.

16.**Pastinen, T., A. Kurg, A. Metspalu, L. Peltonen, and A.C. Syvänen.** 1997. Minisequencing: a specific tool for DNA analysis and diagnostics on oligonucleotide arrays. Genome Res. *7*:606-614.

17.**Pastinen, T., M. Perola, P. Niini, J. Trewilliger, V. Salomaa, E. Vartiainen, L. Peltonen, and A.-C. Syvänen.** 1998. Array-based multiplex analysis of candidate genes reveals two independent and additive genetic risk factors for myocardial infraction in the Finnish population. Hum. Mol. Genet. *7*:1-10.

18.**Phimister, B. (Ed.)** 1999. The Chipping Forecast. Nat. Genet. Suppl. *21*:1-60.

19.**Picoult-Newberg, L., T.E. Ideker, M.G. Pohl, S.L. Taylor, M.A. Donaldson, D.A. Nickerson, and M. Boyce-Jacino.** 1999. Mining SNPs from EST databases. Genome Res. *9*:167-174.

20.**Schena, M., D. Shalon, R.W. Davis, and P.O. Brown.** 1995. Quantitative monitoring of gene expression patterns with a complementary DNA microarray. Science *270*:467-470.

21.**Schena, M., R.A. Heller, T.P. Theriault, K. Konrad, E. Lachenmeier, and R.W. Davis.** 1998. Microarrays: biotechnology's discovery platform for functional genomics. Trends Biotechnol. *16*:301-306.

22.**Shumaker, J.M., A. Metspalu, and C.T. Caskey.** 1996. Mutation detection by solid phase primer extension. Hum. Mutat. *7*:346-354.

23.**Solinas-Toldo, S., S. Lampel, S. Stilgenbauer, J. Nickolenko, A. Benner, H. Dohner, T. Cremer, and P. Lichter.** 1997. Matrix-based comparative genomic hybridization: biochips to screen for genomic imbalances. Genes Chromosomes Cancer *20*:399-407.

24.**Southern, E., K. Mir, and M. Schepinov.** 1999. Molecular interactions on microarrays. Nat. Genet. *21*:5-9.

25.**Syvänen, A.-C., K. Aalto-Setälä, L. Harju, K. Kontula, and H.A. Söderlund.** 1990. Primer-guided nucleotide incorporation assay in genotyping of apolipoprotein E. Genomics *8*:684-692.

26.**Tabor, S. and C.C. Richardson.** 1995. A single residue in DNA polymerases of the Escherichia coli DNA polymerase I family is critical for distinguishing between deoxy- and dideoxyribonucleotides. Proc. Natl. Acad. Sci. USA *92*:6339-6343.

27.**Taillon-Miller, P., E.E. Piernot, and P.Y. Kwok.** 1999. Efficient approach to unique single-nucleotide polymorphism discovery. Genome Res. *9*:499-505.

13 Overview of a Microarray Scanner: Design Essentials for an Integrated Acquisition and Analysis Platform

Trent Basarsky, Damian Verdnik, Jack Ye Zhai, and David Wellis
Axon Instruments, Inc., Foster City, CA, USA

INTRODUCTION

Biological and medical science has entered the Genome Age. With the completion in the next few years of major genome sequences (human, mouse, *Drosophila*, rice, and *Arabidopsis*), the next wave will be functional genomic studies that will rely heavily on microarray technology. DNA chips will have an unprecedented impact and long-lasting influence on basic biology, biomedical research, biotechnology, and health care.

Microarray technology has made it possible to measure various parameters of gene expression for entire genomes reliably and reproducibly (see, for example, Reference 5). Microarrays are used for the acquisition and analysis of the gene expression data, as well as for mutation detection, genotyping, forensics, pathology, and possibly proteomics (the study of proteins and their functions, analogous to genomics). It may also be used in the identification and validation of new drug targets, as well as having widespread medical applications (see reviews in References 2, 4, 6 and 13).

Although glass surface technology is evolving rapidly, DNA microarrays containing tens of thousands of different spots of DNA, RNA, cDNA, or oligonucleotides can actually be prepared on common laboratory microscope slides. Each spot or feature can be used to detect the presence of a unique sequence of messenger RNA, so an array operates as a massively parallel detection device. With a single microarray, an experimenter can check the expression levels of tens of thousands of genes simultaneously. It is the ability to study *complete systems* and consolidated sets of genes at the molecular level and in a largely automated fashion that makes DNA microarrays such powerful analytical tools.

Microarray Biochip Technology
Edited by Mark Schena

One of the keys to the automation of microarray technology is an integrated microarray scanning and analysis system. In most cases these scanners use lasers to illuminate one pixel at a time (each spot of DNA contains many pixels) until all the spots on the DNA chip have been scanned and recorded as a high-resolution image file. The scanned images are analyzed in a data extraction process that measures the absolute and relative fluorescence at two wavelengths, representing test and control mRNA samples, for each spot on the chip. Numerical indicators of the absolute and relative signals for test and control samples at each spot are extracted. This involves quantitative processing to calculate the absolute signals and ratios of the fluorescence intensity at two different wavelengths. With many current scanners, multichannel detection is performed sequentially such that the slide requires a separate scan for each wavelength. These separate images are often saved in one software application and analyzed in another. The modular characteristics of this type of imaging system are not conducive to a high level of automation.

Given the current state of the art in fluorescence imaging, this chapter aims to answer the following question: What is the best design for a microarray scanner and data extraction software analysis system? Having been through the process of evaluating competing technologies while designing and constructing our GenePix 4000A scanner and software system, we are able to provide a survey of the available design options as well as present the logic underlying the instrument that Axon chose to commercialize.

Embedded within the question of superior design features are a host of scientific and business requirements for such a system. Overall, the design goals of a scanning and analysis solution should be set to meet and exceed expectations of sensitivity, speed, ease of use, size, and cost.

- Sensitivity is paramount because researchers desire precise and repeatable measurements of fluorescence originating from genes expressed at extremely low levels (<1:100,000 mRNAs).
- Speed is an important requirement because many first generation ("home grown") instruments required 10 to 40 minutes per channel to scan a microarray (1,11).
- Ease of use is important too because researchers do not want to wade through complicated dialog boxes to complete simple tasks such generating a ratio image.
- Finally, one of the original criticisms of microarray technology was the relatively high cost threshold to enter the field. Affordable instrumentation and analysis tools are desirable to both academic and industrial laboratories.

IMAGING SYSTEM HARDWARE

Because the quality of the raw image is so critical for accurate microarray quantitation, the primary goal when designing an imaging system should be to maximize the signal-to-noise ratio (see below). The importance of the signal-to-noise ratio is readily conveyed by considering a simple example. With a signal-to-noise ratio of 200, a 14% change in signal intensity can be detected with 95% confidence. If the signal-to-noise ratio is reduced to 15, the minimum detectable signal change that can be detected with 95% confidence is reduced to 52% (9). Optimizing the signal-to-noise ratio in an optical system requires sophisticated engineering and prudent selection of the individual components within the system.

Signal-to-Noise Ratio

The signal-to-noise ratio quantifies how well one can resolve a true signal from the noise of the system. It is typically computed by taking the peak signal divided by the variation in the signal. If a microarray scanning system has a poor signal-to-noise ratio, the variation in the signal alone can prevent accurate quantitation of each spot. When discussing noise, there are two noise components to consider: noise present in the absence of light input (dark current noise) and noise generated by the input of light (shot noise).

Dark current noise. The dark current is measured in electrons per second and is introduced by the photomultiplier tubes (PMT), or any photon-detecting device. This noise originates from the thermal emissions from the photosensor such as a photocathode, and/or leakage current through the dynodes of the PMT. In some cases, electronic noise is included in the dark current values.

One can reduce dark current noise to negligible levels using two approaches, both of which are used in the GenePix 4000A:

1. Select PMTs that show very low dark current levels.
2. Minimize the dwell time of the illumination light on each pixel; as a result, the number of dark current electrons generated for each pixel is negligible.

If one uses these two approaches, the only significant noise value to consider in a scanning system is the shot noise.

Shot noise. The shot noise is background that derives from the fundamental particle nature of light; in essence, the shot noise relates to the statistical nature of light. When discussing the shot noise, it is important to realize that its absolute magnitude increases as the signal increases. (More photons results in more variability in the number of photons.) Therefore, as the signal intensity increases, the shot noise also increases. However, since this noise increases as the square root of the signal, the ratio of the signal to the noise actually increases as signal intensity increases.

In a properly designed system, the major contributor to noise is the shot noise of the detected fluorescent emission light. In suboptimal systems, noise may be introduced from many different sources including the electrical processing of the signals such as the analog-to-digital conversion process and the amplification.

Determining the noise of an optical system. In the previous section, we discussed the two sources of noise that are common in optical systems (dark current and the shot noise). These two types of noise are present in all optical measuring systems, although in a properly designed device the shot noise component is dominant. When examining how these sources of noise affect a microarray image, there are two measurement regions to consider:

Noise of the background region: The noise of the background region ultimately determines how well one can resolve whether the fluorescence of a DNA feature is significant above background. This type of noise is easily measured by determining the standard deviation of the signal in a background region. The GenePix software allows you to compute the standard deviation easily. This is reported in the analysis results and in the real-time Pixel Plot. One common method used to establish the signal threshold is to compute the standard deviation (SD) of the background and then add 1 or 2 times this value to the signal (background + 1 SD, or background + 2 SD). This new value becomes a threshold whereby any feature with a mean or median intensity greater than this threshold is likely to represent true signal. When discussing the signal-to-noise ratio, it is typically the background noise that is used for the computation.

Noise of the signal (microarray feature) region: The noise of the signal ultimately determines whether there is a significant change in the intensity of one fluorescent feature relative to another. The variation in signal intensity for a given fluorescent feature arises from two sources:

1. The shot noise of the detected fluorescent signal.
2. The variation of signal intensity due to the heterogeneity in the material that was printed. For example, with certain printing technologies it is possible to have features that show a "doughnut" appearance, with strong fluorescence on the perimeter of the feature and dim fluorescence in the center, or irregular borders.

The standard deviation of the signal intensity of the pixels in a feature is largely due to variation caused by the heterogeneity of the feature. To measure the shot noise of a feature, several measurements of the same feature with multiple scans would have to be taken. The difference between each of these measurements would theoretically reveal the shot noise of the signal.

Illumination and Photodetection

CCD or not CCD? As mentioned above, two key factors to consider when evaluating a microarray scanning system are the quality of the image and the time taken to generate such an image. In general terms, charge-coupled devices (CCDs) offer an advantage over photomultiplier tubes in that they allow simultaneous acquisition of relatively large images (approximately 1 cm^2), rather than scanning and generating a series of pixels to create an image (12). However, CCD systems typically use broad-band xenon bulb technology and spectral filtration for excitation. Spectral filtration is a particularly important issue because most commonly used fluors have a small Stoke's shift (the difference between excitation and emission maxima), which makes the effective separation of excitation light and emission light difficult. A laser-based scanning system uses defined excitation wavelengths, which typically provides, in a cleaner excitation light, many more excitation photons delivered to the sample and many more emission photons generated and collected for each pixel in a given amount of time.

There are limitations to lasers, however. For example, power and wavelength change as a function of temperature in both diode and solid state lasers. To control for temperature effects, the lasers can be stabilized either for temperature or power. The best stabilization solution uses both types of feedback controls on the lasers, which is the design solution implemented in the GenePix 4000A microarray scanner. The lasers are temperature controlled, laser output at the source is continually monitored with photodiodes, and input current is automatically adjusted for any fluctuations detected.

To circumvent the shortcoming of limited light collection, CCD-based systems integrate the signal over a significant amount of time (e.g., 1–60 seconds) to allow collection of enough emitted photons to create an acceptable image. In practice, the integration time of some CCD-based systems may actually exceed the time required for a laser scanning system to capture a comparable area. In addition to scan time, one must also consider the signal-to-noise ratio and the dynamic range of the system. Although the output of many digital CCD cameras is 16-bit, the functional dynamic range is often significantly less because the longer integration times lead to increased dark current noise contribution.

An additional consideration is that although CCDs are able to image a relatively large area, they do not provide single imaging capabilities for an entire microscope slide (25 × 76 mm). Large-format, scientific grade CCD chips typically contain 1600 × 1200 pixels, many fewer than the 2500 × 7600 pixels required to scan a microscope slide at 10-μm resolution. To scan a larger area than that provided by the CCD, smaller images must be "stitched" together in software to obtain a single image of the entire sample. Laser scanning systems such as the GenePix 4000A are not limited in the number of pixels that can be scanned in a single scanning step; in fact, alternate chip formats such as the 50- × 50-mm chip currently used by Protogene Laboratories (Palo Alto, CA, USA; www.protogene.com) are also amenable to laser scanning.

Simultaneous versus sequential scanning. Many microarrays for gene expression applications use two fluorophores, most commonly cyanine 3 (Cy3; green excitation light) and cyanine 5 (Cy5; red excitation light) from Amersham Pharmacia Biotech (Piscataway, NJ, USA). To create a composite microarray image representing both fluors, it is necessary to acquire an image at both the Cy3 and Cy5 wavelengths. To acquire two or more images, there are two different scanning approaches. A *sequential* scanner acquires one image at a time and then builds the ratio image after acquisition is complete. A *simultaneous* scanner, such as the GenePix 4000A, acquires both images at the same time.

Simultaneous scanning, although more difficult to implement, has the following advantages over sequential scanning:

1. Simultaneous scans are fast. With the GenePix 4000A, a 25- × 75-mm slide at 10 µm resolution is scanned in about 5 minutes; a 20- × 20-mm area can be scanned at two wavelengths in 1.5 minutes.
2. There is no need for image alignment after the scanning is completed. Image alignment is often necessary when using a sequential scanning system to compensate for slight movements of the slide during the repetitive scan process and for inherent mechanical variations in stage movements.
3. Given the proper software, it is possible to view either the raw Cy3 or Cy5 data images separately or the composite ratio image while scanning (for examples, see Figure 1). This allows the researcher to obtain instant, on-the-fly analysis results. With the GenePix software, one also has full control of the scan display features including zoom, brightness, and contrast. While scanning, one can view the data in gray scale or in any of a number of pseudocolor representations as well as in a histogram of the image pixel distribution.
4. Because the image is displayed as it is acquired, one can quickly evaluate the slide and reject it immediately if the image is substandard.

The most common potential concern of the simultaneous scanning approach is the possibility of *cross-talk*. Cross-talk is the phenomenon whereby optical signal from one channel contaminates the optical signal in the second channel. The result is an elevated, erroneous fluorescence reading in the contaminated channel. Because of the Stoke's shift, optical cross-talk is more severe from the shorter wavelength channel into the longer wavelength channel than vice versa. Cross-talk can be largely eliminated from a simultaneous scanning system by correct optical design. Axon's patent pending optical design, in combination with custom filters, allows simultaneous scanning without significant cross-talk between the two optical channels. Cross-talk in the GenePix 4000A is <1 part in 1000 (<0.1%) from the green (shorter wavelength) into the red (longer wavelength) channel, and even less than that from the red into the green channel. One can test for the lack of significant cross-talk in the GenePix 4000A by comparing images acquired with both lasers enabled, versus images obtained with only one laser enabled for each scan.

Confocal or not confocal? Confocal microscopes are used to reduce contributions from out-of-focus fluorescent light, in imaging applications in which the objects are significantly thicker than traditional thin sections (9). Most fluorescent microarrays have a planar profile that is nearly two-dimensional relative to the depth of focus. Additional sources of fluorescence arising from regions out of the spot focal plane on the microarray are typically minimal and can be rejected by conventional optical approaches. Therefore, the need for a confocal device to reduce out-of-focus fluorescent light is minimal in most microarray applications. Spurious fluorescence due to dust and nonspecific hybridization can occur on the printed surface of the microarray and because this fluorescence is in the same focal plane as the spot of interest, confocal optics would not reject this fluorescence anyway; furthermore, because photons are collected from a very thin focal depth, a confocal design inherently limits light collection efficiency. In a microarray experiment, it is desirable to collect as much light as possible, particularly when dealing with weak signals. The reduced focal depth of a confocal system can decrease signal quality two ways:

1. The focal depth may not be sufficient to capture all of the photons from each spot.
2. The plane of focus may not be correct in cases in which poor quality substrates are used.

The correct plane of focus is critical when using a confocal device. At present, most microarrays are printed on microscope slides that have inherent variations in the flatness of their surface; as a result, there may not be a single focal plane that allows optimal signal capture over the entire surface. Recent advances in microarray substrate manufacture with optically flat surfaces (http://arrayit.com/microarray-substrates) are essential for minimizing inaccuracies due to uneven surfaces used in conjunction with confocal devices.

Confocal optics have also been shown to reduce microarray image quality in other ways (6). Cheung and colleagues (4) found that the use of a confocal design resulted in an increased noise level and a sizable attenuation of the desired signal, which ultimately produced an image with a lesser signal-to-noise ratio.

Laser power and photobleaching. When evaluating laser power and its effects on photobleaching of the microarray sample, several factors must be considered. First, it is both the power *and* the time duration (dwell time) of sample illumination that causes photobleaching. Potential photobleaching problems with higher power lasers can be minimized by reducing the dwell time on each pixel. More powerful lasers allow faster scans without increasing photobleaching. Second, it is not the power of the laser per se that determines the extent of bleaching, but rather the final power of the light on the microarray surface that matters. In true confocal microscope systems, the width of the laser beam that illuminates the microarray surface is "diffraction-limited" and <1 μm in most cases. In this case, the power density is high, since all the laser power is contained within a small region. In non-confocal systems, it is desirable to match the laser spot size with the resolution of the system. Therefore, in a 10-μm resolution system, a 5- to 10-μm laser beam diameter is optimal. With larger laser beam diameters, the power density is much lower on the microarray surface and photobleaching is very low (see Figure 2). Finally, the microarray that is being analyzed must also be considered. There is currently no standard test microarray, and microarrays from different sources can vary greatly in their intensities, backgrounds, and photobleaching properties.

PMT issues. PMTs are optical components that convert incident photons into electrons via the photoelectric effect. When an incident photon impinges on the active surface of the PMT (the photocathode), an electron is generated. This electron flows through a series of electron multipliers (dynodes) to the anode. The amount of current that flows from the anode is directly proportional to the amount of incident light at the photocathode (10).

The quality of the signal depends on many factors, both biological and technical. For example, the numerical aperture of the lens in the scanner affects both light delivery to the sample and light collection efficiency. After laser excitation has excited a fluorophore such as Cy3 or Cy5, a number of emission photons are generated. These photons are collected and ultimately directed onto the surface of the PMTs. The amount of signal (S) that is generated by the PMT for each pixel is a function of the number of incident emission photons (N) times the quantum efficiency (Q_E) of the PMT, so that $S \propto N \times Q_E$. It is important to realize that N is the result of not only an efficient light collection pathway, but also an efficient laser excitation pathway.

The amount of amplification that a PMT can produce depends on the number of dynodes in the PMT, and the voltage that is applied to the PMT. With PMTs it is possible to achieve a gain of 10^7. One must always ensure that the gain of the PMT is set to a level that allows enough photons to be detected accurately, and that the brightest signals utilize most of the dynamic range of the system. For example, in a 16-bit system (65 536 intensity levels) one should set the PMT voltage such that the brightest signals on the microarray are 80% to 90% of the full 16-bit range. In general, the signal-to-noise ratio is not improved by increasing the PMT voltage because signal and background increase proportionally when the gain is increased.

In nonoptimal systems, electronic noise in the system is a factor. However, electronic noise is typically a constant value and thus is generally more problematic at lower gains. In a nonoptimal system, the signal-to-noise ratio will improve when a single microarray is scanned at high versus low PMT gains because the constant electronic noise will degrade larger values much less than smaller values.

The PMTs used in a microarray imaging system such as in the GenePix 4000A were selected to fulfill two essential criteria for high-quality imaging. First, because different photocathode models have different sensitivities at specific light wavelengths (quantum efficiency), the PMTs should be optimized for the fluorescent wavelengths used most commonly in microarray assays. Second, PMTs should be further selected based on reliability and optimal signal-to-noise performance.

One aspect of using photomultiplier tubes is the linearity of the output of the PMT over a wide range of incident light intensity. For a given control voltage, the gain of a PMT is constant for a wide range of photon fluxes. However, all PMTs exhibit a strong nonlinearity between the gain and different control voltages. For example, adjusting the control voltage on a PMT from 200 to 400 V will result in a different change in gain than when the voltage is adjusted from 400 to 600 V; nevertheless, for a given control voltage the output of the PMT is linear over a wide range of emission signals (10). If the PMT voltages for the red and green channels are different from one another, it is not a problem because the gain will be constant in both cases.

Electronics

Although the optical components of microarray scanners are indispensable for generating an electrical current from the PMTs, the downstream electronic components are just as important. They govern the operation of the PMTs and the conversion of their analog current signals into a digital signal that can be processed by the host computer.

There are several steps between the generation of an electron from an incident photon and the final generation of a series of digital values that represent this photoelectric event. These steps include amplification, processing, and digitization. Axon Instruments has designed and built superior, low-noise electronic instruments for over 15 years. During that period, we have found that digitization performed inside the host computer results in additional noise due to the electrically "noisy" environment inside the computer. Additionally, signal degradation can occur when transferring the analog signal from the scanner to the computer for digitization. The GenePix 4000A has all the digitization circuitry embedded in the scanner, thus avoiding signal degradation that can occur on the way to and inside the host computer. Irrespective of a noisy environment, one should use the lowest noise analog-to-digital converters to prevent image degradation.

Different methods are used to digitize the output of a PMT. One of the most common techniques is to use passive circuitry (an RC circuit) that accumulates charges from the output of the PMT during light collection. Once the dwell time has been achieved, a reading is made that is proportional to the combined charge buildup and passive decay of charge during the dwell time duration. The limitations of this electrical design are twofold. First, if the passive decay is too long, the residual charge left in the circuit can be carried over into the next sample period, producing pixel-to-pixel cross-talk. This can be observed in images in which "tails" appear to trail from very bright spots in the image. This limitation would suggest that a short time constant of decay would be preferred to limit the pixel-to-pixel cross-talk. However, the short time constant would limit the amount of charge read off the circuit at the end of the dwell time, reducing the efficiency of converting photon events into an electrical signal.

In the GenePix scanner, an alternate solution is implemented, one that uses integration circuitry. This type of circuitry is currently implemented and perfected for Axon's ultra-low-noise amplifiers for electrophysiology and greatly extends the dynamic range of signal conversion. The circuitry in the GenePix actually adds all the charges that are generated by the PMT, without any decrease in sensitivity or collection efficiency. The use of a series of three integrators for each PMT channel eliminates pixel-to-pixel cross-talk and obviates the reset transients necessary for integrators.

Mechanical Factors

Microarray scanners are mechanical devices, and one of the concerns with mechanical devices is robustness and durability. Microarray scanners are currently being used in a wide range of environments. In the small academic laboratory, the scanner might only process a few chips a day, whereas a scanner at a biotechnology or pharmaceutical company might be used for 24 hours a day, 7 days a week. The GenePix 4000 scanner is already widely used in all these settings. The mechanics of the GenePix 4000 were designed for optimal robustness and longevity. Although there are a variety of methods for obtaining two-dimensional microarray scans or images, we chose a combination of "tried and true" approaches. The GenePix 4000A scans the microarray slowly in the *y*-direction (the length of the slide) with a standard translation stage and rapidly in the *x*-direction with a driven slider and a robust voice coil mechanism similar in design to that currently used in computer hard disks.

Rectilinear scanning is common to most microarray scanners and has two main advantages over nonlinear scanning mechanisms. First, the need for an expensive digital signal processing (DSP) board in the host computer to convert the nonrectilinear image to a linear form is obviated. Second, potential distortions and data degradation caused by "unwarping" the image are avoided. A potential problem of a rectilinear system is that the optical path length can change with mechanical movement. This concern is avoided when the design uses collimated laser beams, as in the GenePix 4000.

When using a mechanical scanning device, it is important to monitor variability in the performance of the mechanical components. The mechanical performance of the hardware should be monitored in real time during acquisition and corrected. Here again is the advantage of tight integration between the hardware and the software.

Dynamic Range and Sensitivity

In a general sense, the dynamic range of an instrument is the signal width that can be resolved above the instrument noise. The dynamic range is determined by all the components used in a given system. If a PMT has a dynamic range of 10^6 but is processed through an 8-bit analog-to-digital converter, the instrument will exhibit a poor dynamic range. An instrument with a large dynamic range may not be able to use the whole range reliably if the response is nonlinear. When evaluating an instrument, one must consider the range over which the response remains linear and a given dynamic range that is actually being evaluated. For the GenePix 4000A, the dynamic range of the electronics is a full 16 bits (>4 orders of magnitude), and the dynamic range of fluorescent sample detection is 10^4, for an overall linearity of >1000-fold.

The dynamic range is often considered together with the detectivity of the instrument. Detectivity can be defined as the minimum detectable signal above background noise. For the GenePix 4000A, the sensitivity is on the order of 0.1 fluorophores/μm^2 with a signal-to-noise ratio of >3:1.

MICROARRAY SYSTEM SOFTWARE

The fundamental task performed by the scanner software includes full control and monitoring of the hardware as well as precise extraction of data from the image produced by the scanner. Once the microarray substrate is placed in the scanner, the user wants to address both biological and computational questions. One biological question might be: Is gene *x* overexpressed or underexpressed in two-mRNA populations? A computational question is: What confidence do I have in the data? The aim of the software designer should be to provide the answers to these questions as seamlessly and efficiently as possible.

Ideally, the same software that controls and monitors the scanner should acquire and analyze the images. Not only does this produce a streamlined workspace, it allows direct and immediate access to data. The multithreaded GenePix software allows the user to analyze images as they are acquired; in addition, the GenePix workflow options allow the user to concatenate a number of common tasks, which is the first step toward full automation of microarray scanning and analysis.

On the acquisition side, the ability to monitor laser power and mechanical performance continually allows the user to confirm that the instrument is operating properly and within specification. The ability to adjust PMT voltages, brightness, and contrast of the image and to observe the pixel intensity distribution directly during the actual scanning enhances the ability of the user to configure the hardware optimally in real time for each type of microarray scanned.

Whereas the primary constraint on imaging system hardware design is data quality, the primary constraint on software design is data analysis. Data analysis can be grouped into three areas, graphical tools, informatics tools, and numerical methods. One of the design goals is ease of use and thus a unique graphical user interface (GUI) was developed for the software. Instead of developing a standard Windows®-type application, all the functionality of the software is embedded in buttons, more like a true instrument interface. This design, along with linking all functions to "hot keys," allows for an intuitive and powerful interface that is easy to learn for the novice and quick to control by the expert (see Figure 3).

Finally, our design strategy for analysis was based partly on first obtaining the best possible "raw" image. The goal was to avoid compensating for hardware inadequacies by processing the image in software. Image filtering routines were avoided to allow the user to have access to the excellent quality raw data.

Graphical Tools

Ratio Image. Simultaneous scanning with the GenePix system allows the user to view the ratio image of a microarray as it is being read into the computer and therefore allows the user to have instant access to critical information that the microarray provides. What biological information is conveyed by a ratio image? For a gene expression study, the ratio image represents mRNA levels in two different samples. Suppose the reference mRNA is labeled with Cy3 (excited by green laser light at 532 nm) and the test mRNA is labeled with Cy5 (excited by red laser light at 635 nm). Both of the cDNA populations are hybridized to a spotted cDNA microarray on a microscope slide. In a ratio image displayed in a red/green color palette, a red spot indicates that a given gene in the test mRNA sample is overexpressed relative to the reference mRNA population, a green spot indicates that a given gene is repressed relative to the reference mRNA population, and a yellow spot means that there is no significant change in the expression of a given gene in the two samples. A ratio image should also provide a visual summary of the expression levels for the entire microarray (see also References 8 and 9).

Scatter plot. Scatter plots convey the same amount of information as ratio images, but in a different format. Whereas a ratio image displays expression levels by color, a scatter plot displays them as a distance from the diagonal on a graph. The GenePix 4000A software employs two different types of scatter plots. The first, known as the Feature Pixel Plot, depicts pixel intensities from individual features, so that one can evaluate the quality of the data from a single microarray feature. The second type allows the user to depict the extracted data from an entire microarray globally and compare them with a second entire microarray. In the global scatter plot, one can depict mean intensities at the first wavelength against mean intensities at the second wavelength, medians against medians, or raw pixel intensities against raw pixel intensities.

Informatics Tools

Data extraction analysis systems do not include the sophisticated informatics tools that one uses when managing a database assembled from many microarrays (7,8). Analysis systems do need to allow clone tracking information from microplate to spotter to extracted data. The GenePix software reads in plain text files containing gene identifiers (IDs) and names, allowing the user to obtain the gene ID and gene name simply by moving the mouse to a given spot on the microarray.

Another simple yet effective method of tracking microarray experiments through the analysis process uses bar codes. Large corporate and academic labs are labeling microarrays with bar codes as a further step toward the automation of microarray analysis. GenePix includes Axon's own bar code reading software allowing bar codes to be read automatically.

Numerical Methods

Spot finding. Before calculations are performed, a definition of which pixels in the image represent signal and which represent nonsignal (background) must be made. The current state of most microarray fabrication technologies does not allow for precise, exact replicates of microarrays, even from the same printing batch. In addition, spots can move slightly during the printing and processing steps of the microarray. Simply superimposing a regular grid pattern over the microarray to define signal and background for each feature does not allow for slight variations in microarray pattern. The current state of image analysis also does not provide a robust solution for error-free detection of all the spots in a microarray image without some degree of human intervention and rejection of proposed "'finds". We chose to offer both a fully automated and a fully manual solution for microarray quantitation.

In the manual mode, the user employs information about the physical array, such as the number of rows and columns within a subarray and the number of subarrays in the microarray to place an analysis grid easily onto the image. Then, by providing access to each circle, or feature-indicator, the user can manually adjust both the position and size of the feature-indicator to define signal from background precisely for each spot on the array. This manual task is obviously quite time-consuming with arrays that contain thousands of spots, although there are laboratories that believe that careful manual adjustment is rewarded by better quality analysis results. Manual adjustment qualifies as the major bottleneck in the microarray analysis process.

For the automated mode, Axon developed a proprietary algorithm to locate the spots precisely in a wide variety of microarray image types. Human intervention is still required but is limited to either opening a file that describes how the array was printed or using a simple interface to place analysis grids roughly onto the microarray image. Both the size and position of each feature-indicator is automatically adjusted to fit each feature. In our latest testing of the algorithm, the software can automatically find every spot on a yeast genome chip (>6000 spots) in <20 seconds using a standard personal computer. Effectively eliminating additional bottlenecks in the microarray scanning and analysis process will empower the technique further.

To quantify changes in gene expression accurately, it is necessary to extract the intensity of a given feature from an image and to refine these data before a final ratio value is computed. In addition, a number of other parameters can be measured to evaluate the quality of the data for a given feature. During the development of the analysis software, we found that no standard algorithms exist for the microarray field, partly due to the wide inherent variation in the microarrays themselves. Therefore, the strategy we chose was to include multiple methods of ratio calculation. The following section describes some of the data quantities used in GenePix.

Computing feature intensities. For each raw image, the median, mean, and SD of the pixel intensities in each feature-indicator is computed. The pixels used to compute the SD include all the pixels that are completely inside the circular feature-indicator. Any pixel that contacts the circular ring of the feature-indicator is excluded.

Computing background intensities. To reduce the effect of nonspecific fluorescence, such as autofluorescence of the glass slide or nonspecific binding of test and reference cDNA mixtures, GenePix software uses a background subtraction method. For all analysis calculations, the background intensity is subtracted from the feature intensity before any ratio calculations are performed. The method for determining the background intensity is a local background subtraction technique. By this method, different background is computed for each individual feature-indicator, and the mean, median, and SD of the background are reported. For all ratio calculations that require background subtraction, the median local background value is used.

In addition to computing the local background, GenePix software reports the percentage of pixels that exceed a background-related threshold. GenePix software reports the percentage of pixels that are greater than the background plus 1 SD and background plus 2 SDs. These threshold-related values are often useful in evaluating the quality of the data.

Which pixels are used to compute the background? Almost all microarray analysis packages support the determination of "local background" because the background of most microarrays has variable intensity. In the GenePix software, a circular region is drawn that is centered on the feature-indicator. This region has a diameter that is three times the diameter of the current feature-indicator. All the pixels within this area are used to compute the background unless one of the following is true:

1. The pixel resides in a neighboring feature-indicator.
2. The pixel is not wholly outside a 2-pixel-wide ring around a feature-indicator.
3. The pixel is within the feature-indicator of interest.

We have found that 3 diameters wide is optimal for obtaining a representative number of background pixels to be used for correction with a variety of microarray spot densities.

Computing ratios. GenePix software computes a large number of different ratio quantities, each of which provides a different insight into the raw data. The main ratio quantities are described below, and their significance is discussed briefly. They are divided into three different subgroups: those that are ratios of quantities derived from whole features (such as median and mean pixel intensities), those that are derived from pixel-by-pixel ratios of intensities, and extrapolated quantities.

All ratios and derived quantities are corrected for background effects. Signals at a given wavelength are obtained by subtracting the median intensity of background pixels from the same wavelength to produce a corrected intensity.

Ratios of overall derived feature properties: In gene expression studies, one is typically interested in the difference in expression levels between test and reference mRNA populations. This eventually translates into differences in intensity on the two images analyzed by GenePix software, which can be quantified by taking a ratio of image intensities for each feature. Because each feature in the image contains many pixels, a method of finding a representative intensity for each feature is required, as follows:

1. *Ratio of medians.* Median intensities are often preferred to arithmetic mean intensities because they are less susceptible to extreme values at either end of a

distribution. The ratio of medians is the ratio of the background-subtracted median pixel intensity at the second wavelength to the background-subtracted median pixel intensity at the first wavelength.

2. ***Ratio of means.*** Mean intensities are familiar quantities, but they can be significantly skewed by extreme values at either end of a distribution. Calculating the mean does allows one to evaluate the variability of the data. This variability is reported as the standard deviation. The ratio of means is the ratio of the arithmetic mean of the background-subtracted raw pixel intensities at the second wavelength to the arithmetic mean of the background-subtracted raw pixel intensities at the first wavelength.

Quantities derived from pixel-by-pixel ratios: An alternative to computing the ratio from calculated means and medians of each feature is to compute the ratio on a pixel-by-pixel basis and then calculate the arithmetic mean and median of these many ratio values. An advantage of this approach is that nonspecific signal, which appears in both wavelength images, has less of an effect than when the feature is treated as a single entity. Axon has extensive experience in this area of data analysis owing to the development of our cellular imaging products.

1. ***Median of Ratios.*** As is the case with the overall derived feature properties, median intensities are often preferred to mean intensities because they are not as seriously affected by extreme values at either end of a distribution compared with calculation of the arithmetic mean. The median of ratios is the median of the pixel-by-pixel ratios of pixel intensities that have had the median background intensity subtracted.
2. ***Mean of Ratios.*** The *mean of ratios* is the arithmetic mean of the pixel-by-pixel ratios of the raw pixel intensities and can also provide a measure of the variability of this type of ratio calculation.
3. ***Regression Ratio.*** The regression ratio is another method for computing the ratio that does not require rigid definition of the background and signal pixels. The regression ratio is the most objective of the three methods. All the pixels within a circle that have a radius within 2 diameters of the feature of interest are used. The relationship between wavelength 1 and wavelength 2 is determined by computing a linear regression between these two populations of pixels. The slope of the line of best fit (least-squares method) is the regression ratio. The coefficient of determination provides a measure of the level of accuracy of the fit.
4. ***Quality measures.*** The sum of medians and the sum of means are useful measures to interrogate the quality of an individual feature. If the sum of the median or mean is unreasonably low, this indicates that there was very little signal in both channels, suggesting that any ratio derived from this feature should be interpreted with caution. A number of groups are working toward developing a more robust quality measure of the ratios calculated from microarray images. Since no group to date has provided a solution, the GenePix software reports many different ratio calculations. If they are consistent, a researcher can have confidence that the measure is valid. At the 1999 Nature

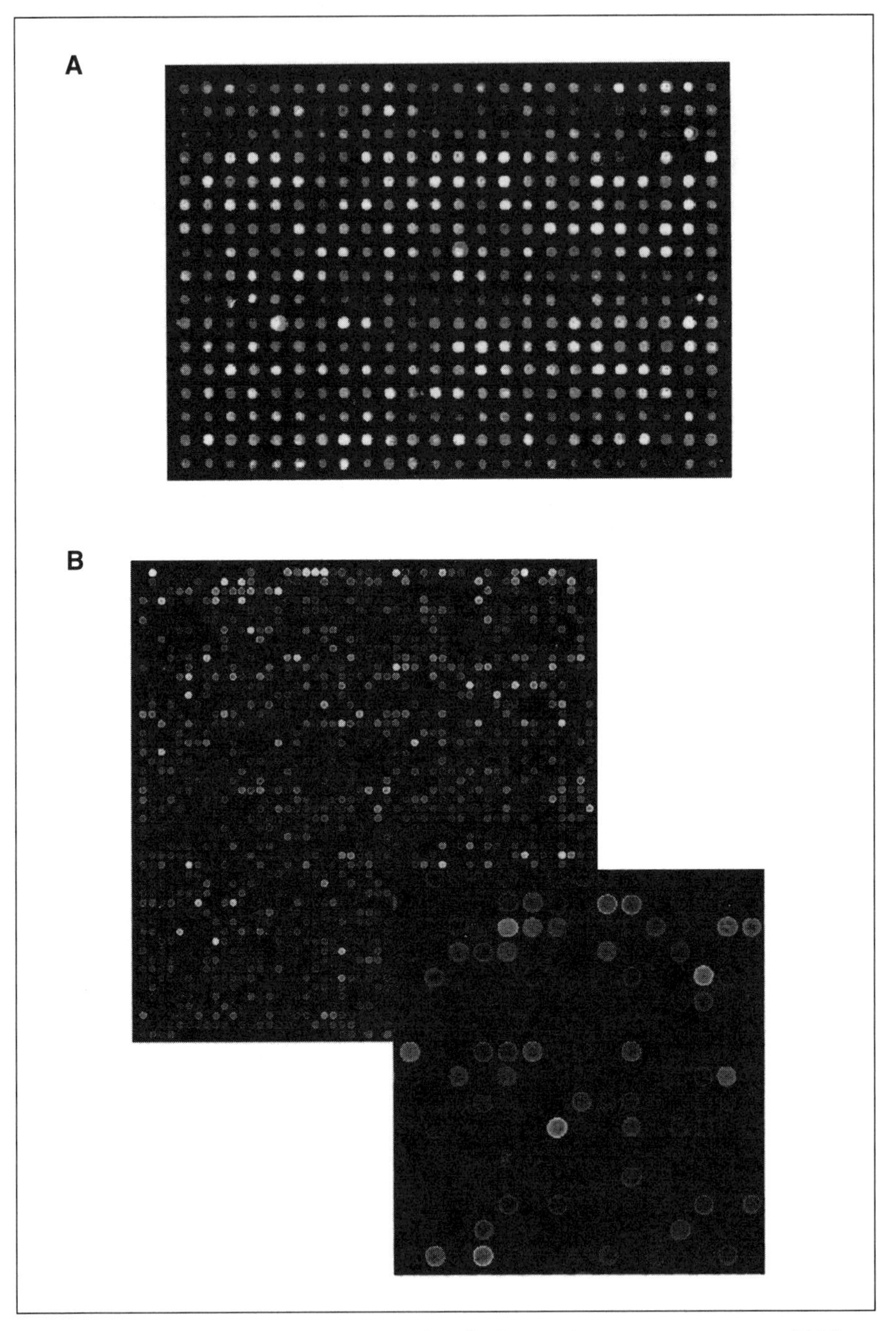

Figure 1. Examples of high-quality images acquired on the GenePix 4000A microarray scanner. (A) Microarray image of a yeast cDNA subarray kindly provided by Joe DeRisi, UCSF. (B) Microarray image of a GEM subarray kindly provided by Incyte Pharmaceuticals. Inset shows an enlarged portion of the subarray. (**See color plate A31.**)

Genetics Microarray meeting, Yidong Chen from the National Cancer Institute presented a promising technique of weighting a number of characteristics of the spot, including size and intensity, to develop a quality score from 0 to 1 for ratio calculations. Clearly this will be an active area of development as microarray technology matures.

5. ***Log Ratio.*** The log ratio is a base-2 logarithm of the ratio of medians. This type of log transformation is often used in gene expression studies as it helps to compare levels of over- and underexpression. For example, for a fourfold increase in expression, the log base 2 ratio is +2. However, if you have a fourfold decrease in the ratio (i.e., a ratio value of 0.25), the value is -2. The log base-2 transform is simply a quicker way to observe the similarity in magnitude between these two conditions.

Data normalization. Comparing the data from different microarray experi-

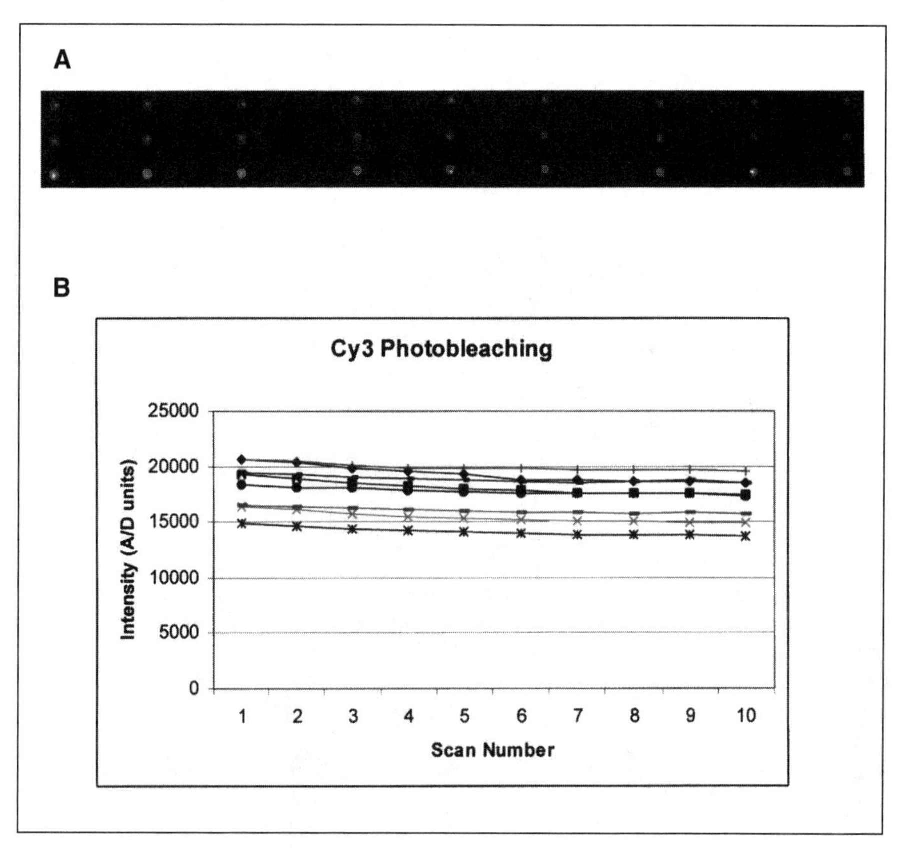

Figure 2. Photobleaching of Cy3 by the 532 nm laser. (A) Image of a portion of a dilution series of Cy3 incorporated into 35-mer oligonucleotides. The array was scanned sequentially 10 times on the GenePix 4000A scanner. The center row of features was used for the analysis of photobleaching shown in B. (B) The features are 200 µm in diameter and placed about 1.5 mm apart. Photobleaching is shown for 10 scans of each of the 9 features analyzed. The average rate of photobleaching was 0.7% per scan. (Microarray kindly provided by Operon.)

ments is a nontrivial task. Small variations in the many steps that produce a microarray image can render comparisons across microarrays problematic. Variability in the data from many microarrays can be corrected during either the analysis or the acquisition stage of an experiment. In the analysis stage, one can improve comparisons across many microarrays by normalizing the data from each microarray. One method of normalization is based on the premise that the arithmetic mean of the ratios from every feature on a given microarray should be equal to 1 (3). If the mean is not 1, a value is computed that represents the amount by which the ratio data should be scaled such that the mean value returns to 1. This value is the normalization factor.

In GenePix software, the user can choose normalization factor computation based on one of the five calculated ratio quantities (ratio of medians, ratio of means, median of ratios, mean of ratios, regression ratio). The type of calculation performed in each case is identical; the only difference is in the source data that are used to compute the normalization factor. For example, if the user selects the ratio of medians method, and the arithmetic mean of all the ratio of medians values is 0.8, a normalization factor of 1.25 would be calculated (1/0.8). This normalization factor is stored in the Results file that is saved from GenePix 4000A and is reported at the bottom of the Results spreadsheet. It is important to note

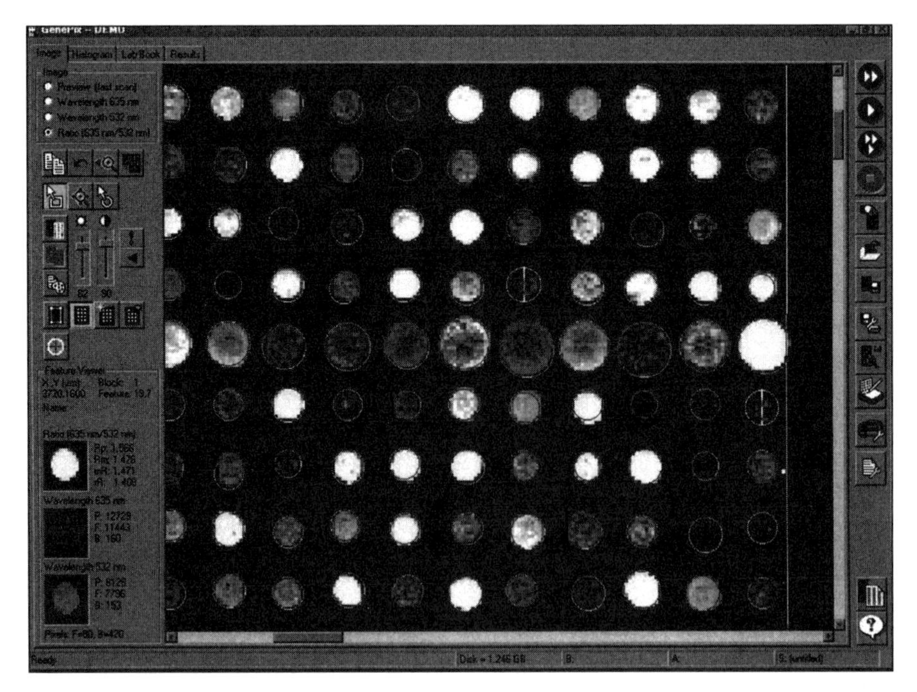

Figure 3. GenePix software interface. This "screen dump" image of the GenePix software shows that the application is designed more like an instrument interface, rather than a standard Windows program with drop-down menus. (**See color plate A32.**)

that GenePix software does *not* change any of the analyzed data, rather, the information is used to aid the user in setting up the hardware correctly.

Given that GenePix software does not normalize the data, what is the usefulness of the normalization factor? GenePix software can use the normalization factor to determine whether the PMT values were set correctly during acquisition. For example, if the user sets the PMT voltage for the red (635 nm) channel too high, the computed ratios will all be shifted to higher values. If the shift causes mean ratios >2 (or <0.5), the user is prompted to rescan the image. Although this shift can be corrected during analysis using normalization techniques, rescanning the image at the correct PMT settings ensures maximum data integrity.

Another option for normalization that does not rely on the above assumption is to embed control features within the array. In this case, these control features can be normalized to each other across all the microarrays that comprise an experiment. As microarrays enter the market from a variety of commercial sources, it is likely that these will include control spots to aid in the normalization and quality control of each microarray experiment.

SUMMARY AND PROSPECTS

Data quality from hardware and data analysis and confidence measures from software form the basis of a well-designed microarray scanner and data extraction software system. Successful hardware design is only possible if one has a deep understanding and experience of optical and electronic technologies, whereas the usability and efficiency of such a system is derived from the tightly integrated communication between hardware and software, including optimized algorithms and a thoughtful and easy-to-use software interface. The final requirement of cost can be met by offering the scanner and multiple copies of the acquisition and analysis software at a value price point attractive to both academia and industry, as accomplished with the GenePix 4000.

The future of microarray scanning and analysis can be summarized in one word: automation. On the hardware side, multiple lasers, improved resolution, and the autoloading of slides will be standard in the next generation of research scanners. On the software side, there is not yet an analysis package that can extract the data from a microarray without human intervention, but existing software is rapidly approaching this point. Analysis of datasets from multiple arrays is already offered (8 for example), but not as a component of a completely integrated system. On both the hardware and software sides, full automation is not far off for integrated scanning and analysis systems.

ACKNOWLEDGMENTS

Axon Instruments, Inc. would like to extend its thanks to Drs. Stephen J.

Smith and Michael B. Eisen of the Stanford University School of Medicine and Dr. Joseph L. DeRisi of the University of California at San Francisco for their creative ideas and constructive feedback during the development of the GenePix hardware and software.

REFERENCES

1. **Bowtell, D.D.L.** 1999. Options available—from start to finish—for obtaining expression data by microarray. Nat. Genet. Suppl. *21*:25-32.
2. **Brown, P.O. and David Botstein.** 1999. Exploring the new world of the genome with DNA microarrays. Nat. Genet. Suppl. *21*:33-37.
3. **Chen, Y., E.R. Dougherty, and M.L. Bittner.** 1997. Ratio-based decisions and the quantitative analysis of cDNA microarray images. J. Biomed. Optics *2*:364-374.
4. **Cheung, V.G., M. Morley, F. Aguilar, A. Massimi, R. Kucherlapati, and G. Childs.** 1999. Making and reading microarrays. Nat. Genet. Suppl. *21*:15-19.
5. **DeRisi, J.L, V.R. Iyer, and P.O. Brown.** 1997. Exploring the metabolic and genetic control of gene expression on a genomic scale. Science *278*:680-686.
6. **Duggan, D., M. Bittner, Y. Chen, P. Meltzer, and J.M. Trent.** 1999. Expression profiling using cDNA microarrays. Nat. Genet. Suppl. *21*:10-14.
7. **Eisen, M.B. and P. O. Brown.** 1999. DNA arrays for analysis of gene expression. Methods Enzymol. *303*:179-205.
8. **Eisen, M.B., P.T. Spellman, P.O. Brown, and D. Botstein.** 1999. Cluster analysis and display of genome-wide expression patterns. Proc. Natl. Acad. Sci. USA *95*:14863-14868.
9. **Inoue, S. and K.R. Spring.** 1997. Video Microscopy. Plenum Press, New York.
10. **Photomultiplier Tube.** 1994. Principle to Application. Hamamatsu Photonics K.K.
11. **Schena, M., D. Shalon, R.W. Davis, and P.O. Brown.** 1995. Quantitative monitoring of gene expression patterns with a complementary DNA microarray. Science *270*:467-470.
12. **Schena, M. and R.W. Davis.** 1999. Parallel Analysis with Biological Chips: PCR Applications, p. 445-456. Academic Press, San Diego.
13. **Schena M. and R.W. Davis.** Genes, genomes and chips, p. 1-16. *In* M. Schena (Ed.), DNA Microarrays: A Practical Approach. Oxford University Press, Oxford.

SELECTED SUPPLIERS

ACLARA Biosciences, Inc.
1288 Pear Avenue
Mountain View, CA 94043 USA
Phone: (650) 210-1200
www.aclara.com

Advanced Labs
182-184 Ferrar Street
South Melbourne, 3205
Victoria, Australia
Phone: (61 3) 9686 5022
Fax: (61 3) 9686 5033
www.advancedlab.com.au

Affymetrix, Inc.
3380 Central Expressway
Santa Clara, CA 95051 USA
Phone: (408) 731-5000
Fax: (408) 481-0422
www.affymetrix.com

Agilent Technologies
3500 Deer Creek Road
P.O. Box #10395
Palo Alto, CA 94304 USA
www.agilent.com

Alpha Innotech Corporation
14743 Catalina Street
San Leandro, CA 94577 USA
Toll Free: (800) 795-5556
Phone: (510) 483-9620
Fax: (510) 483-3227
Email: info@aicemail.com
www.alphainnotech.com

Amersham Pharmacia Biotech, Inc.
800 Centennial Ave.
P.O. Box 1327
Piscataway, NJ 08855-1327 USA
Phone: (732) 457-8000
Fax: (732) 457-0557
www.apbiotech.com

Applied Precision, Inc.
1040 12th Avenue Northwest
Issaquah, WA 98027 USA
Phone: (425) 557-1000
Fax: (425) 557-1055
www.api.com

Arcturus Engineering, Inc.
400 Logue Avenue
Mountain View, CA 94043 USA
Phone: (650) 962 3020
Fax: (650) 962 3039
Email: contact@arctur.com
www.arctur.com

arrayit.com
524 E. Weddell Drive
Suite 3
Sunnyvale, CA 94089 USA
Phone: (408) 744-1331
Fax: (408) 744-1711
Email: arrayit@arrayit.com
www.arrayit.com

Asper Ltd.
3 Oru St.
51014 Tartu
Estonia
Phone: +372 55 466 98
Fax: +372 7 422 168
Email: hpavel@ebc.ee
web.ee/virtual/est/e-asper.htm

Axon Instruments, Inc.
1101 Chess Drive
Foster City, CA 94404-1102 USA
Phone: (650) 571-9400
Fax: (650) 571-9500
Email: sales@axon.com
www.axon.com

Beckman Coulter, Inc.
4300 N. Harbor Boulevard
P.O. Box 3100
Fullerton, CA 92834-3100 USA
Phone: (714) 871-4848
Fax: (714) 773-8283
www.beckman.com

Beecher Instruments
P.O. Box 8704
Silver Spring, MD, USA
Phone: (301) 585-6621
www.beecherinstuments.com

BioDiscovery, Inc.
11150 W. Olympic Blvd.
Suite 805
Los Angeles, CA 90064 USA
Phone: (310) 966-9366
Fax: (310) 966-9346
Email: info@biodiscovery.com
www.biodiscovery.com

Biodot Ltd.
1 Home Farm Court
Diddington
Cambs, PE18 9XU
United Kingdom
Phone: 44 (0) 1480 810846
Fax: 44 (0) 1480 811311
Email: biodot@compuserve.com
www.biodot.com

Bio Medical Science Company, Ltd.
BMS Building
1617-55
Seocho-dong
Seocho-ku, Seoul
Korea
Phone: 82-2-3471-6500
Fax: 82-2-3472-1211
Email: bmskorea@chollian.net

Bio Robotics, Inc.
12 Walnut Hill Park
Woburn, MA 01801 USA
Phone: (781) 376 9791
Fax: (781) 376 9792
Email: info@biorobotics.com
www.biorobotics.com

BM Equipment Company, Ltd.
25-4 Hongo 3-Chome
Bunkyo-Ku, Tokyo 113
Japan
Phone: 03-3818-5091
Fax: 03-3818-5530

Cartesian Technologies, Inc.
17781 Sky Park Circle
Irvine, CA 92614 USA
Phone: (949) 440-3680
Fax: (949) 440-3694
www.cartesiantech.com

Cellomics, Inc.
635 William Pitt Way
Pittsburgh, PA 15238 USA
Phone: (412) 826-3600
Fax: (412) 826-3850
www.cellomics.com

Cepheid
1190 Borregas Avenue
Sunnyvale, CA 94089-1302 USA
Phone: (408) 541-4191
Fax: (408) 541-4192
www.cepheid.com

Ciphergen
490 San Antonio Road
Palo Alto, CA 94306 USA
Phone: (650) 496-3770
Fax: (650) 424-1651
www.ciphergen.com

Clinical Micro Sensors, Inc.
126 West Del Mar Blvd.
Pasadena, CA 91105 USA
Phone: (626) 584-5900
Fax: (626) 584-0909

CLONTECH Laboratories
1020 East Meadow Circle
Palo Alto, CA 94303 USA
Phone: (650) 424-8222
Fax: (650) 424-1088
Email: products@clontech.com
www.clontech.com

Corning, Inc.
Science Products Division
45 Nagog Park
Acton, MA 01720 USA
Phone: (800) 492-1112
www.corning.com

Diamond Multimedia Systems, Inc. (formerly Micronics)
S3 Incorporated
2841 Mission College Blvd.
Santa Clara, CA 95054-1838 USA
Phone: (408) 588-8000
Phone: (408) 588-8585
Fax: 408-980-5444
www.diamondmm.com

Display Systems Biotech, Inc.
1260 Liberty Way, Suite B
Vista, CA 92083 USA
Toll Free: (800) 697-1111
Phone: (760) 599-0598
Fax: (760) 599-9930
Email: info@displaysystems.com
www.displaysystems.com

Double Twist, Inc.
1999 Harrison Street, Suite 1100
Oakland, CA 94612 USA
Phone: (510) 628-0100
Fax: (510) 628-0108
Email: info@doubletwist.com
www.doubletwist.com

Epicentre Technologies Corporation
1402 Emil Street
Madison, WI 53713 USA
Phone: (608) 258-3080
Fax: (608) 258-3088
Email: techhelp@epicentre.com
www.epicentre.com

Gene Company Ltd.
Unit A, 8/F, Shell Industrial Building
12 Lee Chung Street
Chai Wan
Hong Kong
Phone: (852) 2896-6283
Fax: (852) 2515-9371

Gene Logic, Inc.
708 Quince Orchard Road
Gaithersburg, MD 20878 USA
Phone: (301) 987-1700
Fax: (301) 987-1701
Email: hyang@genelogic.com
www.genelogic.com

Gene Machines
935 Washington Street
San Carlos, CA 94070 USA
Phone: (650) 508-1634
Fax: (650) 508-1644
Email: info@genemachines.com
www.genemachines.com

Gene Trace Systems, Inc.
1401 Harbor Bay Parkway
Alameda, CA 94502 USA
Phone: (510) 748-6000
Fax: (510) 748-6001
www.genetrace.com

Genetic MicroSystems, Inc.
34 Commerce Way
Woburn, MA 01801 USA
Phone: (781) 932-9333
Fax: (781) 932-9433
Email: info@geneticmicro.com
www.geneticmicro.com

Genetix Limited
63-69 Somerford Road
Christchurch
Dorset, BH23 3QA
United Kingdom
Phone: 44 (0) 1202 483900
Fax: 44 (0) 1202 480289
Email: sales@genetix.co.uk
www.genetix.co.uk

Genisphere, Inc.
3 Fir Court
Oakland, NJ 07436 USA
Phone: (877) 888-3DNA
Phone: (201) 651-3100
Fax: (201) 651-3122
Email: info@genisphere.com
www.genisphere.com

Genome Systems, Inc.
4633 World Parkway Circle
St. Louis, MO 63134-3156 USA
Toll Free: (800) 430-0030
Phone: (314) 577-2733
Fax: (314) 427-3324
www.genomesystems.com

Genometrix
3608 Research Forest Drive
Suite B7
The Woodlands, TX 77381 USA
Phone: (281) 465-5000
Fax: (281) 465-5002
Email: info@genometrix.com
www.genometrix.com

Genomic Solutions Inc.
4355 Varsity Drive, Suite E
Ann Arbor, MI 48108 USA
Phone: (734) 975-4800
Fax: (734) 975-4808
Email: info@genomicsolutions.com
www.genomicsolutions.com

Genosys Biotechnologies, Inc.
1442 Lake Front Circle, Suite 185
The Woodlands, TX 77380 USA
Phone: (800) 2345-DNA
Fax: (281) 363-2212
Email: genosys@genosys.com
www.genosys.com

GENPAK Limited
Science Park Square
Falmer, Brighton BN1 9SB
United Kingdom
Phone: +44 1273 704470
Fax: +44 1273 626213
Email: info@genpakdna.com
www.genpak.co.uk

Genset SA
24, rue Royale
75008 Paris
France
Phone: +33 1 55 04 59 00
Fax: +33 1 55 04 59 29
www.genxy.com

GeSiM GmbH
Bautzener Landstrasse 45
D - 01454 Grosserkmannsdorf
Germany
Phone: +49-(0) 351-2695-322
Fax: +49-(0) 351-2695-320
Email: gesim@gesim.de
www.gesim.de

GSI Lumonics
500 Arsenal Street
Watertown, MA 02472-2888 USA
Phone: (617) 924-1010
Fax: (617) 924-7327
www.gsilumonics.com

Hewlett-Packard
3000 Hanover Street
Palo Alto, CA 94304-1185 USA
Phone: (650) 857-1501
Fax: (650) 857-5518
www.hp.com

Hitachi Genetic Systems
1201 Harbor Bay Parkway
Alameda, CA 94502 USA
Toll Free: (800) 624-6176
Phone: (510) 337-2000
Fax: (510) 337-2099
www.hitsoft.com

Hyseq Inc.
670 Almanor Avenue
Sunnyvale, CA 94086 USA
Phone: (408) 524-8100
Fax: (408) 524-8141
Email: cd@sbh.com
www.hyseq.com

Incyte Pharmaceuticals, Inc.
3160 Porter Drive
Palo Alto, CA 94304 USA
Phone: (650) 855-0555
Fax: (650) 855-0572
Email: sales@incyte.com
www.incyte.com

Informax, Inc.
6010 Executive Blvd., 10th Floor
N. Bethesda, MD 20852 USA
Phone: (301) 984-2206
Fax: (240) 223-0025
Email: sales@informaxinc.com
www.informaxinc.com

Intelligent Automation Systems
149 Sidney Street
Cambridge, MA 02139 USA
Phone: (617) 354-3830
Fax: (617) 547-9727
www.ias.com

INTERACTIVA The Virtual Laboratory
Sedanstr.10
D-89007 Ulm
Germany
Phone: +49 (0) 731-93579-290
Fax: +49 (0) 731-93579-291
www.interactiva.de

JOUAN, Inc.
170 Marcel Drive
Winchester, VA 22602 USA
Phone: (540) 869-8623
Fax: (540) 869-8626
Email: info@JouanInc.com
www.jouaninc.com

Life Technologies, Inc.
9800 Medical Center Drive
Rockville, MD 20850 USA
Phone: (301) 610-8000
www.lifetech.com

LION Bioscience AG
Im Neuenheimer Feld 515-517
D-69120 Heidelberg, Germany
Phone: 49 (0) 6221-40 38-0
Fax: 49 (0) 6221-40 38-101
Email: contact@lionbioscience.com
www.lion-ag.de

Luminex Corp.
12212 Technology Blvd.
Austin, TX 78727 USA
Toll Free: (800) 219-8020
Phone: (512) 219-8020
Fax: (512) 258-4173
Email: info@luminexcorp.com
www.luminexcorp.com

Meyer Instruments, Inc.
1304 Langham Creek, Suite 235
Houston, TX 77084 USA
Phone: (281) 579-0342
Fax: (281) 579-1551
Email: Meyer@MeyerInst.com
www.meyerinst.com

MJ Research, Inc.
149 Grove Street
Watertown MA 02172 USA
Phone: (617) 923-8000
Fax: (617) 923-8080
www.mjr.com

Molecular Dynamics
928 East Arques Avenue
Sunnyvale, CA 94086-4520 USA
Phone: (408) 773-1222
Fax: (408) 773-1493
Email: info@mdyn.com
www.mdyn.com

Molecular Probes, Inc.
4849 Pitchford Avenue
Eugene, OR 97402 USA
Phone: (541) 465-8300
Fax: (541) 344-6504
Email: webmaster@probes.com
www.probes.com

Monsanto
800 N. Lindbergh Blvd.
St. Louis, MO 63167 USA
Phone: (314) 694-1000
www.monsanto.com

Motorola, Inc.
1303 E. Algonquin Road
Schaumburg, IL 60196 USA
Phone: (847) 576-5000
Fax: (847) 576-325
www.mot.com

NEN Life Science Products
549 Albany Street
Boston, MA 02118 USA
Toll Free: (800) 551-2121
Phone: (617) 482-9595
Fax: (617) 482-1380
Email: webmaster@nenlifesci.com
www.nenlifesci.com

Nanogen
10398 Pacific Center Court
San Diego, CA 92121 USA
Phone: (858) 410-4600
Fax: (858) 410-4848
www.nanogen.com

Operon Technologies, Inc.
1000 Atlantic Avenue
Suite 108
Alameda, CA 94501 USA
Toll Free: (800) 688-2248
Phone: (510) 865-8644
Fax: (510) 865-5225
Email: dna@operon.com
www.operon.com

Packard Instrument Company
800 Research Parkway
Meriden, CT 06450 USA
Phone: (203) 238 2351
Fax: (203) 639 2172
Email: webmaster@packardinstrument.com
www.packardinstrument.com

Perkin-Elmer Corporation
761 Main Avenue
Norwalk, CT 06859-0001 USA
Phone: (203) 762-1000
Fax: (203) 762-6000
Email: info@pe-corp.com
www.perkinelmer.com

Phoretix International
Tyne House, 26 Side
Newcastle upon Tyne
NE1 3JA
United Kingdom
Phone: +44 (0) 191-230-2121
Fax: +44 (0) 191-230-2131
Email: web@phoretix.com
www.phoretix.com

ProtoGene Laboratories, Inc.
1454 Page Mill Road
Palo Alto, CA 94304 USA
Phone: (650) 842-7100
Fax: (650) 842-7110
www.protogene.com

QIAGEN, Inc.
28159 Avenue Stanford
Valencia, CA 91355 USA
Phone: (800) 426-8157
Fax: (800) 718-2056
www.qiagen.com

Research Genetics
2130 Memorial Parkway
Huntsville, AL 35801 USA
Phone, US or Canada:
(800) 533-4363
Phone, UK: 0-800-89-1393
Fax, US or Canada: (256) 536-9016
Fax, Worldwide: 001-256-536-9016
Email: info@resgen.com
www.resgen.com

Rosetta Inpharmatics
12040-115th Ave., NE
Kirkland, WA 98034 USA
Phone: (425) 820-8900
Fax: (425) 821-5354
Email: info@rii.com
www.rii.com

Savant Instruments, Inc.
100 Colin Drive
Holbrook, NY 11741-4306 USA
Phone: (516) 244-2929
Fax: (516) 244-0606
www.savec.com

Scanalytics
8550 Lee Highway, Suite 400
Fairfax, VA 22031-1515 USA
Phone: (703) 208-2230
Fax: (703) 208-1960
Email: info@scanalytics.com
www.scanalytics.com

SciMatrix
2300 Englert Drive, Suite B
Durham, NC 27713 USA
Phone: (919) 544-5558
Fax: (919) 572-6889
Email: info@scimatrix.com
www.scimatrix.com

SEQUENOM
11555 Sorrento Valley Road
San Diego, CA 92121-1331 USA
Phone: (858) 350-0345
Fax: (858) 350-0344
www.sequenom.com

Silicon Genetics
935 Washington Street
San Carlos, CA 94070 USA
Phone: (650) 591-4459
Fax: (650) 591-5574
Email: info@sigenetics.com
www.sigenetics.com

Stratagene
11099 North Torrey Pines Road
La Jolla, CA 92037 USA
Phone: (800) 424-5444
www.stratagene.com

Synteni, Inc.
6419 Dumbarton Circle
Fremont, CA 94555 USA
Toll Free: (510) 739-2100
Fax: (510) 739-2200
Email: sales@incyte.com
www.synteni.com

Tecan U.S., Inc.
4022 Stirrup Creek Drive, Suite 310
Durham, NC 27703 USA
Toll Free: (800) 338-3226
Phone: (919) 361-5200
Fax: (919) 361-5201
www.tecan-us.com

TeleChem International, Inc.
524 E. Weddell Drive, Suite 3
Sunnyvale, CA 94089 USA
Phone: (408) 744-1331
Fax: (408) 744-1711
Email: arrayit@arrayit.com
arrayit.com

Third Wave Technologies, Inc.
502 South Rosa Road
Madison, WI 53719 USA
Toll Free: (888) 898-2357
Phone: (608) 273-8933
Fax: (608) 273-8618
www.twt.com

V&P Scientific, Inc.
9823 Pacific Heights Blvd.
Suite T
San Diego, CA 92121 USA
Phone: (800) 455-0643
Fax: (858) 455-0703
www.vp-scientific.com

Virtek Vision, Inc.
300 Wildwood Avenue
Woburn, MA 01801-9998 USA
Phone: (800) 933-9011
Fax: (781) 933-3461
www.virtek.ca

Vysis Inc.
3100 Woodcreek Drive
Downers Grove, IL 60515-5400 USA
Toll Free: (800) 553-7042
Phone: (630) 271-7000
Fax: (630) 271-7138
www.vysis.com

INDEX

E

F

G

H

Q

R

S

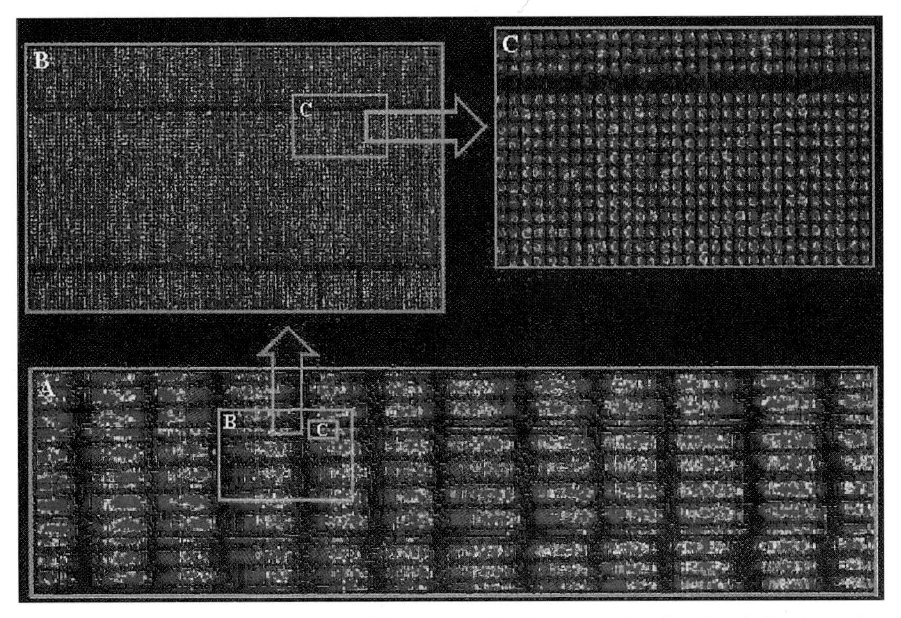

Figure 3. 82-K Microarray. TeleChem's Stealth printing technology was used to deposit 0.4-nL volumes (approximately 100-μm diameter spots) of Cy3-labeled oligonucleotide at 120 μm spacing on an aldehyde-derivatized glass substrate (CEL). The subarray configuration of 36 × 36 × 72 gives a total of 82 944 features. The areas represented by regions A, B, and C correspond to 12.96, 0.25, and 0.07 cm^2, respectively. (See Chapter 1.)

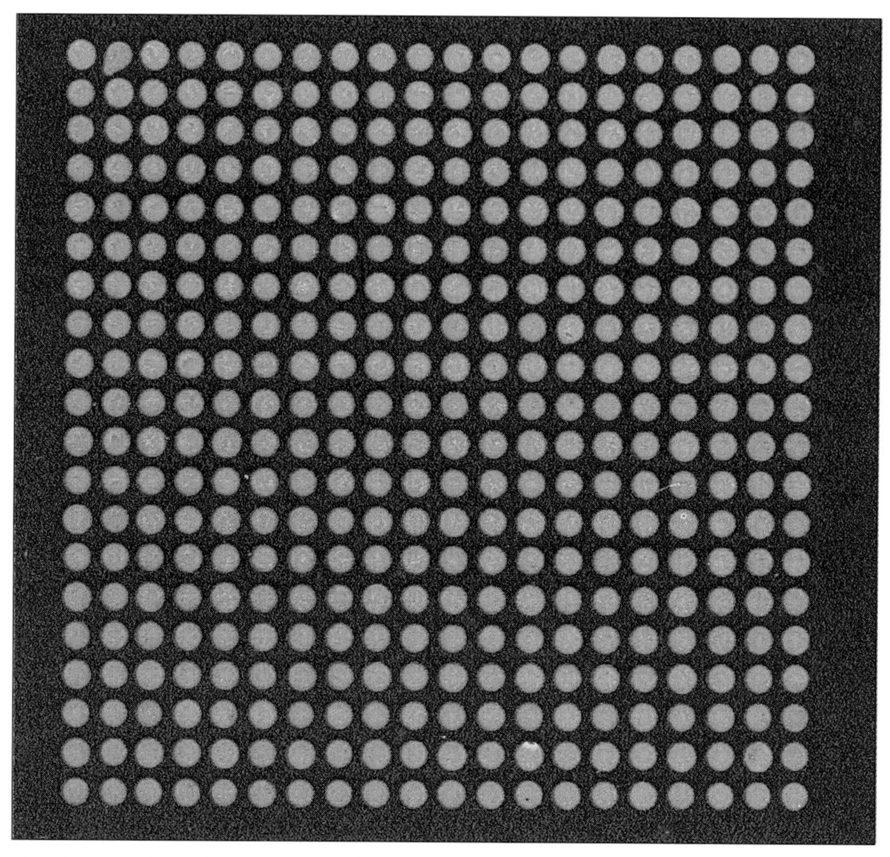

Figure 3. Array of 4.2-nL spots of Cy3 made using a syringe-solenoid type of dispenser. Spot spacing is 500 μm with spot size approximately 325 μm. (See Chapter 2.)

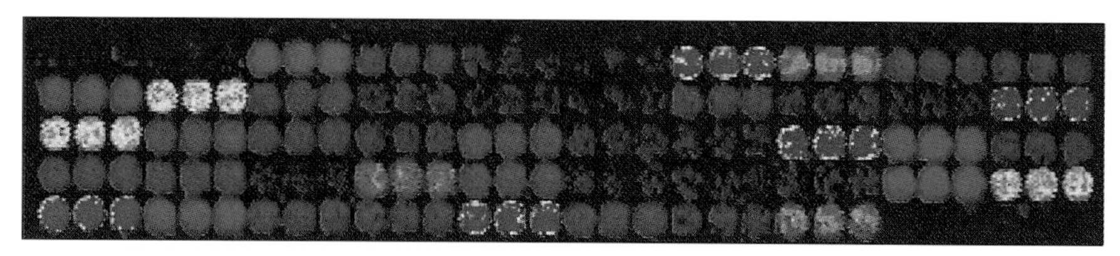

Figure 5. Example of a printed and hybridized microarray made with TeleChem's microspotting pin technology (ChipMaker 3 pins). Spot spacing, 140 μm; spot size, approximately 125 μm, with targets spotted in triplicate. Fluorescent image generated with a ScanArray 3000 (GSI Lumonics). (See Chapter 2.)

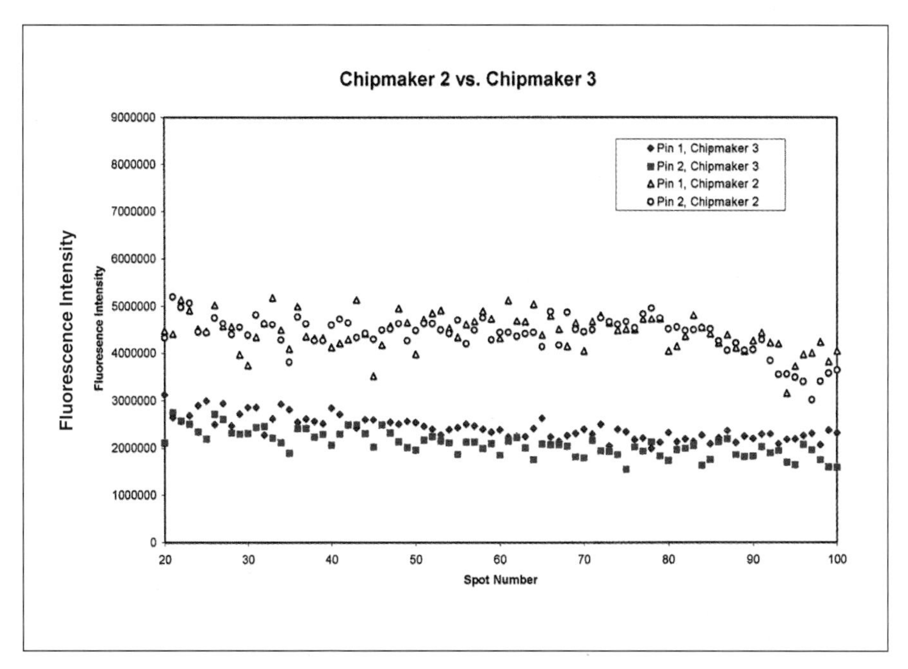

Figure 8. Spot size as a function of spot number for two different size TeleChem pins (ChipMaker 2 and 3). Experimental conditions were the same as in Figure 7, except that 30 preprints were done prior to printing. (See Chapter 2.)

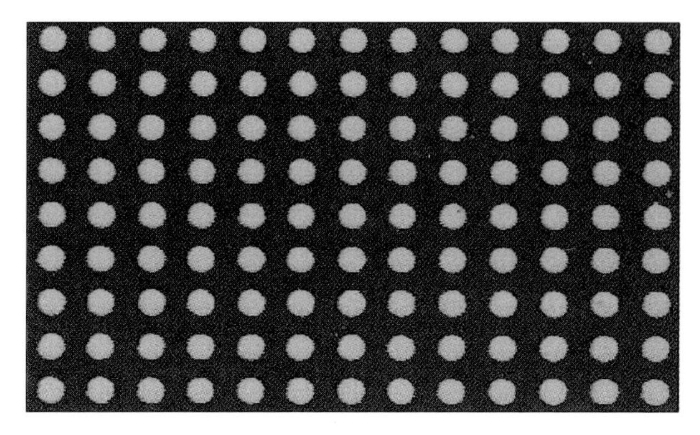

Figure 6. Microarray printed with the Pin and Ring technology. Plant DNA printed as 150-μm spots on polylysine-coated slides stained with eosin. (Data courtesy of Ray Samaha, EOS Biotechnology.) (See Chapter 3.)

Figure 7. Pin and Ring printing on nitrocellulose. Shown is a transmission light micrograph of DNA spotted directly onto a nitrocellulose membrane. (Data courtesy of Dr. Bertrand Jordan.) (See Chapter 3.)

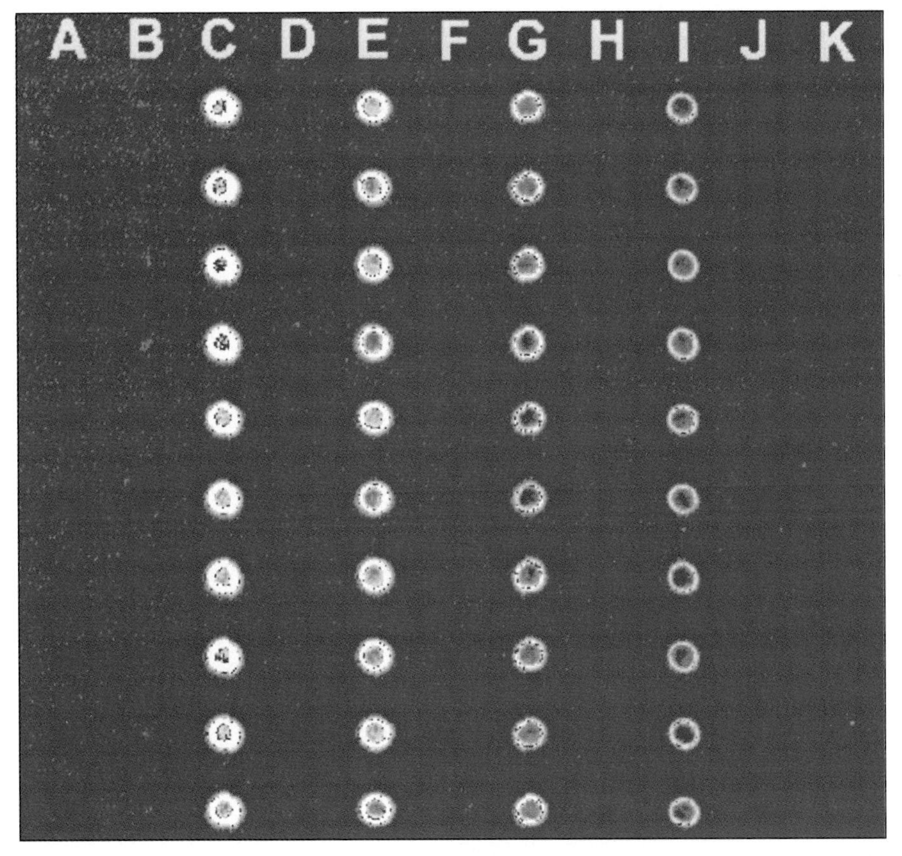

Figure 8. Southern blot microarrays made with the Pin and Ring. (A–K) A dilution series of oligonucleotides interspersed with a nonhomologous sequence was printed onto a nitrocellulose-coated glass microscope slide, and then hybridized with a Cy5-labeled target. No sample carryover is detected. (See Chapter 3.)

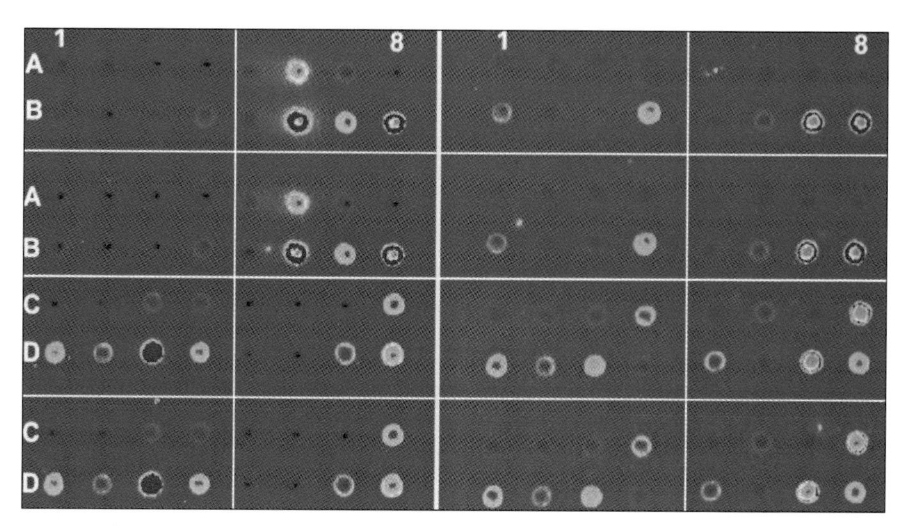

Figure 9. Oligonucleotide arrays made with the Pin and Ring. Shown is a composite image of oligonucleotide microarrays printed on nitrocellulose and hybridized with PCR products from two different individuals (separated by vertical yellow line). Each array contains 16 unique probes with duplicate rows of spots indicated by capital letters. Specificity, repeatability, and signal to background definition are well suited for single-nucleotide polymorphism (SNP) and expression analyses. (See Chapter 3.)

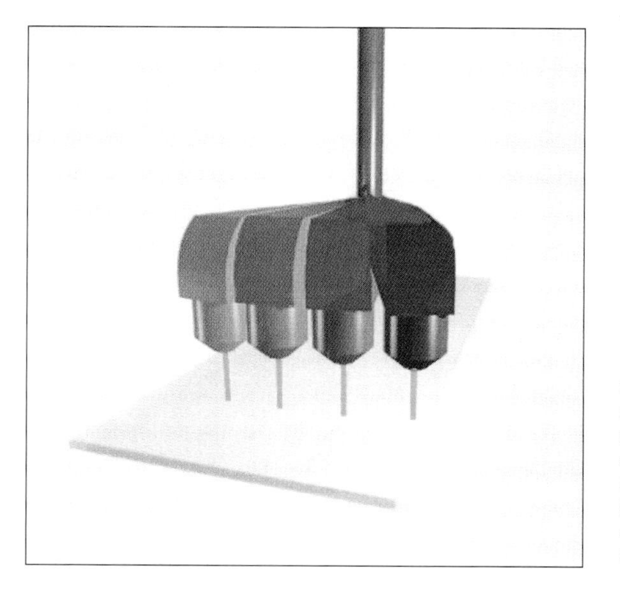

Figure 13. Rotary architecture for microarray detection. A stroboscopic depiction of the rotating arm moving over the entire width of a slide is shown. The light beam is shown as discrete rays, but it scans a continuous path from side to side. (See Chapter 3.)

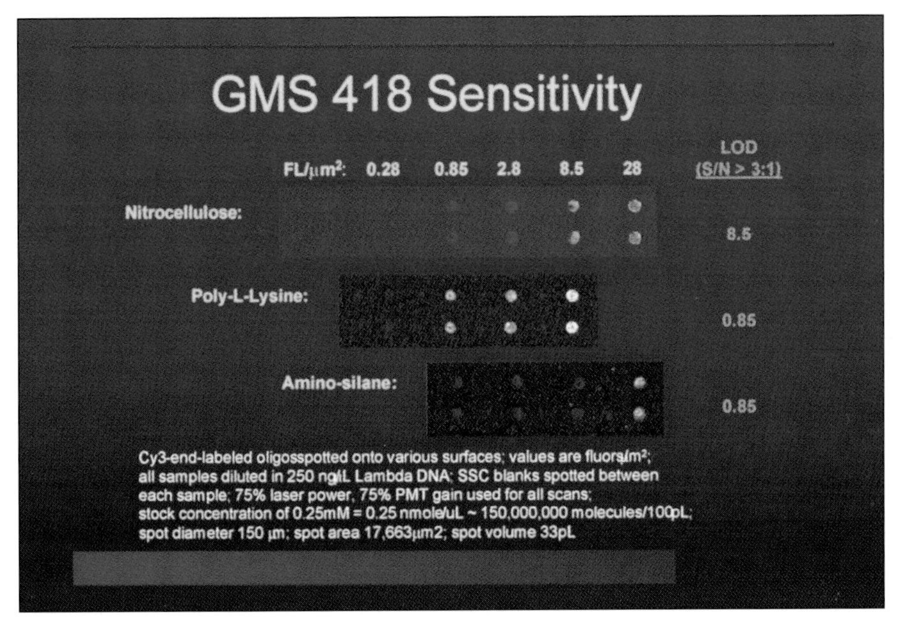

Figure 15. Sensitivity of the GMS 418 scanner. (See Chapter 3.)

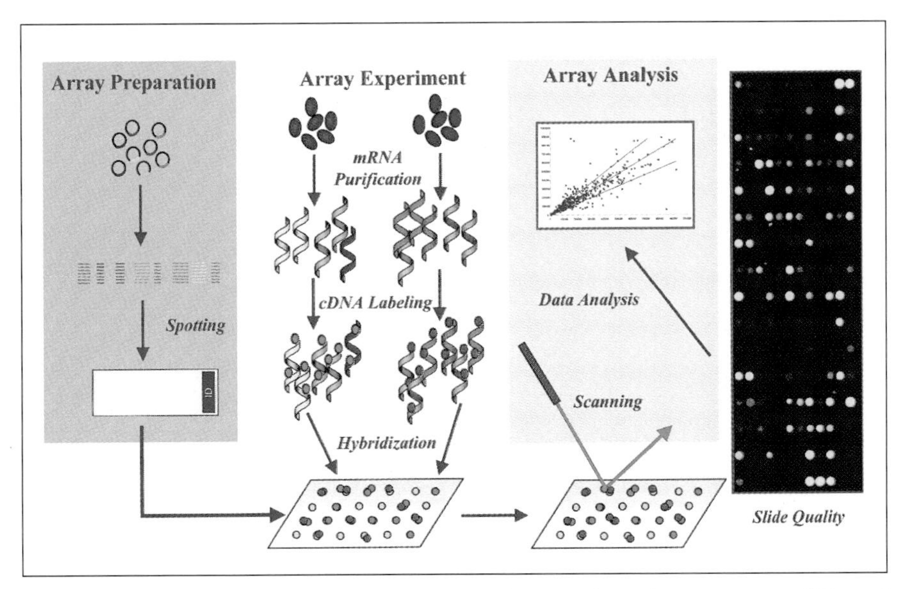

Figure 1. Flow chart of two-color microarray experiments. The major phases of array preparation, differential gene-expression experiment, and analysis are listed, with the topics discussed in this chapter labeled in italics. (See Chapter 4.)

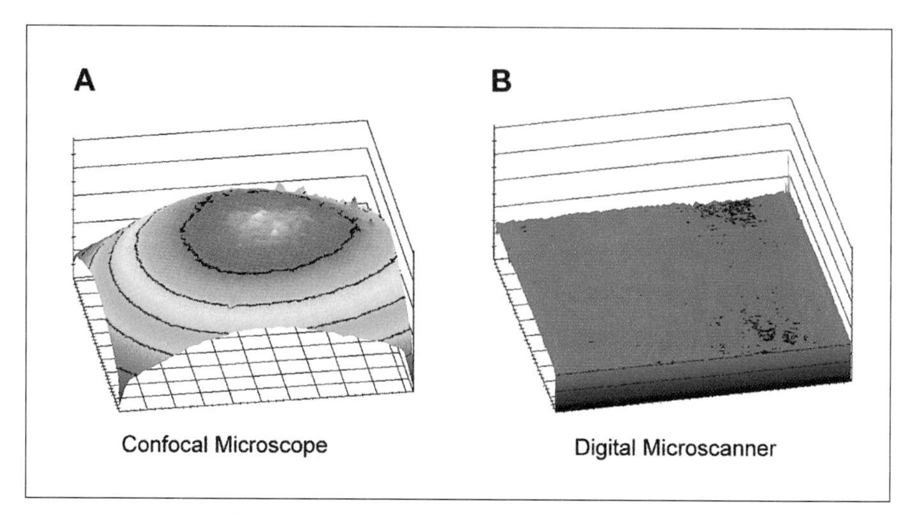

Figure 4. Comparison of field uniformity. (A) Confocal microscope. (B) Molecular Dynamics Microarray Scanner. A thin uniform fluorescent substrate was scanned in both a confocal microscope (4× objective) and the microarray scanner. In both figures the *x*- and *y*-axes map the field of view, and intensity in arbitrary units is plotted on the *z*-axis. [Reprinted with permission from D.K. Hanzel et al. (8)]. (See Chapter 4.)

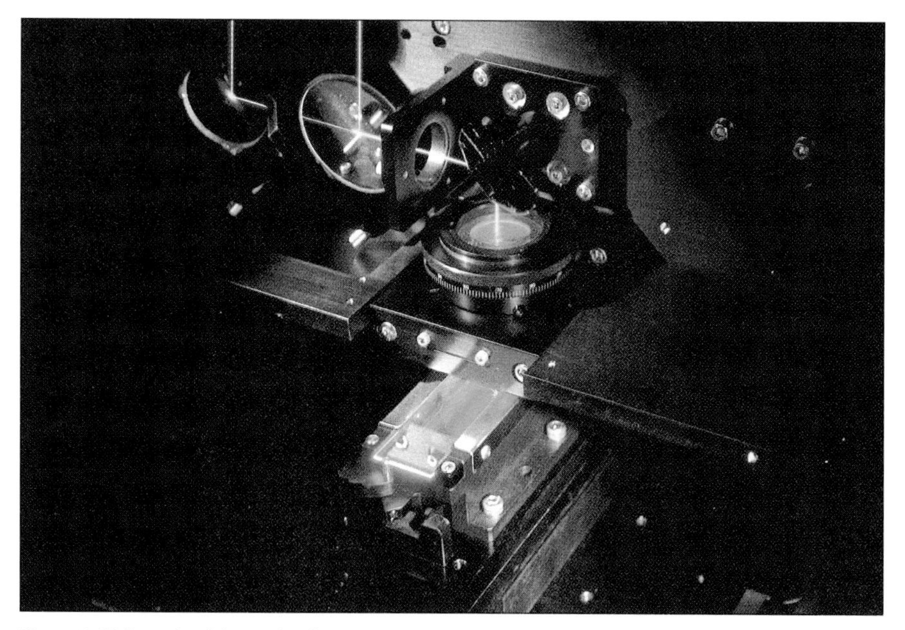

Figure 5. Light path of the Molecular Dynamics Microarray Scanner. The green laser beam reflects off a dichroic beam splitter and is scanned by a galvo-controlled mirror through the wide-field objective lens. Fluorescent light emitted from the surface of the slide is collected by the lens and "descanned" by the mirror. The fluorescent light (indicated by the orange line) passes through the dichroic beam splitter and is reflected by a mirror toward the confocal pinhole and the PMT detector. (See Chapter 4.)

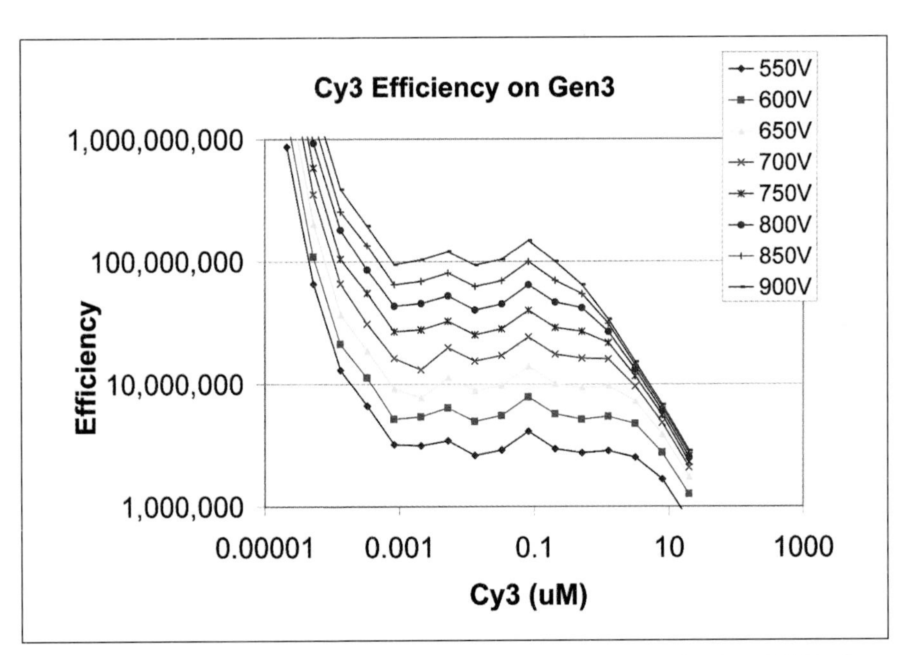

Figure 6. Efficiency of detection of spotted Cy3 on the microarray scanner as a function of PMT voltage. The flat regions of an efficiency plot reveal the linear dynamic range of the system. (See Chapter 4.)

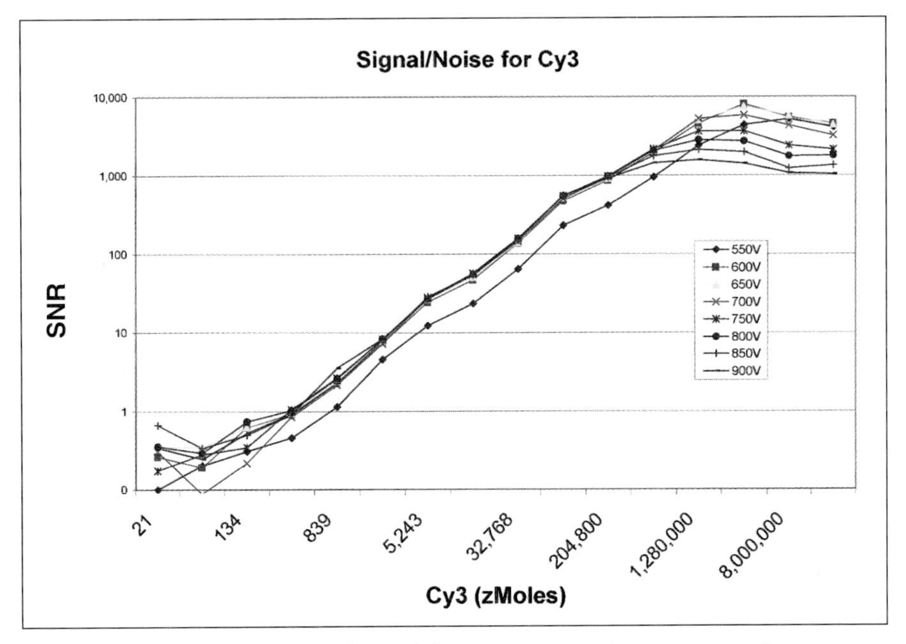

Figure 7. Signal-to-noise ratio (SNR) of spotted Cy3 on a prototype microarray scanner. Scans were conducted at a series of PMT voltages. These plots are used to determine the limit of detection (LOD) of dye fluorescence and the acceptable values for PMT voltage. 10^3 zmole = 1 amole. (See Chapter 4.)

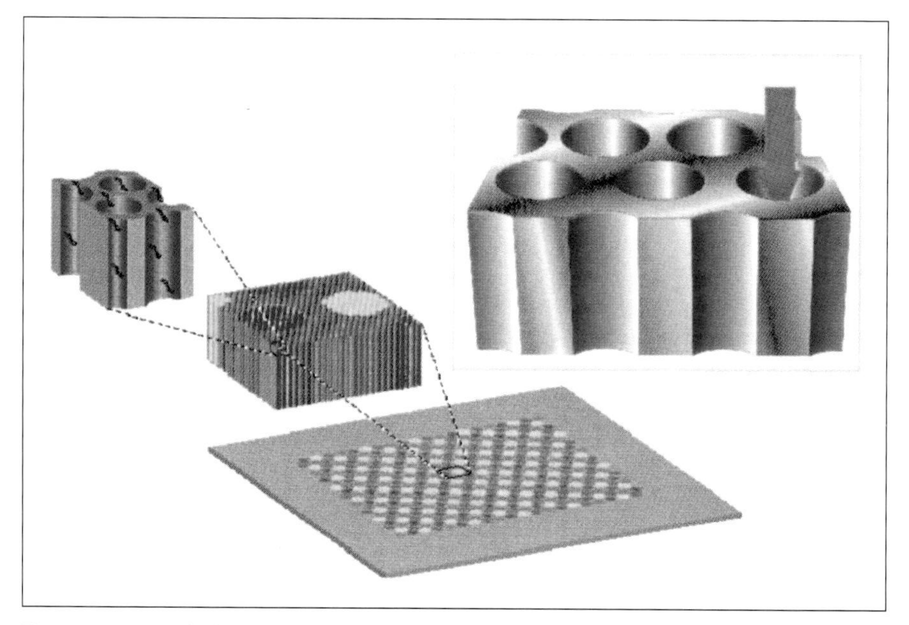

Figure 2. Conceptual schematic of the Flow-Thru Chip. The chip is composed of an ordered array of microscopic channels that transverse the thickness of the substrate. Capture elements (DNA probes) are deposited on the chip in spots. Each spot incorporates several individual microchannels. (See Chapter 5.)

Figure 1. The HuGeneFL array. It contains probes representing more than 5000 full-length human genes. It is shown hybridized with human brain RNA. (See Chapter 6.)

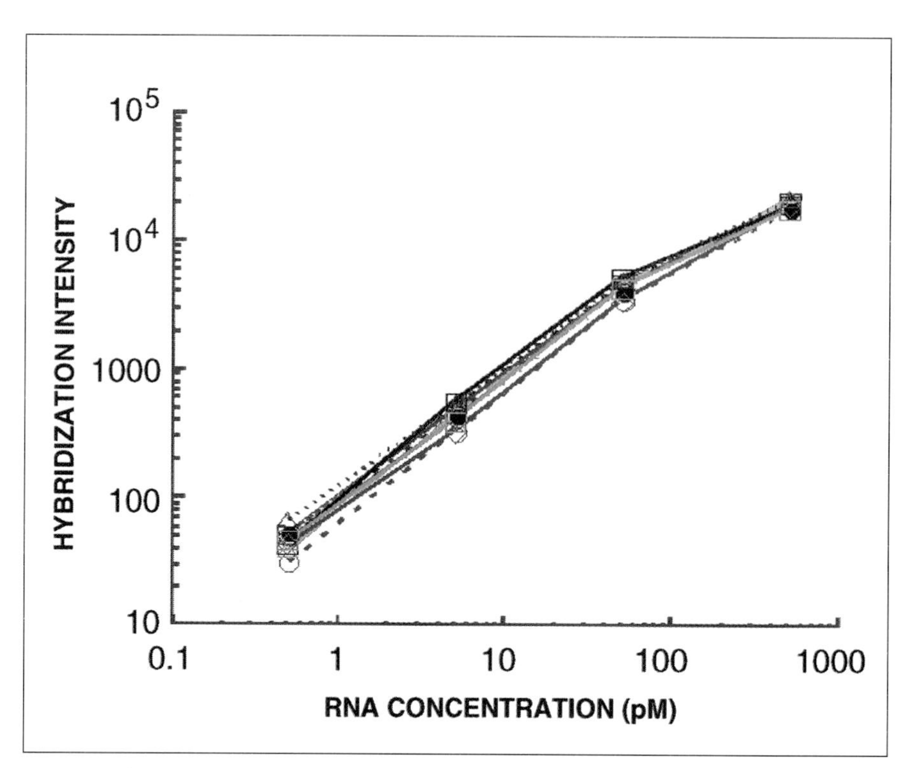

Figure 11. The relationship between hybridization signals and RNA concentration is linear. Log-log plot of the average hybridization intensities versus concentration for 11 different RNA targets spiked into labeled T10 RNA at 0.5, 5.0, 50, and 500 pM. (See Chapter 6.)

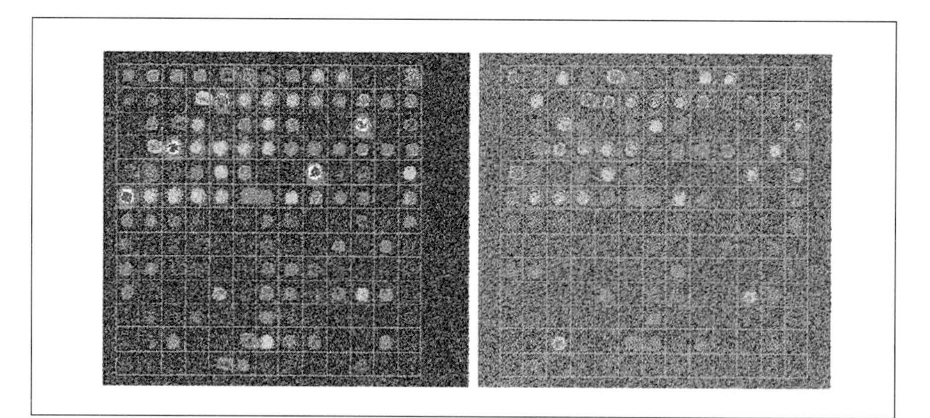

Figure 3. All arrays are gridded for analysis. Cy3 (right) and Cy5 (left) scans. Identical grids and element locations are calculated for the two images. Element identification occurs through a series of steps based on size and estimated location of each cDNA deposited within the grid. (See Chapter 7.)

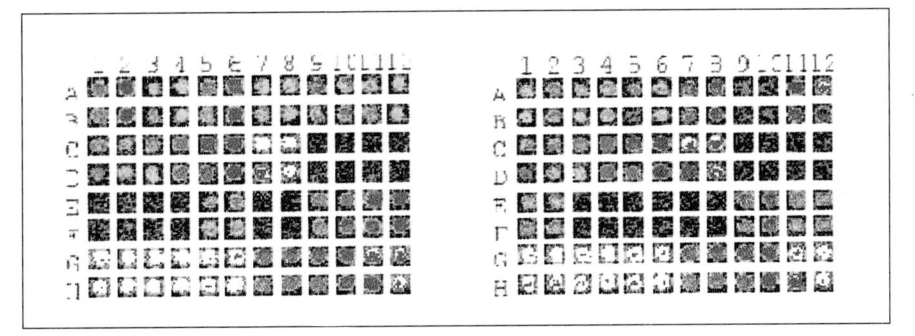

Figure 4. Every GEM contains a 96-well plate with controls for labeling and hybridization. In the Cy3 (right) and Cy5 (left) images shown here, the 96 wells are divided so that each control element is arrayed in quadruplicate. Intergenic yeast fragments are located in rows A–F. RNA sample quality is tested in rows G and H, where a heterogeneous population of DNA and individual housekeeping genes is arrayed. (See Chapter 7.)

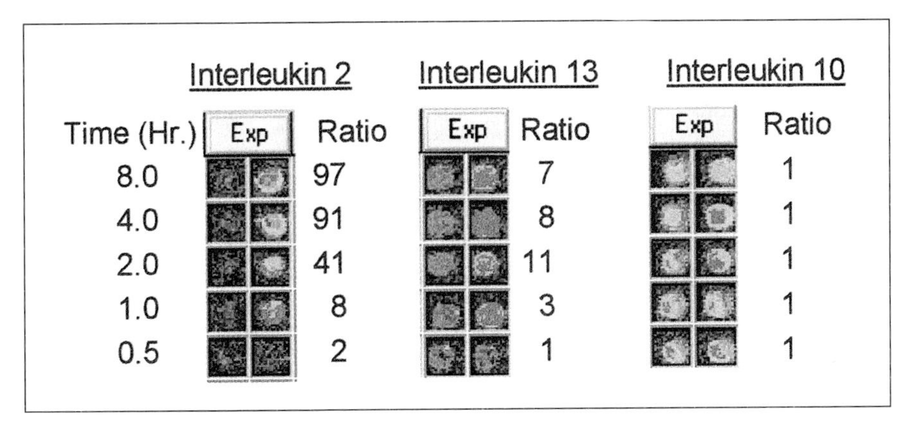

Figure 11. Differential responses in three interleukin genes. Responses are viewed in Incyte's GEMTools software over five time points. The control (left column) is the 0 treatment sample for all three genes. As time after PMA treatment increases, the three interleukins show markedly different responses (right columns). (See Chapter 7.)

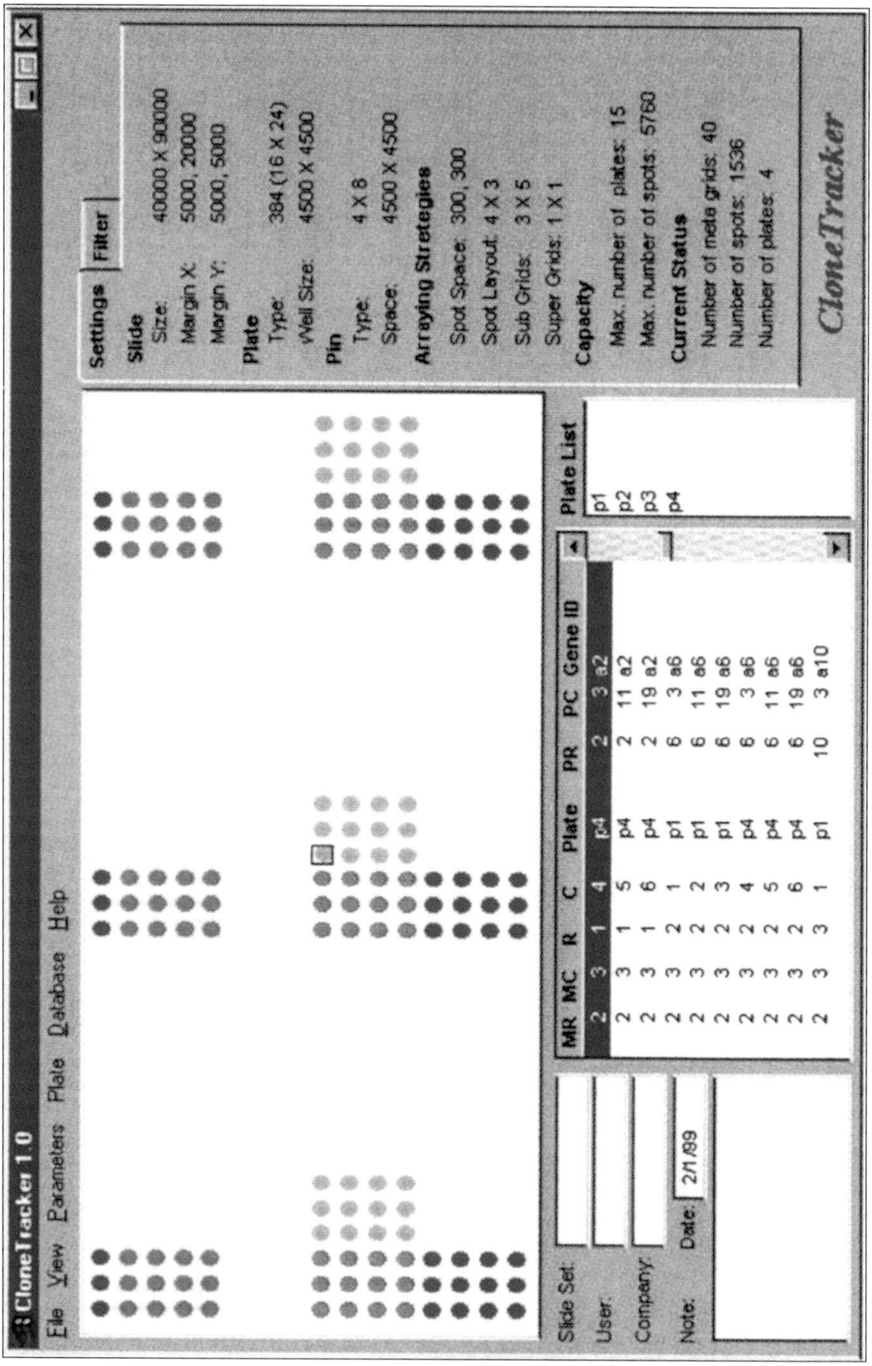

CloneTracker 1.0

File View Parameters Plate Database Help

Slide Set:
User:
Company:
Note: Date: 2/1/99

MR	MC	R	C	Plate	PR	PC	Gene ID
2	3	1	4	p4	2	3	a2
2	3	1	5	p4	2	11	a2
2	3	1	6	p4	2	19	a2
2	3	2	1	p1	6	3	a6
2	3	2	2	p1	6	11	a6
2	3	2	3	p1	6	19	a6
2	3	2	4	p4	6	3	a6
2	3	2	5	p4	6	11	a6
2	3	2	6	p4	6	19	a6
2	3	3	1	p1	10	3	a10

Plate List
p1
p2
p3
p4

Settings Filter

Slide
Size: 40000 X 90000
Margin X: 5000, 20000
Margin Y: 5000, 5000
Plate
Type: 384 (16 X 24)
Well Size: 4500 X 4500
Pin
Type: 4 X 8
Space: 4500 X 4500
Arraying Stretegies
Spot Space: 300, 300
Spot Layout: 4 X 3
Sub Grids: 3 X 5
Super Grids: 1 X 1
Capacity
Max. number of plates: 15
Max. number of spots: 5760
Current Status
Number of meta grids: 40
Number of spots: 1536
Number of plates: 4

CloneTracker

Figure 4. Array fabrication application user interface. (See Chapter 8.)

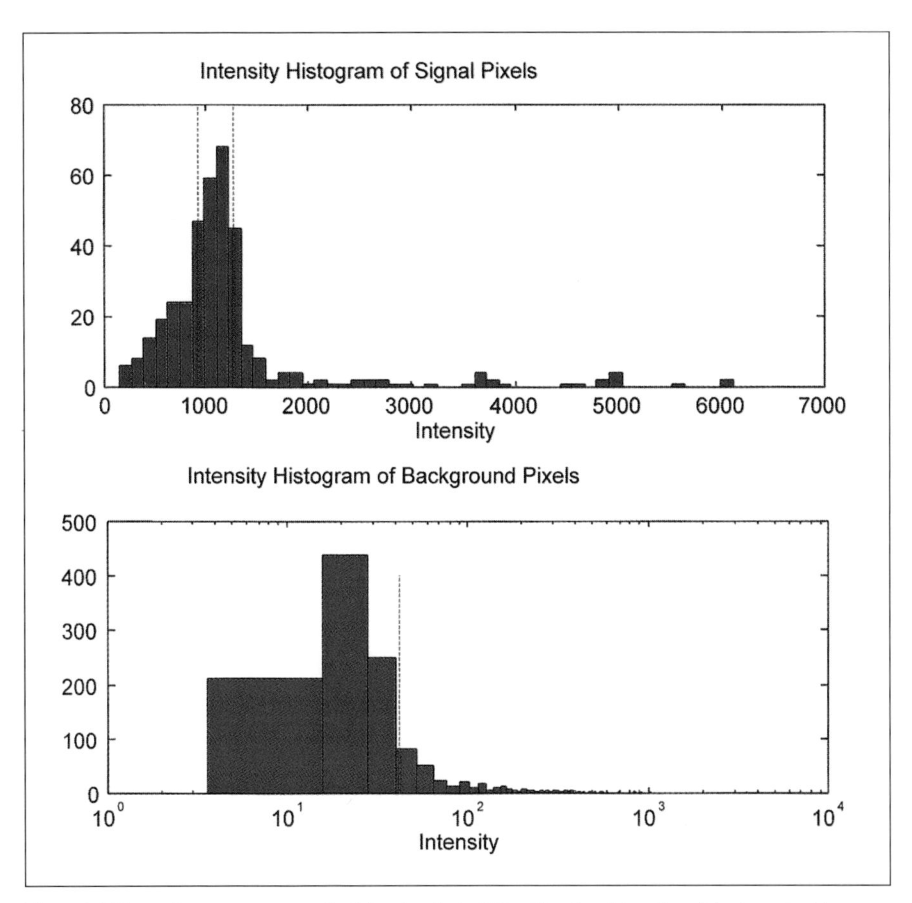

Figure 9. Trimmed measurement method for signal pixel identification. Top, plot of the intensity histogram of the pixels inside the target circle. Thirty percent of the pixels are trimmed off at the lower intensity side (left of the left vertical line). Twenty percent of the pixels are trimmed off the high-intensity side (right of the right vertical line). Bottom, plot of the histogram of the pixels outside the target circle. Thirty percent of pixels are trimmed off at the high-intensity side of the histogram (right of the vertical line). The abscissa is a log scale to show the detail on the lower intensity side. The mean intensities are 1100 for signal and 20 for background. They differ by 32% and -29% from the true value. (See Chapter 8.)

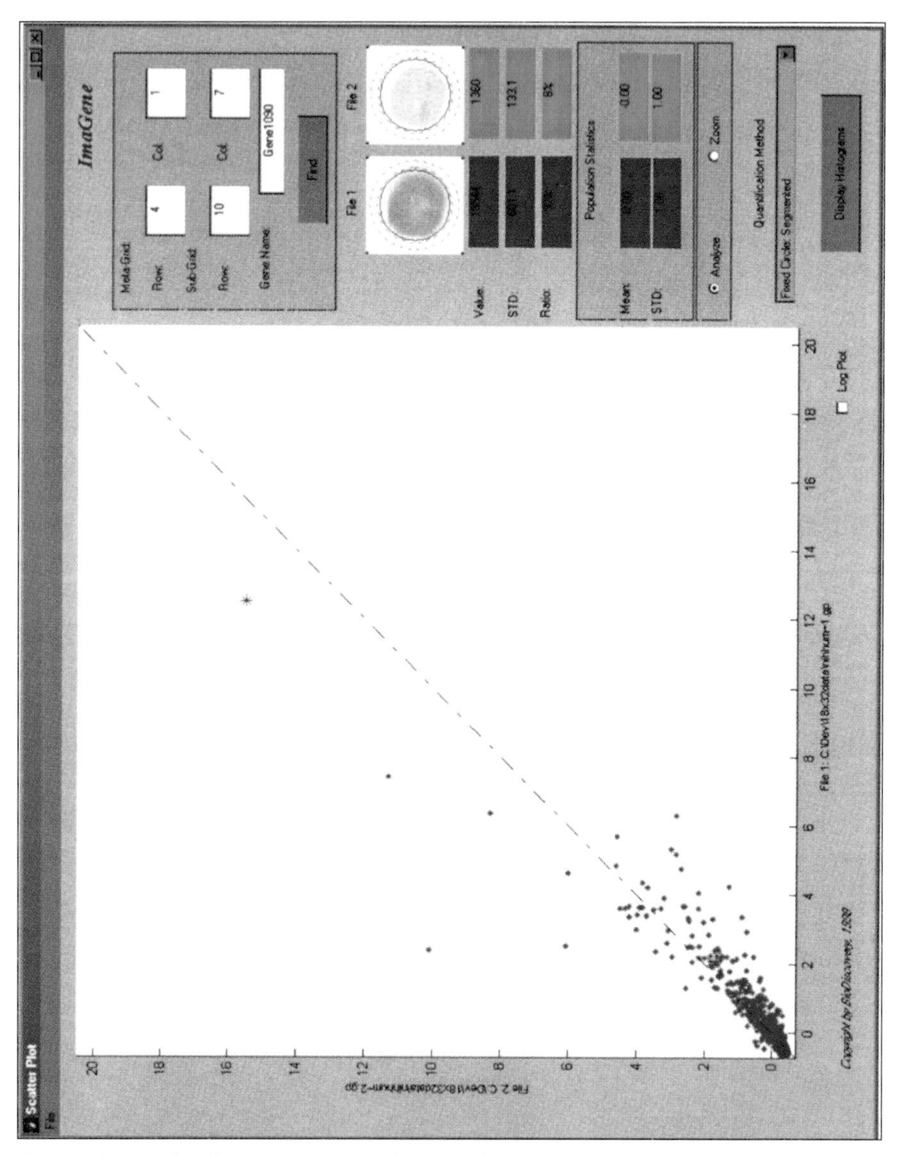

Figure 11. Scatter plot of two experiments. Each point in the plot represents a single gene. The position of the point on the graph reveals the absolute expression level of the gene, such that highly expressed genes lie far from the origin. Genes that are expressed at the same level in two samples fall along the identity (45° angle) line. (See Chapter 8.)

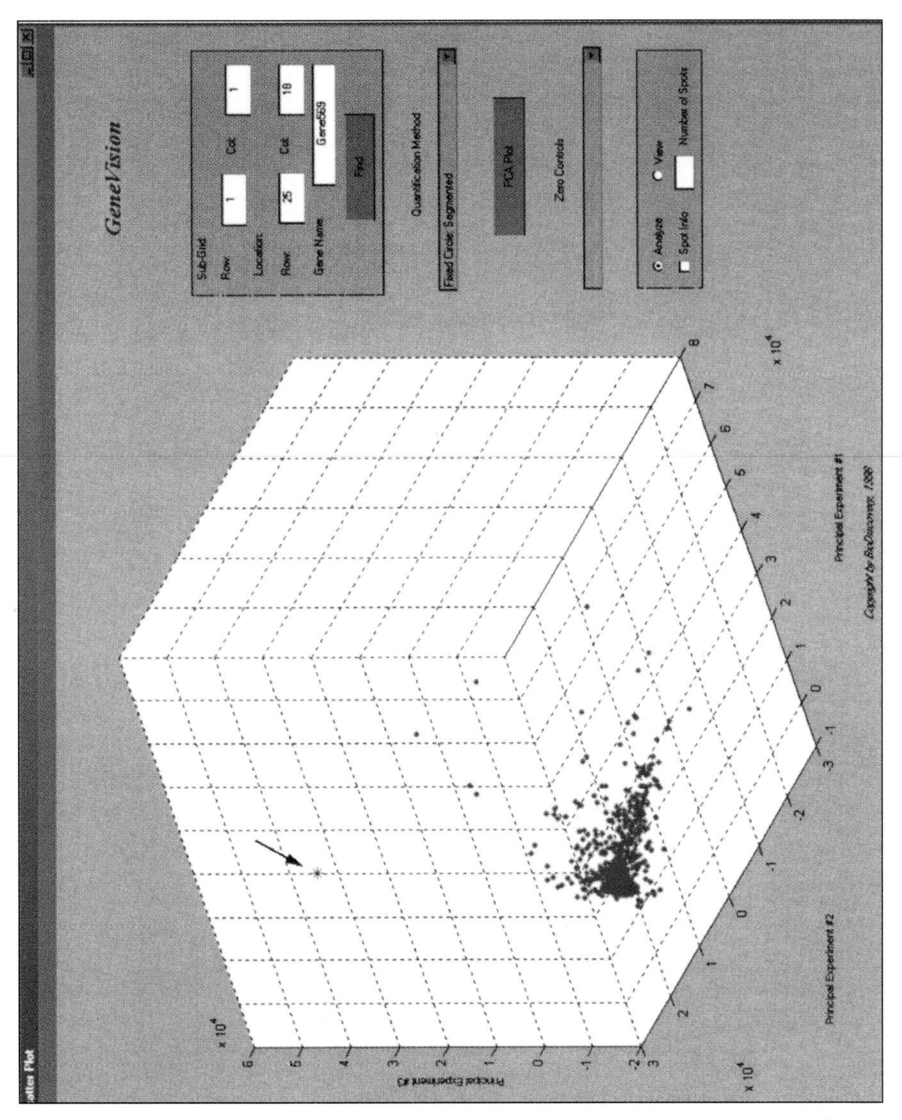

Figure 12. Principal component analysis on 600 genes across 21 experiments. Gene scores are plotted on the first three principal components. The arrow indicates a possible outlier. (See Chapter 8.)

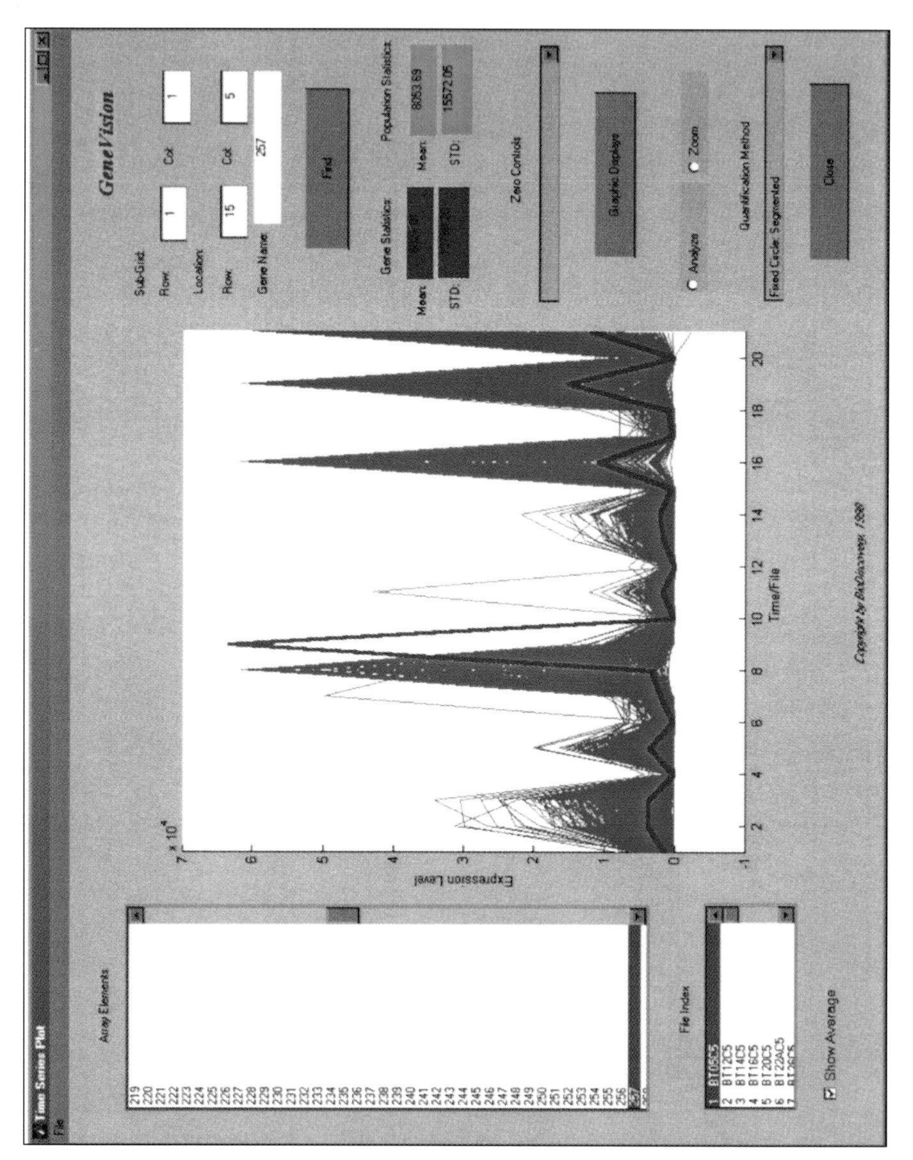

Figure 13. The parallel coordinate plot. The plot displays the expression levels of all genes across all experiments/files in the analysis. On the *x*-axis, the experiments or experimental files are ordered. The *y*-axis reveals the expression level of all genes across all experiments. (See Chapter 8.)

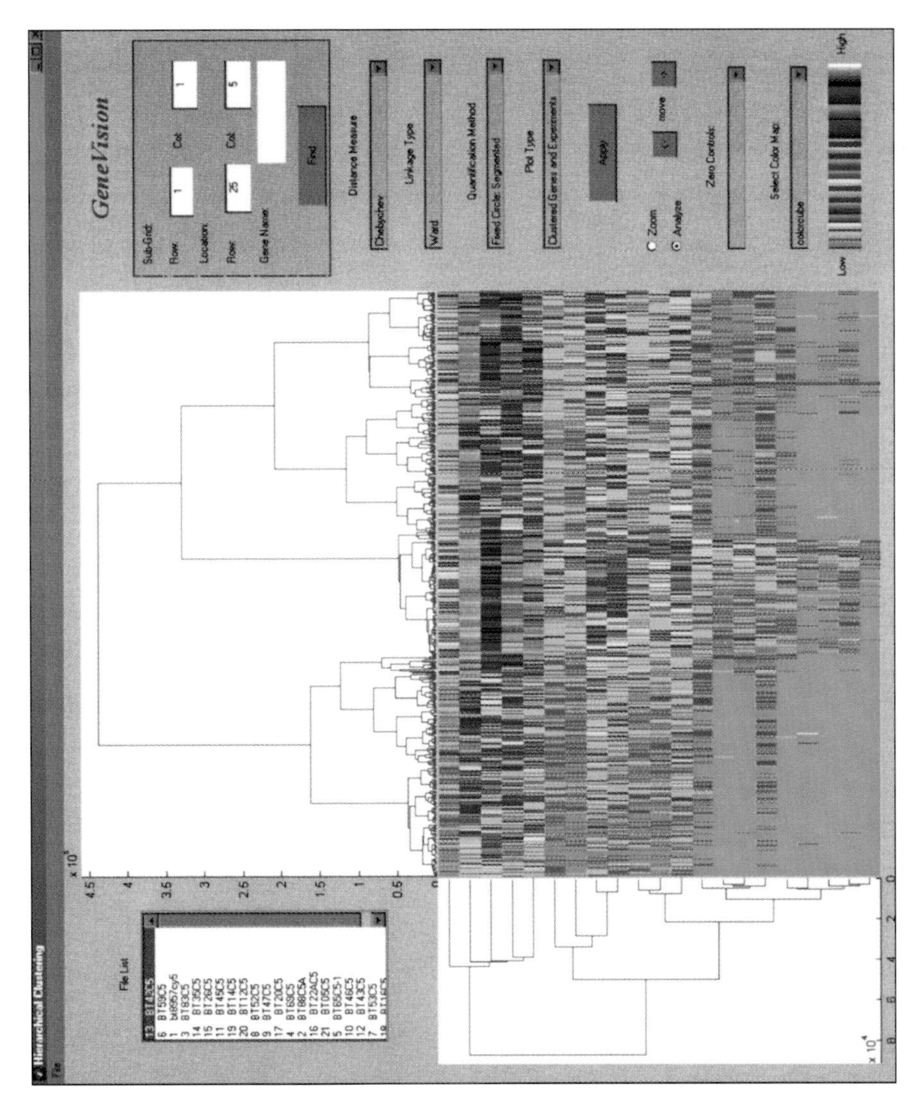

Figure 14. Color-coded gene expression values. Values are given for 600 genes (displayed horizontally) over 21 experiments (displayed vertically). Simultaneous clustering of genes and experiments is visualized by the top and left side dendrograms, respectively. (See Chapter 8.)

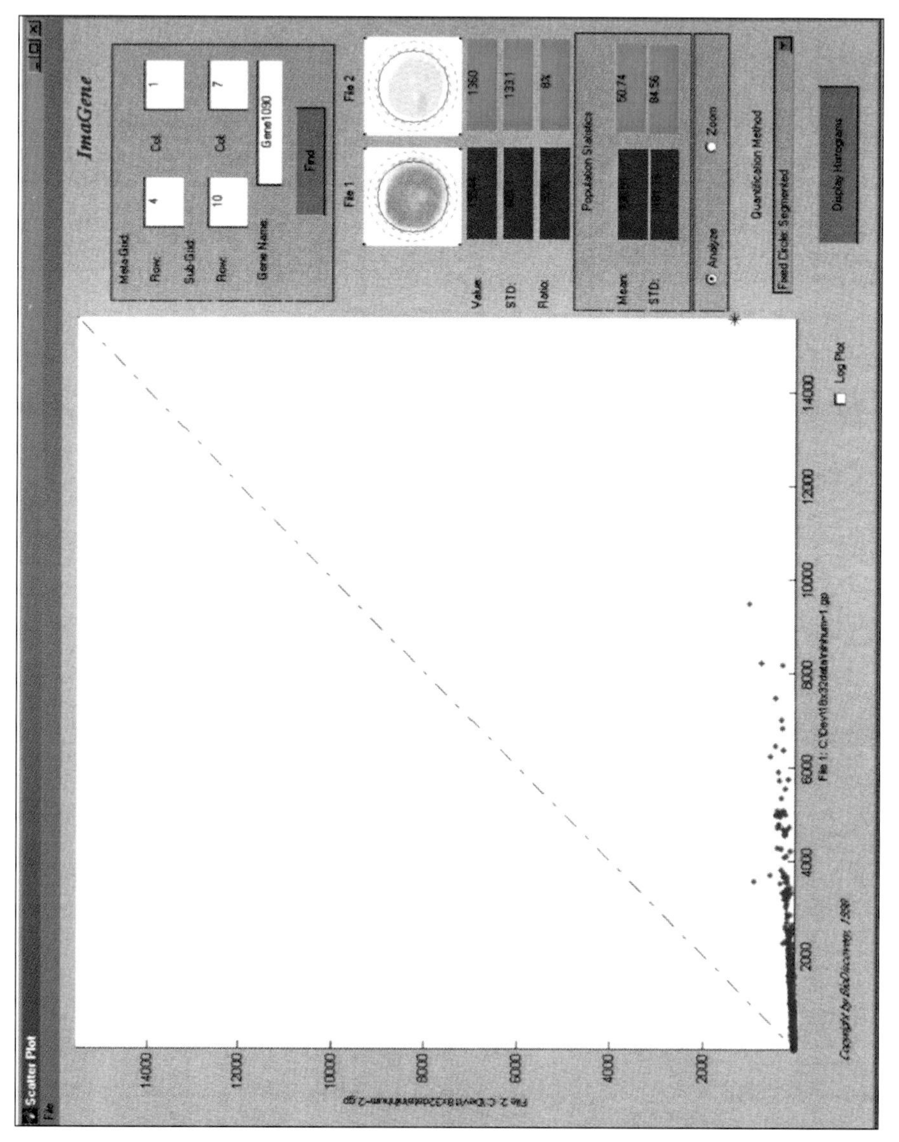

Figure 15. Comparison of fluorescent signals. Scatter plot of two experiments in which the overall fluorescent signal for one file is significantly stronger than the other. (See Chapter 8.)

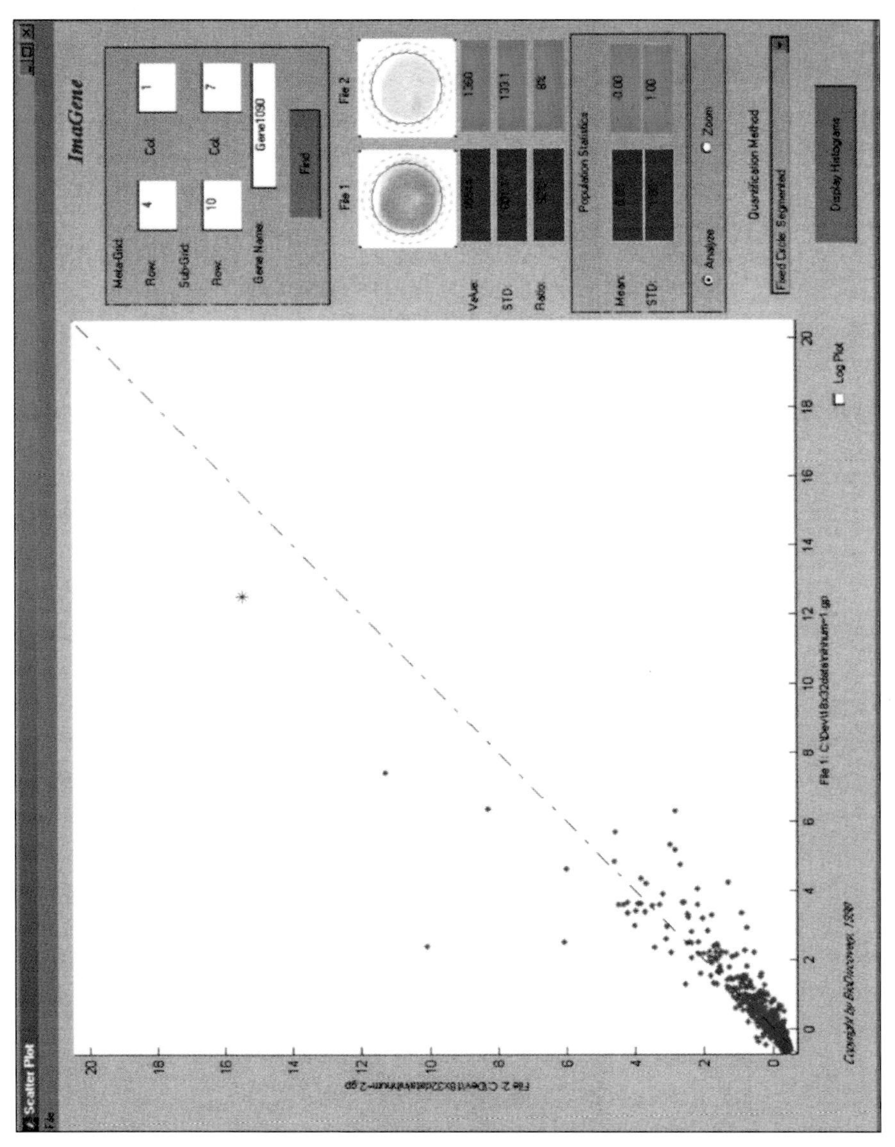

Figure 16. The same data as in Figure 15, but after normalization. (See Chapter 8.)

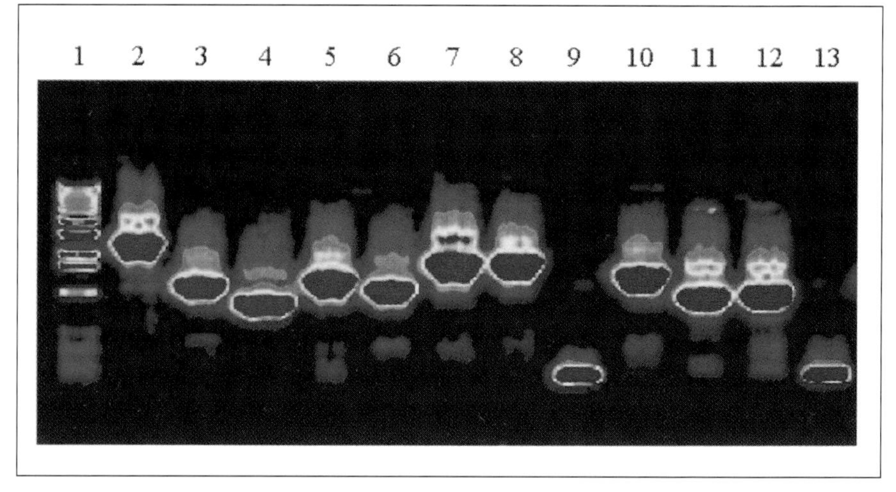

Figure 2. Purified PCR products. A TeleChem PCR Purification Kit was used to purify PCR products. One-tenth of the yield from a 100-µL PCR was analyzed by agarose gel electrophoresis. Lane 1 is a molecular weight marker, and lanes 2 to 13 contain the PCR products. (See Chapter 9.)

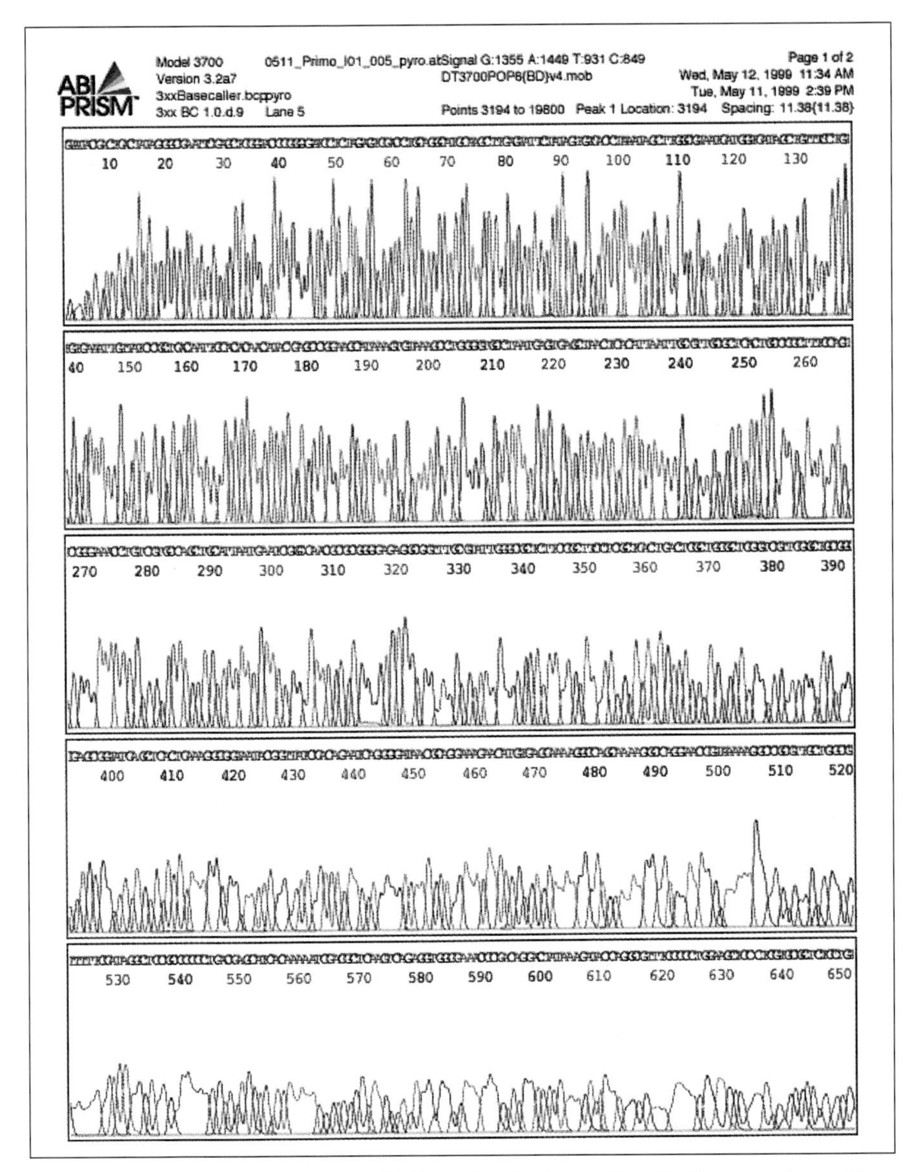

Figure 3. **Sequencing purification kits.** A TeleChem Dye Terminator Clean-Up Kit was used to purify sequencing products. Sequencing was done on a 3700 machine from PE Applied Biosystems. (Data courtesy of Primo Baybayan's Laboratory.) (See Chapter 9.)

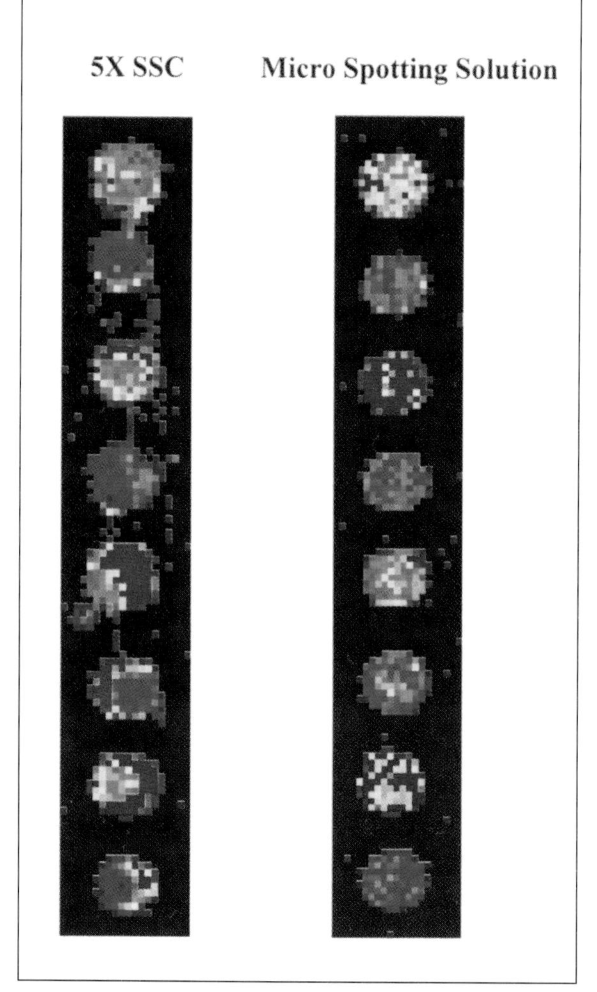

Figure 4. Micro Spotting Solution. Samples print more evenly with Micro Spotting solution, which increases the viscosity and surface tension of the sample and allows uniform drying. Samples were printed with TeleChem's Micro Spotting technology at 150-μm spacing, and hybridized with a Cy3-labeled probe; chips were then scanned with a ScanArray 3000 (GSI Lumonics). (See Chapter 9.)

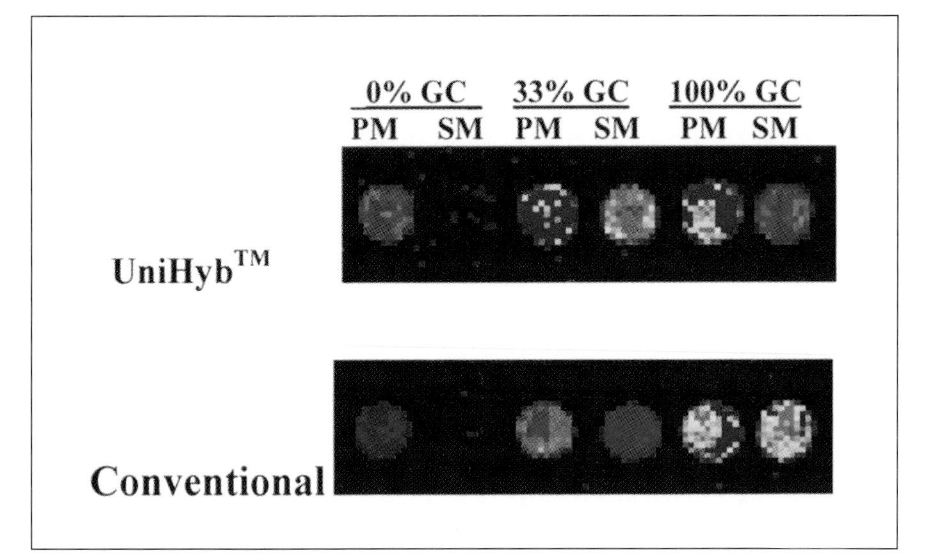

Figure 5. UniHyb Hybridization Solution. Scanned images of oligonucleotide microarrays printed with TeleChem's ChipMaker 3 microspotting device. Spacing is 150 µm center to center on aldehyde slides. The oligonucleotides are three pairs of amino-linked 15-mers with a perfect match (PM) or a single base mismatch (SM) relative to a Cy3-labeled probe. Hybridizations were performed for 4 hours at 42°C with 0.2 pmol/µL probe in Unihyb Hybridization Solution (right) or 5× SSC + 0.2% SDS (bottom). Fluorescent detection was performed using the ScanArray 3000 (GSI Lumonics). Improvements in hybridization signal and specificity are easily observed with Unihyb Hybridization Solution compared with a conventional hybridization buffer. Point mutations are easily identified in sequences spanning the entire range of AT and GC base composition. (See Chapter 9.)

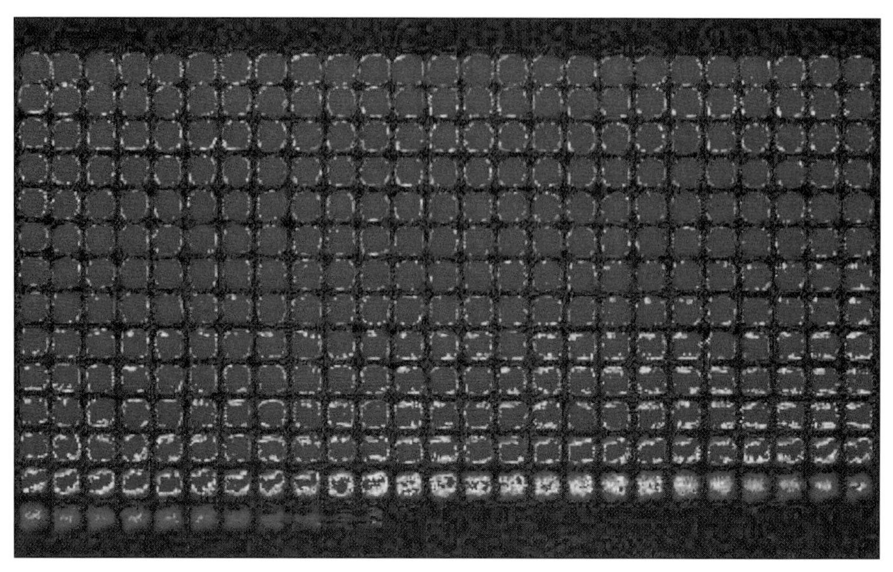

Figure 8. Printed microarray. Stealth 3 pins on a Cartesian PixSys 5500 robot printing at 140-μm spacing. A single loading with a Stealth 3 pin is sufficient to print approximately 250 spots. (See Chapter 9.)

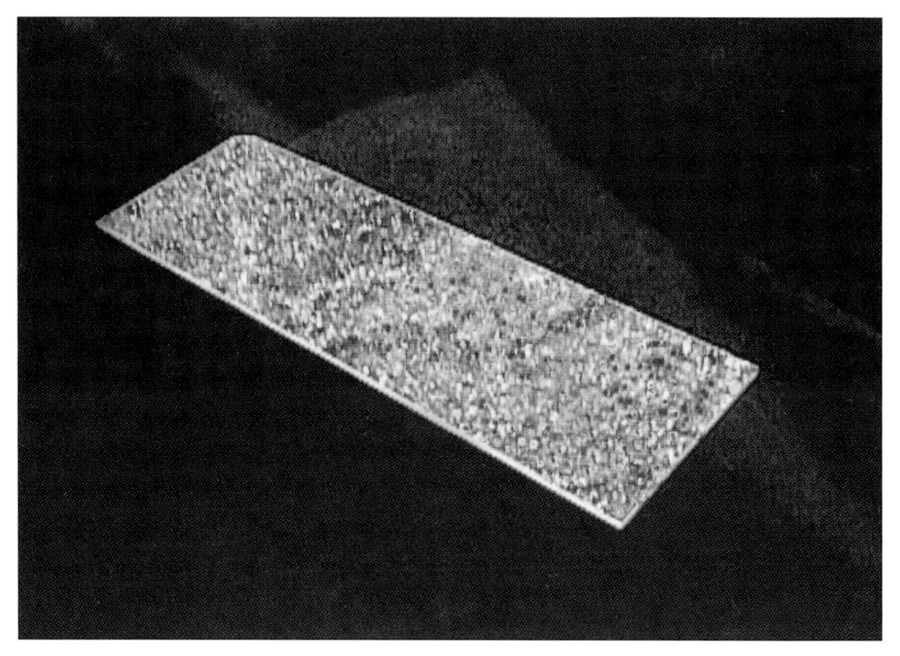

Figure 14. Custom microarray services. Artist's rendition of a custom microarray. (See Chapter 9.)

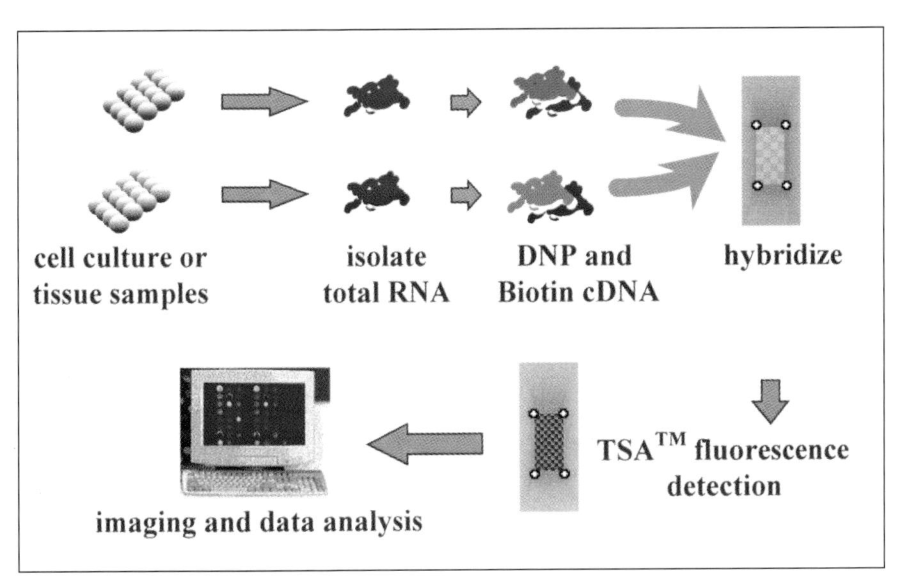

Figure 1. MICROMAX differential gene expression analysis. Two populations of total RNA are isolated from cells or tissues and labeled by incorporating dinitrophenol (DNP) or biotin, respectively, into cDNA. The cDNA is hybridized to a microarray and stained with Tyramide Signal Amplification (TSA) technology. Cyanine 3 and cyanine 5 signals are acquired for the two samples, and the data are analyzed in software. (See Chapter 10.)

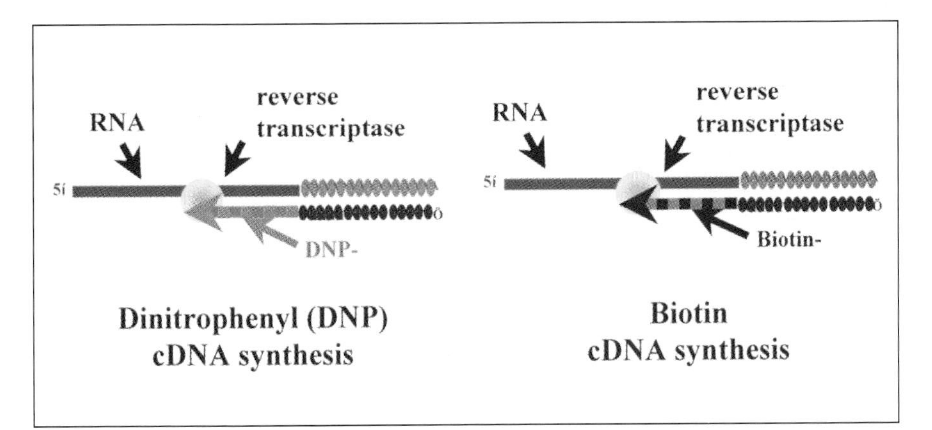

Figure 2. MICROMAX cDNA synthesis. Two samples of total RNA are primed, and reverse transcriptase is used to incorporate dinitrophenol (DNP)- and biotin-modified nucleotides, respectively, into two cDNA samples. (See Chapter 10.)

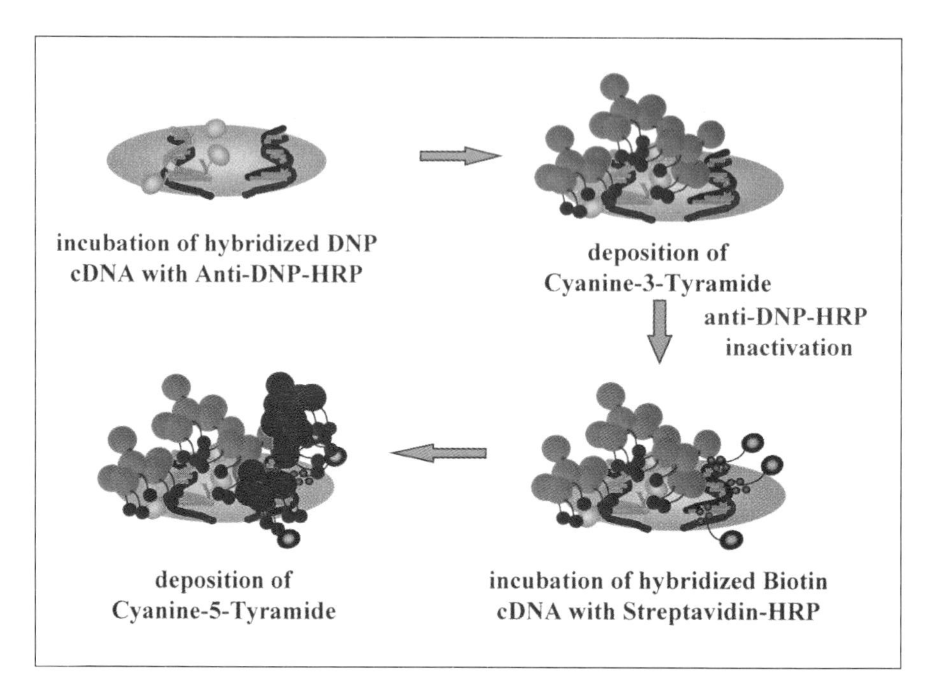

Figure 3. Fluorescence deposition via sequential TSA detection steps. The hybridized microarray is reacted with an α-DNP/HRP conjugate and mixed with a cyanine 3-tyramide reagent. In a second step, the microarray is reacted with a streptavidin/HRP conjugate and mixed with a cyanine 5-tyramide reagent. The cyanine 3 and cyanine 5 dyes are then detected by fluorescence emission. (See Chapter 10.)

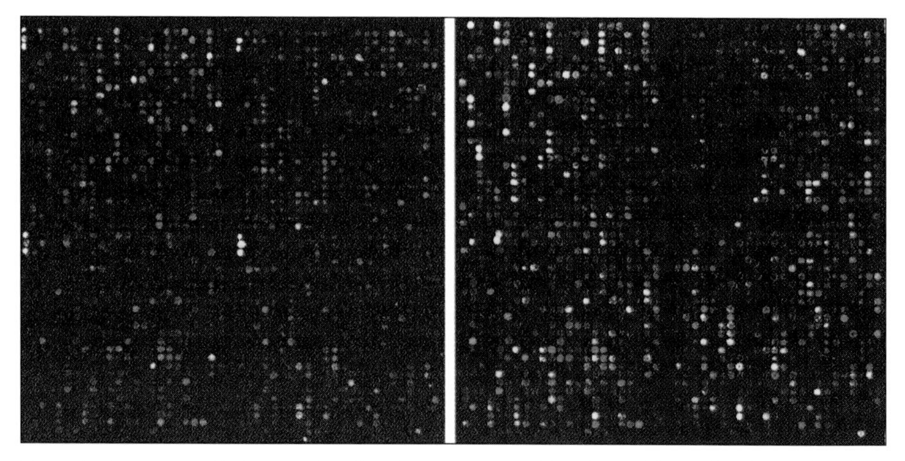

Figure 4. Microarray analysis comparing direct fluorescent and TSA. Overlay bitmap files from a direct fluorescence method with 120 μg total LNCaP and PC3 RNA after cDNA synthesis and hybridization (left), versus the TSA fluorescence method with 2 μg of total RNA input (right). Color overlays are composite images of cyanine 3 and cyanine 5 fluorescence signals: red represents cyanine 3 > cyanine 5, green represents cyanine 5 > cyanine 3, and yellow represents cyanine 3 = cyanine 5. (See Chapter 10.)

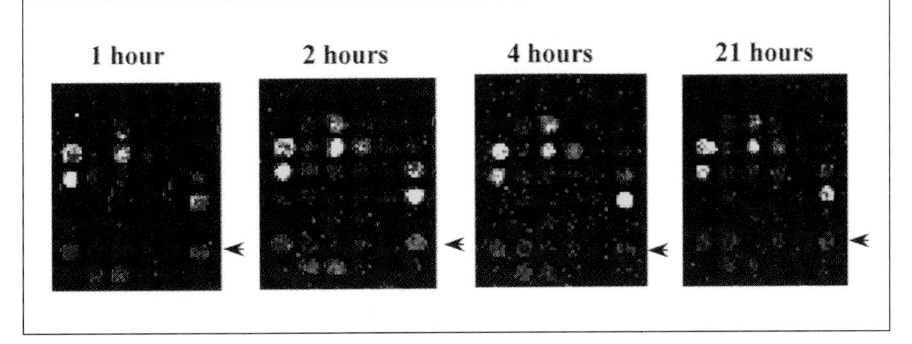

Figure 5. Gene expression time course obtained with MICROMAX. Transcription factor ETR101 induction (arrowhead) is measured in a time course experiment at various times (1 to 21 hours) after stimulation of Jurkat cells with phorbol myristate acetate. (See Chapter 10.)

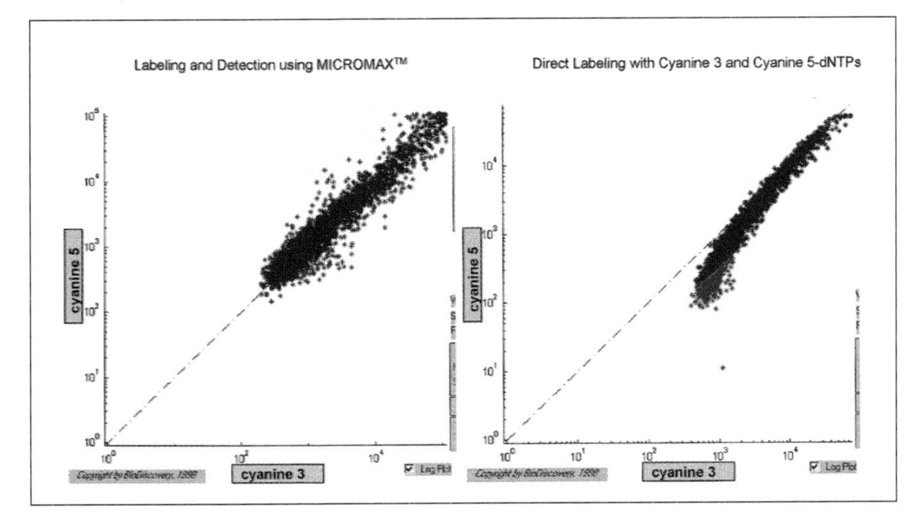

Figure 6. Scatter plots comparing TSA and direct labeling. Log versus log plots of a single source of RNA compared with two different double-labeling methods. A single sample of mRNA from Jurkat cells was divided into four equal aliquots and labeled with cyanine 3 (*x*-axis) and cyanine 5 (*y*-axis) using TSA fluorescence (left) or direct fluorescence (right). Blue dots represent gene sequences that have fluorescent signals within twofold and pink dots represent gene sequences that show more than twofold difference. (See Chapter 10.)

A

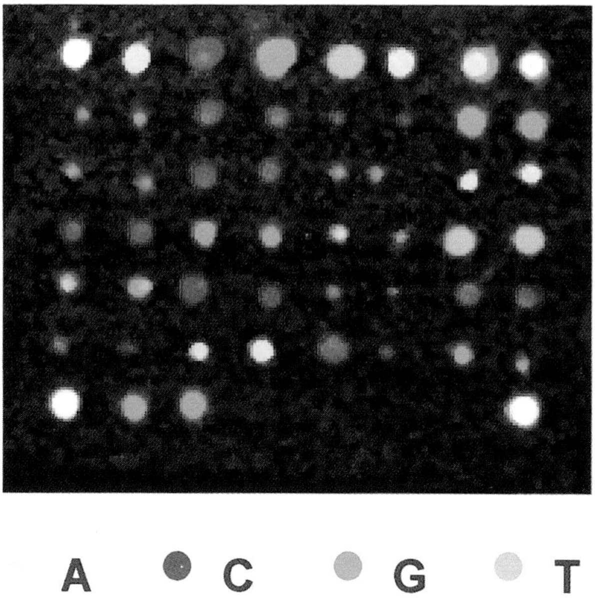

A C G T

B

	S 25mer	S 20mer	As 25mer	As 20mer	S 25mer	S 20mer	As 25mer	As 20mer
1	N	A	C	G	T	AC	AG	N
2	IVS-I-110 G to A	IVS-I-110 G to A	IVS-I-110 C to T	IVS-I-110 C to T	Codon 39 C to T	Codon 39 C to T	Codon 39 G to A	Codon 39 G to A
3	IVS-I-1 G to A	IVS-I-1 G to A	IVS-I-1 C to T	IVS-I-1 C to T	IVS-I-6 T to C	IVS-I-6 T to C	IVS-I-6 A to G	IVS-I-6 A to G
4	IVS-II 745 C to G	IVS-II 745 C to G	IVS-II 745 G to C	IVS-II 745 G to C	Codon 6 A to G	Codon 6 A to G	Codon 6 T to C	Codon 6 T to C
5	IVS-II 1 G to A	IVS-II 1 G to A	IVS-II 1 C to T	IVS-II 1 C to T	IVS-I-5 G to A	IVS-I-5 G to A	IVS-I-5 C to T	IVS-I-5 C to T
6	Codon 5 C to T	Codon 5 C to T	Codon 5 A to G	Codon 5 A to G	-87 C to G	-87 C to G	-87 G to C	-87 G to C
7	N	Codon 5 G	Codon 5 G					N

Figure 5. β-Thalassemia APEX with a wild-type DNA sample. (A) Color code of the incorporated nucleotides is shown below the image. (B) Key code of the array. Seven rows and eight columns of oligonucleotides were arrayed onto the slide. Two oligonucleotides (20- and 25-mer) for the sense strand (s) and two for the antisense strand were used for each mutation site. Upper row and four corners are self-elongating marker oligonucleotides. Wild-type signals and signals expected from mutations are indicated at the bottom of each cell. (See Chapter 12.)

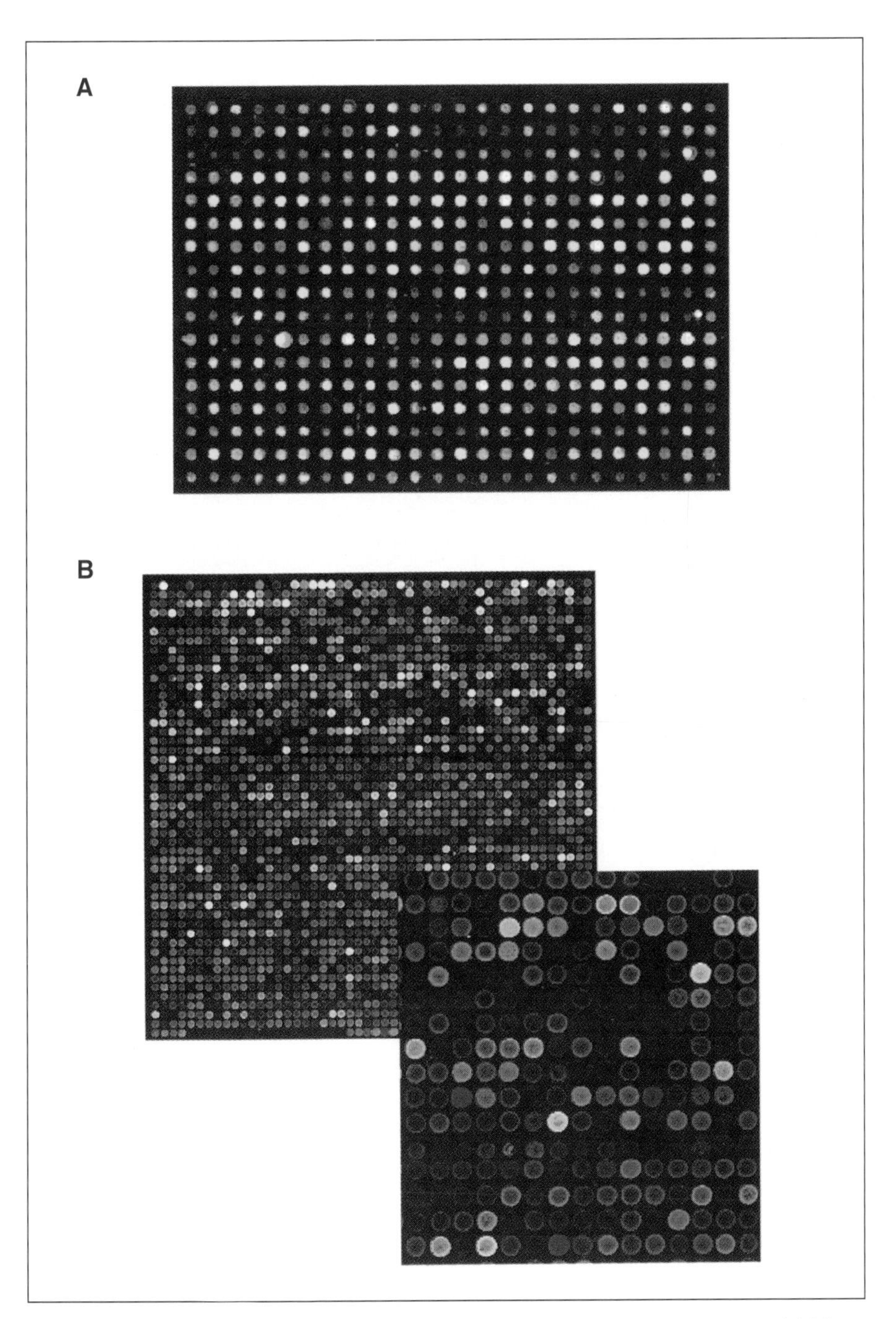

Figure 1. Examples of high-quality images acquired on the GenePix 4000A microarray scanner. (A) Microarray image of a yeast cDNA subarray kindly provided by Joe DeRisi, UCSF. (B) Microarray image of a GEM subarray kindly provided by Incyte Pharmaceuticals. Inset shows an enlarged portion of the subarray. (See Chapter 13.)

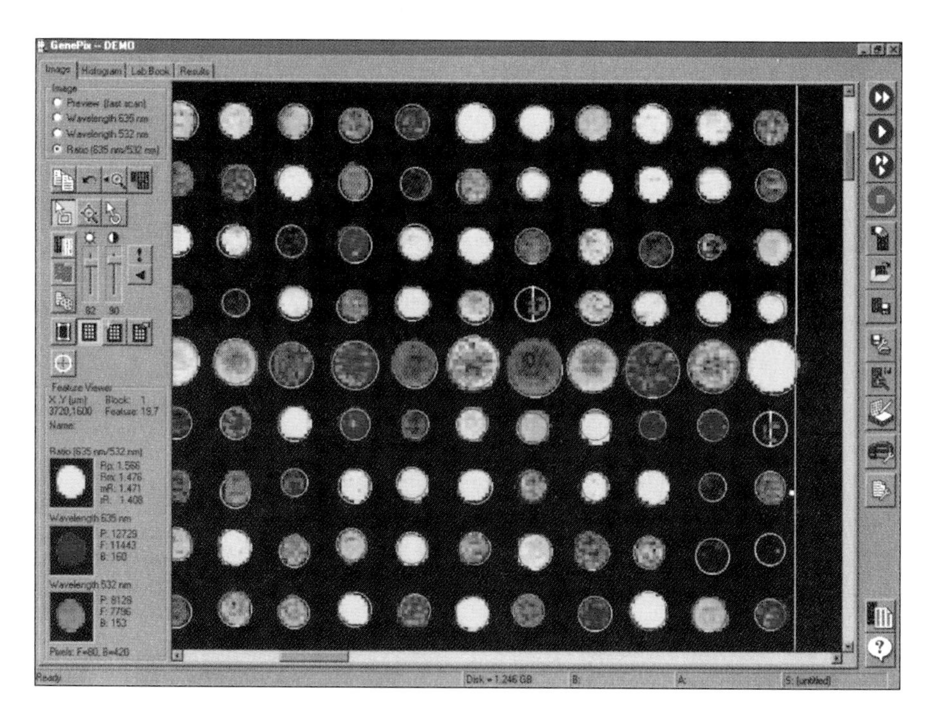

Figure 3. GenePix software interface. This "screen dump" image of the GenePix software shows that the application is designed more like an instrument interface, rather than a standard Windows program with drop-down menus. (See Chapter 13.)